AN INTRODUCTION TO STATISTICAL PHYSICS

ELLIS HORWOOD SERIES IN
MATHEMATICS AND ITS APPLICATIONS

Series Editor: Professor G. M. BELL, Chelsea College, University of London

The works in this series will survey recent research, and introduce new areas and up-to-date mathematical methods. Undergraduate texts on established topics will stimulate student interest by including present-day applications, and the series can also include selected volumes of lecture notes on important topics which need quick and early publication.

In all three ways it is hoped to render a valuable service to those who learn, teach, develop and use mathematics.

MATHEMATICAL THEORY OF WAVE MOTION
G. R. BALDOCK and T. BRIDGEMAN, University of Liverpool.
MATHEMATICAL MODELS IN SOCIAL MANAGEMENT AND LIFE SCIENCES
D. N. BURGHES and A. D. WOOD, Cranfield Institute of Technology.
MODERN INTRODUCTION TO CLASSICAL MECHANICS AND CONTROL
D. N. BURGHES, Cranfield Institute of Technology and A. DOWNS, Sheffield University.
CONTROL AND OPTIMAL CONTROL
D. N. BURGHES, Cranfield Institute of Technology and A. GRAHAM, The Open University, Milton Keynes.
TEXTBOOK OF DYNAMICS
F. CHORLTON, University of Aston, Birmingham.
VECTOR AND TENSOR METHODS
F. CHORLTON, University of Aston, Birmingham.
TECHNIQUES IN OPERATIONAL RESEARCH
VOLUME 1: QUEUEING SYSTEMS
VOLUME 2: MODELS, SEARCH, RANDOMIZATION
B. CONNOLLY, Chelsea College, University of London
MATHEMATICS FOR THE BIOSCIENCES
G. EASON, C. W. COLES, G. GETTINBY, University of Strathclyde.
HANDBOOK OF HYPERGEOMETRIC INTEGRALS: Theory, Applications, Tables, Computer Programs
H. EXTON, The Polytechnic, Preston.
MULTIPLE HYPERGEOMETRIC FUNCTIONS
H. EXTON, The Polytechnic, Preston
COMPUTATIONAL GEOMETRY FOR DESIGN AND MANUFACTURE
I. D. FAUX and M. J. PRATT, Cranfield Institute of Technology.
APPLIED LINEAR ALGEBRA
R. J. GOULT, Cranfield Institute of Technology.
MATRIX THEORY AND APPLICATIONS FOR ENGINEERS AND MATHEMATICIANS
A. GRAHAM, The Open University, Milton Keynes.
APPLIED FUNCTIONAL ANALYSIS
D. H. GRIFFEL, University of Bristol.
GENERALISED FUNCTIONS: Theory, Applications
R. F. HOSKINS, Cranfield Institute of Technology.
MECHANICS OF CONTINUOUS MEDIA
S. C. HUNTER, University of Sheffield.
GAME THEORY: Mathematical Models of Conflict
A. J. JONES, Royal Holloway College, University of London.
USING COMPUTERS
B. L. MEEK and S. FAIRTHORNE, Queen Elizabeth College, University of London.
SPECTRAL THEORY OF ORDINARY DIFFERENTIAL OPERATORS
E. MULLER-PFEIFFER, Technical High School, Ergurt.
SIMULATION CONCEPTS IN MATHEMATICAL MODELLING
F. OLIVEIRA-PINTO, Chelsea College, University of London.
ENVIRONMENTAL AERODYNAMICS
R. S. SCORER, Imperial College of Science and Technology, University of London.
APPLIED STATISTICAL TECHNIQUES
K. D. C. STOODLEY, T. LEWIS and C. L. S. STAINTON, University of Bradford.
LIQUIDS AND THEIR PROPERTIES: A Molecular and Macroscopic Treatise with Applications
H. N. V. TEMPERLEY, University College of Swansea, University of Wales and D. H. TREVENA, University of Wales, Aberystwyth.
GRAPH THEORY AND APPLICATIONS
H. N. V. TEMPERLEY, University College of Swansea.

AN INTRODUCTION TO STATISTICAL PHYSICS

W. G. V. ROSSER, M.Sc., Ph.D., F.Inst.P
Reader in Electromagnetism
Department of Physics
University of Exeter

ELLIS HORWOOD LIMITED
Publishers · Chichester

**Halsted Press: a division of
JOHN WILEY & SONS**
New York · Chichester · Brisbane · Toronto

First published in 1982 by
ELLIS HORWOOD LIMITED
Market Cross House, Cooper Street, Chichester, West Sussex, PO19 1EB, England

The publisher's colophon is reproduced from James Gillison's drawing of the ancient Market Cross, Chichester.

Distributors:

Australia, New Zealand, South-east Asia:
Jacaranda-Wiley Ltd., Jacaranda Press,
JOHN WILEY & SONS INC.,
G.P.O. Box 859, Brisbane, Queensland 40001, Australia

Canada:
JOHN WILEY & SONS CANADA LIMITED
22 Worcester Road, Rexdale, Ontario, Canada.

Europe, Africa:
JOHN WILEY & SONS LIMITED
Baffins Lane, Chichester, West Sussex, England.

North and South America and the rest of the world:
Halsted Press: a division of
JOHN WILEY & SONS
605 Third Avenue, New York, N.Y. 10016, U.S.A.

© **1982 W. G. V. Rosser/Ellis Horwood Ltd.**

British Library Cataloguing in Publication Data
Rosser, W. G. V.
An introduction to statistical physics. —
(Ellis Horwood series in physics in medicine and biology)
1. Statistical mechanics
I. Title
530.1'3   QC174.8

Library of Congress Card No. 81–4139 AACR2

ISBN 0–85312–272–5 (Ellis Horwood Ltd., Publishers – Library Edn.)
ISBN 0–85312–357–8 (Ellis Horwood Ltd., Publishers – Student Edn.)
ISBN 0–470–27241–4 (Halsted Press – Library Edn.)
ISBN 0–470–27242–2 (Halsted Press – Student Edn.)

Typeset in Great Britain by Preface, Salisbury.
Printed in the USA by The Maple-Vail Book Manufacturing Group, New York.

Table of Contents

Author's Preface

This book is based on a course given by the author to Honours Physics and Combined Honours students in their second year course at Exeter University. Nowadays, solid state physics and related topics form an important part of the third year honours course at British Universities. This means that students should preferably do an introductory course on statistical physics in their second year. At this stage the students are already familiar with Newtonian mechanics, and may have done a course on classical equilibrium thermodynamics, but they will probably only just have started their first comprehensive course on quantum mechanics. Clearly with this restricted background, in the initial stages, a second year course on statistical physics requires a far more simplified approach than can be adopted in a third year course. The author has tried to write the book at the level the average second year student can understand. To achieve this, the first presentation of new material is by means of simple numerical examples. All the mathematical steps are given in full. The basic principles of statistical physics are illustrated by simple numerical examples in Chapter 2. The reader can always return to these numerical examples, if he has any difficulties with the basic formulae used in later chapters. From Chapter 3 onwards the subject is developed axiomatically from three main postulates, namely (i) the existence of discrete quantum states, whose energies are the energy eigenvalues of the appropriate Schrödinger N particle equation; (ii) the principle of equal *a priori* probabilities, according to which, at thermal equilibrium, all the accessible microstates of a closed isolated system are equally probable; (iii) the law of conservation of energy. These axioms are applied to the case of two systems separated by a partition and surrounded by adiabatic walls so that they make up a closed isolated system. (This corresponds to the microcanonical ensemble.) The Boltzmann distribution (canonical ensemble) and the grand canonical distribution are developed later as special cases. For the benefit of more advanced readers, extra topics are discussed in sections marked with a star (*). These sections can be omitted in a first reading. The treatment is generally far more concise in these sections. Some of these sections, for example Chapter 7*, are designed to make the more advanced reader really think about the subject. Other starred sections are meant as an introduction to more advanced text books. Whenever appropriate, references are given for more advanced reading.

The book has three main aims:

Aim 1: To interpret the laws of classical equilibrium thermodynamics in terms of statistical mechanics. This is done mainly in Chapter 3. An interpretation of heat, work and cycles in terms of statistical mechanics is given later in Chapter 7 for the benefit of advanced readers. Ideally, the students should have done an introductory course on classical equilibrium thermodynamics. However, a review of the basic principles of classical equilibrium thermodynamics is given in Chapter 1 in a form suitable for comparison later with the approach to thermodynamics based on statistical mechanics. After covering the basic principles from both points of view, students who have not done an introductory course on classical equilibrium thermodynamics, should be able to follow the application of these principles to practical thermodynamic examples in any of the standard texts on thermodynamics.

Aim 2: To develop the Boltzmann distribution in Chapter 4, and the approach to thermodynamics based on the partition function in Chapter 5. The properties of the Helmholtz free energy F and the Gibbs free energy G are developed and interpreted in terms of both classical equilibrium thermodynamics and statistical mechanics in Chapter 6.

Aim 3: The application of statistical mechanics to quantum phenomena. Planck's radiation law is developed in Chapter 9. This is followed in Chapter 10 by a discussion of the Einstein and Debye theories of heat capacities, leading up to the concept of a phonon. In Chapter 11, the grand canonical distribution is developed and used to derive the Fermi–Dirac and Bose–Einstein distribution functions. Some introductory applications of the Fermi–Dirac and Bose–Einstein distributions are given in Chapter 12.

The author has consulted many text books during the development of the course on which this book is based. He wishes to acknowledge a particular debt to *Statistical Physics* by F. Reif and *Thermal Physics* by C. Kittel, which were used originally as the main course texts. The early influence of *Basic Concepts of Physics* by C. W. Sherwin will be apparent in Chapter 2. I would like to thank Mrs M. Madden and Mrs M. Cornish for typing the manuscript. Finally, I would like to acknowledge my debt to the students at Exeter University, whose questions and enthusiasm made the giving of the course, on which this book is based, such a pleasure.

W. G. V. ROSSER

Chapter 1

A Review of Classical Equilibrium Thermodynamics

1.1 INTRODUCTION

This Chapter gives a brief review of the main principles of classical equilibrium thermodynamics, presented in a form convenient for comparison later in Chapters 2 and 3 with the microscopic approach to thermodynamics based on statistical mechanics. This Chapter should serve as a refresher course for readers already familiar with classical thermodynamics. Probably the best approach for readers, who have not yet done a full course on classical thermodynamics, is to move on fairly quickly to the microscopic approach presented in Chapters 2 and 3. They can then follow the axiomatic approach to classical thermodynamics given by Callen [1]. Alternatively, having developed the basic principles of thermodynamics in this book, they can read about the practical applications of thermodynamics using a text book which follows the traditional approach to classical thermodynamics, for example, Adkins [2], Pippard [3], Sears and Salinger [4], and Zemansky [5]. As another alternative, they could go on to read a book in which the microscopic and the macroscopic approaches to thermodynamics are combined, as for example in Reif [6]. A reader wanting to do supplementary reading for this Chapter, is referred to any of the standard textbooks on classical thermodynamics such as Adkins [2], Pippard [3], Sears and Salinger [4] and Zemansky [5].

1.2 THERMODYNAMIC SYSTEMS

Classical equilibrium thermodynamics developed, mainly in the Nineteenth Century, as a series of laws relating the macroscopic thermodynamic state variables of a system, such as pressure and volume. This was before the development of detailed atomic models in the Twentieth Century. In the practical applications of thermodynamics, by a system we mean that portion of the matter of the universe which is bounded by a closed surface, as shown in Figure 1.1(a). The rest of the universe forms the surroundings of the system. The boundaries of the system are not necessarily fixed. For example, if a gas expands, the volume of the gas increases and the boundary of the gas system changes. [An example, is shown later in Figure 1.2(a)]. If changes are made to the surroundings of a system, after a period of time the macroscopic thermodynamic variables of a homogeneous system reach new **constant values**. The system is then said to be in a state of internal thermodynamic equilibrium.

As the general case of thermodynamics, it will be convenient to consider the idealised example of two subsystems, labelled 1 and 2, making up a composite closed system, as shown for example in Figures 1.1(b) and 1.1(c). The two subsystems are surrounded by rigid **adiabatic** walls. (An adiabatic wall is a wall which does not conduct heat, whereas a **diathermic** wall allows heat to flow through it). If the outer walls in Figure 1.1(b) are

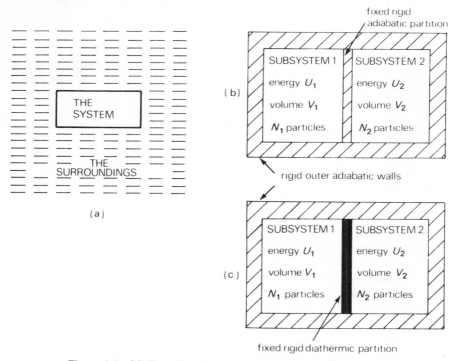

Figure 1.1—(a) Example of a system in thermal equilibrium with its surroundings. (b) An idealised example of two subsystems separated by a fixed rigid adiabatic partition. The two subsystems are inside rigid, outer adiabatic walls and form a closed system of fixed total energy, fixed total volume and fixed total number of particles. (c) The partition in this case is a fixed rigid diathermic partition.

rigid, the total volume $(V_1 + V_2)$ of the subsystems does not change. The total energy $(U_1 + U_2)$ and the total number of particles $(N_1 + N_2)$ in Figure 1.1(b) are also constant. The composite system made up from subsystem 1 plus subsystem 2 is a **closed system**, that is an isolated system of fixed total energy, fixed total volume and a fixed total number of particles.

Subsystems 1 and 2 in Figure 1.1(b) are separated by a partition. If the partition is a fixed, solid, rigid, adiabatic partition, it is an internal constraint of the composite system which prevents the exchange of heat between subsystems 1 and 2. The partition also prevents changes in the volumes of subsystems 1 and 2 and the exchange of particles between the two subsystems. If the partition is a fixed, solid, rigid, diathermic partition, as shown in Figure 1.1(c), heat can flow from one subsystem to the other, but the partition is still an internal constraint of the composite system, preventing changes in the volumes of the subsystems and the exchange of particles between the subsystems.

If the diathermic partition in Figure 1.1(c) is free to move, the volumes of the subsystems can change and one subsystem can do mechanical work on the other. If there are holes in the partition, the subsystems can exchange particles.

To correspond with Figure 1.1(a), we can treat subsystem 1 in Figures 1.1(b) and 1.1.(c) as the surroundings and subsystem 2 as the system. As a special case, we can assume that, in Figure 1.1(c), subsystem 1 is very much bigger than subsystem 2, so that subsystem 1 acts as a heat reservoir for subsystem 2. (A heat reservoir is a system whose heat capacity is so very much bigger than the heat capacity of the system in thermal contact with it, that heat flow from or to the heat reservoir does not change the temperature of the heat reservoir significantly.)

1.3 MACROSCOPIC AND MICROSCOPIC PHYSICS

There is no need in classical equilibrium thermodynamics to make any assumptions about atomic structure. The laws of thermodynamics were developed before the development of detailed atomic models in the Twentieth Century. The laws of thermodynamics lead to general relations *between* thermodynamic variables, enabling us to predict the value of one thermodynamic variable from the values of other thermodynamic variables. The laws of thermodynamics cannot predict the actual magnitudes of individual quantities directly from an atomic model. Our aim in Chapter 3 will be to interpret the laws of classical thermodynamics using a microscopic theory. Statistical mechanics can also be used to predict the values of individual macroscopic quantities. This approach will be developed from Chapter 4 onwards.

Macroscopic variables, such as pressure and volume, can be appreciated directly by our senses. Our sense of sight enables us to visualise, qualitatively, macroscopic changes in volume. Our sense of touch responds to changes in pressure. Atomic theory goes beyond the realm of direct perception. An important quantity relating microscopic and macroscopic quantities is Avogadro's constant N_A, which is equal to the number of $^{12}_{6}C$ atoms in 0.012 kilogramme of the isotope $^{12}_{6}C$. Avogadro's constant was first determined accurately by Perrin. (One of his methods is outlined in Problem 4.12). The experimental value of Avogadro's constant N_A is $(6.022\ 5 \pm 0.000\ 3) \times 10^{23}\ mol^{-1}$. A **mole** of any substance is the amount of the substance which contains as many elementary units as there are $^{12}_{6}C$ atoms in 0.012 kilogramme of $^{12}_{6}C$. The elementary unit must be specified and may be an atom, a molecule, an ion, an electron etc., or a group of such entities. For example, a mole of electrons consists of $6.022\ 5 \times 10^{23}$ electrons.

Avogadro's constant is an extremely large number (see Problem 1.1). The unaided eye can resolve about ten lines per millimetre at a distance of 25 cm, and should just about see a cube of side 0.1 mm and volume $10^{-12}\ m^3$ held at a distance of 25 cm. X-ray analysis has shown that the separation of the atoms in a solid is typically of the order of $10^{-10}\ m$, so that each atom should occupy a volume of the order of $10^{-30}\ m^3$. Thus a cube of side 0.1 mm made from a solid would contain of the order of 10^{18} atoms, which is a *very* large number. As a typical example of a *small* macroscopic system, we shall therefore take a system of 10^{18} atoms, corresponding to the smallest cube the unaided eye could see.

To illustrate a typical relation between a microscopic and a macroscopic quantity, consider the definition of mass density as the mass per unit volume. Though the diameter of an atom is typically of the order of $10^{-10}\ m$, most of the mass of each atom is in the atomic nucleus, which has a diameter of the order of $10^{-14}\ m$. On the atomic (microscopic) scale there are enormous fluctuations in mass density in distances of the order of $10^{-10}\ m$. The macroscopic mass density can be defined as

$$\rho = \Delta m/\Delta V \qquad (1.1)$$

where Δm is the total mass in a volume element ΔV, which is large on the atomic scale but small on the laboratory scale. For example, a cube of side 0.1 μm of a solid would contain about 10^9 atoms. The fluctuations in mass density average out in such a volume element, which is still small on the laboratory scale. Hence in a macroscopic theory, the mass density, defined by equation (1.1), can be treated as a smooth continuous function of position.

For purposes of discussion, assume that the system in Figure 1.1(a) is a gas. The **macrostate** of the system can be specified by any three of the macroscopic variables p, V, T and n, where p is the pressure, V is the volume of the system, T the temperature and n the number of moles in the system. (In statistical mechanics, instead of the number of moles, we shall generally use N, the number of particles in the system, where N is equal to the product of the number of moles n and Avogadro's constant N_A.) The equation of state of a real gas must be determined experimentally, or using atomic theory and statistical mechanics. On the basis of Boyle's law and Charles' law, it was suggested that the experimental equation of state of a gas was

$$pV = nRT \ . \tag{1.2}$$

In equation (1.2), R is the gas constant for one mole and T is the thermodynamic temperature. Equation (1.2) is not accurate for real gases, except in the limit of very low pressures. It is convenient to postulate an *ideal gas* which obeys equation (1.2) at all pressures and temperatures.

To specify the **microscopic** state (the **microstate**) of an *ideal* monatomic gas system in quantum theory, one would have to specify the single particle quantum states occupied by each monatomic gas atom. Each single particle state can be specified by four quantum numbers, one of which specifies the component of the spin of the particle. [Reference: Section 2.1.4] Thus, to specify the microstate of the N particle ideal gas system requires $4N$ quantum numbers, which is roughly 2.4×10^{24} quantum numbers per mole. This is an impossible task.

It is important to distinguish two types of macroscopic thermodynamic variable. An **extensive variable** is one whose value depends on the mass of the system, whereas an **intensive variable** is one whose value is independent of the mass of the system. If two identical homogeneous systems which have the same mass density, temperature and pressure are placed in thermal contact, the mass and volume of the composite system are both double the mass and volume of either original system, but we shall find that, according to the laws of thermodynamics the pressure, temperature and mass density of the composite system are the same as the pressure, temperature and mass density of either of the original systems. Provided the surface energy is negligible, the internal energy of the composite system is double the internal energy of either original system. The surface energy is negligible in what is known as the **thermodynamic limit**, in which the volume V and the number of particles N in the system both tend to infinity, but the ratio (N/V), the number of particles per unit volume, is kept constant. Thus in the thermodynamic limit, the internal energy is an extensive variable.

Summarising, the mass, volume and internal energy of a system are extensive variables whereas the pressure, temperature and density of a homogeneous system in internal thermodynamic equilibrium are intensive variables.

1.4 THE ZEROTH LAW OF THERMODYNAMICS

According to the **zeroth law of thermodynamics, if two systems A and B are in thermal equilibrium with a third system C, then A and B must also be in thermal equilibrium with each other.** In classical equilibrium thermodynamics the term that two systems are in thermal equilibrium is often applied to two separated systems, which are in such thermodynamic states that, if they were connected through a rigid diathermic wall, no heat would flow from one system to the other and the two systems would be in thermal equilibrium immediately after they were placed in thermal contact. The zeroth law forms the basis of thermometry. For example, if system C is a thermometer of negligible heat capacity, and if system C gives the same reading when it is in thermal contact with system A as it does when it is in thermal contact with system B, according to the zeroth law of thermodynamics systems A and B should be in thermal equilibrium. This means that, if A and B were placed in thermal contact, no heat would flow either from A to B or from B to A. The zeroth law leads to the ideas of empirical temperature and equations of state. Scales of empirical temperature can be defined in terms of arbitrary fixed points. For example, the fixed points on the Celsius, or centigrade scale, which was in common use before 1954, were the ice and the steam points which were defined to be exactly 0°C and 100°C respectively. Since 1954, the Kelvin scale has been adopted. On this scale the temperature of the triple point of water is defined to be *exactly* 273.16K, where the symbol K stands for degrees on the Kelvin scale. The temperature of the triple point of water is that unique temperature at which ice, water and water vapour can coexist in equilibrium.

If X is the value of the property used to measure temperature at the empirical temperature θ, and X_3 is the value of the property at the triple point of water, then on the Kelvin scale

$$\theta = 273.16(X/X_3) \; . \tag{1.3}$$

The values of empirical temperature determined using equation (1.3) depend on the type of thermometer used. For a constant volume gas thermometer, X would stand for p, the pressure of the gas in the thermometer. Real gases approximate closely to ideal gases in the limit of zero pressure. In practice, one approximates to an ideal gas thermometer by making measurements with various amounts of gas in the constant volume gas thermometer and extrapolating to zero pressure. The temperature on the ideal gas (Kelvin) scale is then given by

$$T = 273.16 \underset{p \to 0}{\text{Limit}} \, p/p_3 \; . \tag{1.4}$$

In equation (1.4), p is the pressure of the gas in the constant volume gas thermometer at the unknown temperature T, and p_3 is the pressure of the same gas at the temperature of the triple point of water.

In classical equilibrium thermodynamics, the thermodynamic scale of temperature is developed after the second law of thermodynamics is introduced. It is then shown that the ideal gas scale agrees with the thermodynamic scale of temperature. It was for this reason that the symbol T was used for temperature in equations (1.2) and (1.4).

Equation (1.3) is not very convenient for thermometers other than ideal gas thermometers. For example, if equation (1.3) were applied to a thermocouple, the empirical temperatures would depart very significantly from the ideal gas scale. It is better to calibrate such thermometers by direct comparison with an ideal gas thermometer or by using a number of standard points, which have been determinded accurately using ideal gas thermometers. Low temperatures, below 1K, are often determined using a thermodynamic relation such as the Clausius–Clapeyron equation. (Reference: Adkins [2].)

The Celsius scale is now defined by the equation

$$\theta°C = T - 273.15 \qquad (1.5)$$

where T is the temperature on the Kelvin scale, and θ is the temperature on the Celsius scale.

1.5 THE FIRST LAW OF THERMODYNAMICS

According to the **first law of thermodynamics**

$$\Delta U = \Delta W + \Delta Q \ . \qquad (1.6)$$

In equation (1.6), ΔU is the total increase in the internal energy U of the system. It is equal to the sum of ΔW, the total work done *on* the system, and ΔQ, the total heat supplied *to* the system. Equation (1.6) expresses the law of conservation of energy. The total work done on the system, ΔW, and the total heat supplied to the system, ΔQ, both depend on the way the system is taken from its initial equilibrium state to its final equilibrium state. The individual values of ΔW and ΔQ cannot be determined from the initial and final equilibrium states. According to the first law of thermodynamics, ΔU, the total change in the internal energy, which is equal to $(\Delta W + \Delta Q)$, depends only on the initial and final equilibrium states of a system, so that the internal energy U is a function of the thermodynamic state of the system, and, for infinitesimal changes, dU is a perfect differential. [These conclusions of classical equilibrium thermodynamics will be interpreted in Section 7.2.2* of

Chapter 7.] For infinitesimal changes, the first law of thermodynamics is generally written in the form

$$dU = đW + đQ \ . \tag{1.7}$$

Classical equilibrium thermodynamics developed in the 19th Century, before detailed atomic models were developed in the 20th Century. It is not necessary to introduce any precise atomic model for the internal energy U into classical equilibrium thermodynamics. [The internal energy U will be interpreted as the sum of the total kinetic energy and the total potential energy of all the atoms in the system in Section 3.4.]

The mechanical work done on a hydrostatic system depends on the pressure and the change of volume of the system. [Reference: equation (1.8).] The heat supplied to a system is associated with a difference of temperature between the system and its surroundings, and can take place by thermal conduction, thermal convection and radiative transfer. The heat added to a system can be measured by calorimetric methods. The term heat should only be applied to energy in transit. The flow of heat to the system and the performance of work on the system are different ways of increasing the internal energy of the system. One cannot say afterwards which part of the internal energy of the system was supplied as heat and which part was due to the work done on the system. A reader interested in a full discussion of the concepts of heat and internal energy in the context of classical equilibrium thermodynamics is referred to Adkins [2] and to Pippard [3]. [The differences between the effects of adding heat to a system and doing work on the system will be developed from the viewpoint of quantum theory in Chapter 7*.]

In an **adiabatic** process, no heat is added to or taken from the system. In an **isothermal** process, the temperature of the system is kept constant by placing the system in thermal contact with a heat reservoir. In an isothermal process heat flows into or out of the system to keep the temperature of the system constant.

1.6 THERMODYNAMIC REVERSIBILITY

For a thermodynamic process to be **reversible**, it must be possible to reverse the direction of the process by making infinitesimal changes in the applied conditions. As an example of a reversible process, consider a gas contained in a cylinder by a frictionless, tightly fitting piston, as shown in Figure 1.2(a). Let p be the pressure of the gas and p_0 the external pressure. If there is no friction and if p is infinitesimally bigger than p_0, the piston will move outwards in Figure 1.2(a). If p_0 is infinitesimally bigger than p, the piston will move inwards. In this case an infinitesimal change in external

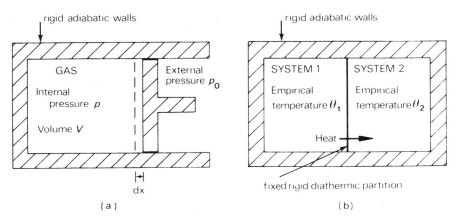

Figure 1.2—(a) The reversible expansion of a gas. (b) The flow of heat is reversible, if θ_1 is only infinitesimally bigger than θ_2.

pressure reverses the direction of the process. Reversibility in this case implies the absence of friction.

Let the external pressure p_0 be infinitesimally greater than p, the pressure of the gas, so that the piston moves inwards a small distance dx, at a very slow speed, until the pressure of the gas is equal to p_0. The work done *on* the gas is equal to the applied force times the distance moved by the piston, which is $p_0 A\, dx$, where A is the area of cross section of the piston. Since $A\, dx$ is equal to $(-dV)$, where dV is the increase in the volume of the gas, and $p_0 = p$, the mechanical work done *on* the system in the reversible process is

$$đW_{\text{rev}} = -p\, dV \ . \qquad (1.8)$$

If the gas is compressed, the increase in volume dV is negative, so that in this case $đW_{\text{rev}}$, the work done in compressing the gas, is positive. (The work done by the gas on the surroundings is equal to $+p\, dV$.)

If the decrease in the volume of the gas is finite, to be reversible the change in volume must be quasi-static. To achieve this, the external pressure must be increased by infinitesimal amounts and the decrease in volume allowed to take place very slowly. Each successive state of the system is then an equilibrium state, in which the pressure of the gas has reached the same value in all parts of the gas.

If the difference between the external pressure p_0 and the pressure of the gas p in Figure 1.2(a) were finite, the piston would accelerate, and the gas would be driven into motion, which would die away due to viscous damping. Such a change would not be reversible and the work done on the system, namely $(-p_0\, dV)$, would be greater than $(-p\, dV)$.

To transfer heat reversibly from one sub-system to the other in the example shown in Figure 1.2(b), the difference in the temperatures of the two subsystems must be infinitesimal. (It will be pointed out in Section 1.7 that, according to the second law of thermodynamics, if the empirical temperature θ_1 of subsystem 1 is infinitesimally greater than the empirical temperature θ_2 of subsystem 2, heat will flow from subsystem 1 to subsystem 2, whereas if θ_2 is infinitesimally greater than θ_1 heat will flow from subsystem 2 to subsystem 1.)

1.7 THE SECOND LAW OF THERMODYNAMICS

There are several ways of introducing the second law of thermodynamics based on the experimental observations that certain processes, allowed by the first law of thermodynamics, equation (1.6), do not take place in practice. For example, let a copper block at an empirical temperature of 350K be placed in a beaker of water, which is at an empirical temperature of 300K. It is consistent with the first law of thermodydnamics for the copper block to gain energy from the water and becoming red hot, and for the water to lose an equal amount of energy and eventually freezing. In practice, this process is never observed, but the copper block and water reach thermal equilibrium at a temperature somewhere between 300K and 350K. This type of experimental result is summarised in the Clausius statement of the second law of thermodynamics:–

No process is possible whose sole result is the transfer of heat from a colder to a hotter body.

The term **sole result** is important, because it is possible to transfer heat from a cold system to a hotter system using a refrigerator, but in this case external work must be done on the working substance.

Another historical approach to the second law of thermodynamics developed from the experimental result that there is a strict upper limit to the efficiency of a heat engine. Though it would be consistent with the first law of thermodynamics to do so, it is not possible in practice to convert all the heat taken from a heat reservoir into work during a *cyclic* process which brings the working substance back to its original thermodynamic state. This is illustrated in terms of the Carnot cycle in Appendix 2. This type of experimental result is summarised in the **Kelvin–Planck** statement of the second law of thermodynamics, which is:

No process is possible whose sole result is the absorption of heat from a reservoir and the conversion of all of this heat into work.

It is possible to add a quantity of heat Q to a system and allow the system to do external work equal to, or even greater than Q, but in this case the working substance is not back in its initial state. The Kelvin-Planck statement of the second law shows that there must be a fundamental difference between the effect of heat and work on a system. (This point will be interpreted in Chapter 7.)

It is shown in text books on thermodynamics that the truth of either the Clausius or the Kelvin–Planck statements of the second law of thermodynamics is a necessary and sufficient condition for the truth of the other.

The traditional way of developing the concepts of thermodynamic temperature and entropy from the second law of thermodynamics is based on the application of these statements to heat engines. The main steps will now be summarised. For fuller details the reader is referred to a text book on thermodynamics such as Adkins [2], Pippard [3], Sears and Salinger [4] or Zemansky [5]. The Carnot cycle is usually described as an example of a reversible engine. [An account of the Carnot cycle is given in Appendix 2.] The second law is used in the way outlined in Appendix 2, Section A2.2, to derive Carnot's theorems:

(i) **No engine can be more efficient than a reversible engine working between the same limits of temperature.**

(ii) **All reversible engines working between the same two limits of temperature have the same efficiency.**

Thus the efficiency of a Carnot engine is independent of the working substance, and depends only on the temperatures of the hot reservoir and cold sink. Kelvin defined the thermodynamic scale of temperature using the relation

$$Q_1/Q_2' = T_1/T_2 \tag{1.9}$$

where Q_1 is the heat taken in by the Carnot engine from the hot reservoir and Q_2' is the heat given up by the Carnot engine to the cold reservoir. According to equation (1.9), the ratio of the two temperatures T_1 and T_2 on the thermodynamic scale is equal to the ratio of the heats absorbed and rejected by a Carnot engine working between reservoirs at these temperatures. The internationally agreed fixed point on the **Kelvin thermodynamic scale** is the temperature of the triple point of water, which by definition is exactly 273.16K. The Kelvin thermodynamic scale is an absolute scale since, according to Carnot's theorem (ii), the ratio Q_1/Q_2' is independent of the working substance. If we used a different working substance, the values of Q_1 and Q_2' would be different, but for all reversible engines the ratio of Q_1 to Q_2' would be the same, provided the temperatures of the reservoir and sink were unchanged.

Substituting from equation (1.9) into equation (A2.2), we find that the efficiency of a Carnot engine is

$$\eta = (T_1 - T_2)/T_1. \tag{1.10}$$

Equation (1.10) is an important restriction on the maximum possible efficiency of a heat engine. For example, if T_1 is 400K and T_2 is 300K, the maximum possible efficiency of the Carnot engine is 25%. According to Carnot's theorems, no heat engine can be more efficient than this, if it is working between the same two limits of temperature.

If the convention is adopted that the heat given to a system is positive and the heat given up by the system is negative, then the heat *gained* by the system, when the system gives the heat Q_2' to the cold reservoir, is $Q_2 = -Q_2'$. Adopting this convention, for a reversible Carnot cycle, equation (1.9) becomes

$$(Q_1/T_1) + (Q_2/T_2) = 0 \ . \tag{1.11}$$

In equation (1.11) T_1 and T_2 are the thermodynamic temperatures of the hot reservoir and cold sink respectively. Equation (1.11) can be extended to any arbitrary *reversible* cycle to give

$$\oint đQ_{rev}/T = 0 \ . \tag{1.12}$$

This is Clausius' theorem. According to equation (1.12), if a system is taken around a **reversible** cycle, and if the heat $đQ_{rev}$ added reversibly to the system at each point is divided by the thermodynamic temperature of the system at that point, the sum of all such quotients is zero. If we put

$$dS = đQ_{rev}/T \tag{1.13}$$

equation (1.12) gives

$$\oint dS = 0 \ . \tag{1.14}$$

In equation (1.13), dS is the change in the entropy S of the system, when a quantity of heat $đQ_{rev}$ is added to the system in a reversible change from one equilibrium state to another equilibrium state.

Consider the two reversible paths from A to C, shown on the pV indicator diagram in Figure 1.3. From equation (1.14)

$$\int_{ABC} dS + \int_{CDA} dS = 0 \ .$$

Rearranging, $$\int_{ABC} dS = -\int_{CDA} dS = \int_{ADC} dS \ . \tag{1.15}$$

According to equation (1.15), the line integral of the entropy is the same along both the *reversible* paths ABC and ADC between the equilibrium states A and C in Figure 1.3. Generalising, the difference in entropy between the equilibrium states A and C can be calculated along any *reversible* path using the equation

$$S_C - S_A = \int_A^C đQ_{rev}/T \ . \tag{1.16}$$

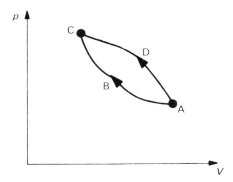

Figure 1.3—Two reversible paths ABC and ADC between equilibrium states A and C are shown on a pV indicator diagram.

The difference in the entropies of the equilibrium states A and C depends only on the initial and final equilibrium states of the system. Hence the entropy of a system is a function of the thermodynamic state of the system. The entropy is only defined for states in internal thermodynamic equilibrium (equilibrium states). Only the difference in the entropy between such equilibrium states can be calculated using equation (1.16).

The total quantity of heat supplied to a system, when the system is taken from one equilibrium state to another depends on the reversible path taken between the initial and the final equilibrium states. Though $đQ_{rev}$ is not a perfect differential, the change in the entropy dS, which is given by equation (1.13), is a perfect differential. The quantity $1/T$ is the integrating factor for $đQ_{rev}$.

To summarise the steps in the development of equation (1.13): the second law of thermodynamics is used in the way described in Section A2.2 of Appendix 2 to derive Carnot's theorem that all reversible engines have the same efficiency; this leads to the definition of thermodynamic temperature by equation (1.9) and to Clausius' theorem, equation (1.12); Clausius' theorem leads to the introduction of a new function of the state of the system, namely the entropy S, which is defined by equation (1.13). This summary illustrates how equation (1.13) depends on the second law of thermodynamics.

Recently the approach due to Carathéodory has become increasingly popular. According to **Carathéodory's principle**, in the immediate neighbourhood of every equilibrium state of a system there exist states which are inaccessible (which cannot be reached) by a reversible adiabatic process. Using an argument based on Pfaffian forms, Carathéodory was able to show that this simple statement implies the existence of an entropy function and a thermodynamic temperature. A reader interested in a simplified approach is referred to Zemansky [5].

1.8 THE THERMODYNAMIC IDENTITY

According to the first law of thermodynamics, equation (1.7),

$$dU = đQ + đW \tag{1.7}$$

where dU is the increase in the internal energy of the system, $đQ$ is the heat supplied to the system and $đW$ is the work done on the system. For an infinitesimal reversible path connecting the initial and the final equilibrium states, according to equation (1.13).

$$đQ_{rev} = T \, dS \; . \tag{1.13}$$

For a reversible change in the volume of a hydrostatic system, such as a gas, according to equation (1.8)

$$đW_{rev} = -p \, dV \tag{1.8}$$

where p is the pressure of the system. Substituting in equation (1.7), we have

$$dU = T \, dS - p \, dV \; . \tag{1.17}$$

Equation (1.17) is sometimes called the **thermodynamic identity**. It is a combination of the first law, equation (1.7) and the second law in the form of equation (1.13). Equation (1.17) relates the initial and final equilibrium states of the system. All the variables in equation (1.17), namely U, T, S, p and V are functions of the thermodynamic state of the system. Equations (1.13) and (1.8) are only valid separately for reversible changes. Equation (1.17) is valid for any irreversible change between the initial and the final *equilibrium* states. For example, if in Figure 1.2(a) the external pressure p_0 were significantly greater than p the pressure of the gas, the work done on the gas $(-p_0 \, dV)$ would be greater than $(-p \, dV)$. If the walls of the system and the piston in Figure 1.2(a) were adiabatic walls, then $đQ$ would be zero in this process. The gas would be accelerated by the motion of the piston. Eventually the gas motions would die away and the system would reach a new equilibrium state. Since equation (1.17) relates the initial and the final equilibrium states, we would have

$$dU = -p_0 \, dV = -p \, dV + T \, dS \; .$$

To calculate the change of entropy in this irreversible process, one would have to choose an alternative reversible path between the same initial and final equilibrium states of the system and apply equation (1.13) for this path. For example, one could compress the gas slowly and reversibly until it reached the final volume and then add heat $đQ$ to the gas in a reversible way, keeping the volume of the gas constant, such that the gas reached the final equilibrium temperature. The change in entropy would then be equal to $đQ/T$.

In an adiabatic change, no heat is supplied to or taken away from the system, so that for an adiabatic change đQ is zero in equation (1.7). According to equation (1.8), for a **reversible adiabatic change** of volume

$$đW_{rev} = -p \, dV \ . \tag{1.8}$$

Hence for a **reversible adiabatic change**, equation (1.7) becomes

$$dU = -p \, dV \ . \tag{1.18}$$

Comparing equation (1.18) with the thermodynamic identity, equation (1.17), it can be seen that, for a **reversible adiabatic change**, $T \, dS$ is zero. Hence, the entropy of a system is constant in a reversible adiabatic change.

The entropy S can be expressed in terms of other thermodynamic state variables. It will be shown in Chapters 2 and 3 that, for our microscopic approach to thermodynamics based on statistical mechanics, the natural variables for expressing S are the extensive variables: U, the internal energy of the system, V, the volume of the system and N, the number of particles in the system. It is assumed in equation (1.17) that the total number of particles in the system is constant. If the volume V is also constant, so that dV is zero in equation (1.17), we have

$$T = (\partial U/\partial S)_{V,N} \ . \tag{1.19}$$

(An account of partial differentiation is given in Section A1.2 of Appendix 1.) If the internal energy U is constant, dU is zero in equation (1.17) which then gives

$$p = T(\partial S/\partial V)_{U,N} \ . \tag{1.20}$$

Alternatively, if S is constant

$$p = -(\partial U/\partial V)_{S,N} \ . \tag{1.21}$$

If the fundamental relation, giving S in terms of U, V and N is known, the thermodynamic temperature and pressure of the system can be calculated using equations (1.19) and (1.20) respectively.

If the system can exchange particles with its surroundings, equation (1.17) must be extended to

$$dU = T \, dS - p \, dV + \mu \, dN \ . \tag{1.22}$$

In equation (1.22), dN is the increase in the number of particles in the system and μ is the chemical potential per particle. Equation (1.22) will be developed in Section 3.7.5. It follows from equation (1.22) that

$$\mu = -T(\partial S/\partial N)_{U,V} = (\partial U/\partial N)_{S,V} \ . \tag{1.23}$$

So far we have only considered the mechanical work done on hydrostatic systems, that is fluids. Other examples of external work are the work done in magnetising a system, the work done in changing the electric polarisation of a system, and the work done against surface tension.

1.9 IRREVERSIBLE PROCESSES

In Section 1.7 the long path leading up to the introduction of an extra function of state, the entropy, was outlined. At this point the reader could justifiably ask, what is the point of all this effort just to introduce an abstract quantity called entropy? The importance of entropy is that it can be used to specify the direction in which a physical process will go.

Consider the two subsystems in thermal contact, shown previously in Figure 1.2(b). Let the initial thermodynamic temperatures of subsystems 1 and 2 be $T_1 = 350K$ and $T_2 = 300K$ respectively. According to the Clausius statement of the second law of thermodynamics, heat will only flow from subsystem 1 to subsystem 2, never the other way round. This flow of heat is an irreversible process. Let one joule of heat be transferred from subsystem 1 to subsystem 2 through the diathermic partition, which is then changed back immediately to an adiabatic partition, so that each subsystem can reach a new state of internal thermodynamic equilibrium. It will be assumed that the changes in the thermodynamic temperatures of subsystems 1 and 2 are small enough to be neglected. According to equation (1.13), the change in the entropy of subsystem 1 is $\Delta S_1 = -1/350$ J K^{-1}, and the change in the entropy of subsystem 2 is $\Delta S_2 = +1/300$ J K^{-1}. The total change in entropy is

$$\Delta S_1 + \Delta S_2 = -(1/350) + (1/300) = (1/2100) \text{ J K}^{-1} .$$

The total entropy of subsystem 1 plus subsystem 2 increases. If one joule of heat flowed the other way from subsystem 2 to subsystem 1, the total entropy would go down by $(1/2100)$ J K^{-1}. This latter process is never observed in practice. Hence the direction of heat flow is given by

$$(\Delta S_1 + \Delta S_2) > 0.$$

If $T_1 = 300.001K$ and $T_2 = 300.000K$, according to the second law of thermodynamics heat would still flow from subsystem 1 to subsystem 2. In this case

$$\Delta S_1 + \Delta S_2 = -(1/300.001) + (1/300.000)$$
$$= +0.000\ 000\ 011 \text{ J K}^{-1} .$$

In the limit when $T_1 = T_2$, the change would be reversible and $(\Delta S_1 + \Delta S_2)$ would be zero.

If the diathermic partition in Figure 1.2(b) were left in position, heat would continue to flow from subsystem 1 to subsystem 2 until at thermal equilibrium

$$T_1 = T_2$$

and

$$(S_1 + S_2) \text{ is a maximum .} \tag{1.24}$$

As another example of an irreversible process, consider two subsystems of equal volumes V, as shown in Figure 1.4(a). Subsystem 1 consists of a mole of ideal gas molecules, whereas subsystem 2 is a vacuum. If the partition in Figure 1.4(a) is removed, the gas expands to fill the whole volume $2V$, as shown in Figure 1.4(b). This is an irreversible process, since,

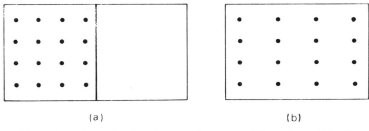

(a)   (b)

Figure 1.4—Example of an irreversible process. When the partition in (a) is removed the ideal gas expands from a volume V to occupy a volume $2V$. The entropy of the gas increases.

after the partition has been removed, one would never find all the gas molecules back in the left hand side half of the composite system in Figure 1.4(b). It can be shown using classical thermodynamics or using statistical mechanics (Reference: Problem 5.9 of Chapter 5) that, in the example illustrated in Figures 1.4(a) and 1.4(b), the increase in the entropy per mole is

$$\Delta S = R \log 2$$

where $\log 2$ is the natural logarithm of 2 to the base e. The direction of the irreversible process is again given by the condition that the total entropy of an isolated system increases in an irreversible process. It can be shown, using the general form of Clausius' theorem, that for the general case of a system interacting with its surroundings, illustrated in Figure 1.1(a),

$$(\Delta S_{\text{system}} + \Delta S_{\text{surroundings}}) \geq 0 . \tag{1.25}$$

The equality sign holds for a reversible change. (Reference: Adkins [2] and Pippard [3].)

1.10 THE THIRD LAW OF THERMODYNAMICS

Integrating equation (1.13) from the absolute zero to a thermodynamic temperature T we have

$$S = \int (đQ_{rev}/T) + S_0 \tag{1.26}$$

where S_0 is the entropy of the system at the absolute zero of temperature. The value of S_0 cannot be determined using only the second law of thermodynamics.

In 1906, on the basis of experiments carried out at temperatures well above liquid helium temperatures, **Nernst** proposed that, **for pure condensed substances at temperatures close to the absolute zero, the change in entropy associated with any change in the external parameters tends to zero.** In 1911, on the basis of his researches in statistical mechanics, **Planck** proposed that **the entropy of every pure condensed substance, in internal thermodynamic equilibrium, is zero at the absolute zero, so that, for this class of substance, according to Planck's proposal, S_0 should be zero in equation (1.26).** These proposals by Nernst and Planck constitute the third law of classical thermodynamics. (A reader interested in a full discussion of the third law of classical thermodynamics is referred to Sears and Salinger [4] or to Wilks [7].)

The **third law of thermodynamics** is best developed and interpreted using statistical mechanics. When developing this latter approach in Section 3.9*, it will be important for us to bear in mind two features of the third law, namely:

(i) **The entropy of a pure solid or liquid in internal thermodynamic equilibrium should be zero at the absolute zero.**

This is a statement of what should happen at $T = 0$.

(ii) **The Nernst form of the third law is often used in practice to predict the asymptotic laws of physics at very low temperatures and in some cases these predictions are valid experimentally at temperatures as high as 1K or more.**

As an example, consider the heat capacity of a solid, which is defined by the equation

$$C_V = (đQ/dT)_V \tag{1.27}$$

where dT is the increase in the temperature of the solid when a quantity of heat $đQ$ is added to the solid, the volume of the solid remaining constant. Using equation (1.17), with dV equal to zero, equation (1.27) can be rewritten in the form

$$C_V = (\partial U/\partial T)_V = T(\partial S/\partial T)_V = (\partial S/\partial \log T)_V . \tag{1.28}$$

As $T \to 0$, according to the third law of thermodynamics $\Delta S \to 0$. However, $\Delta \log T$ remains finite since $\log T \to -\infty$ as $T \to 0$. Hence, according to the third law of thermodynamics C_V should tend to zero as $T \to 0$. Experimentally the heat capacities of insulating solids do tend to zero at temperatures of the order of 1K. In a typical case, the molar heat capacity of an insulating solid at 1K is about 10^{-5} times the classical value of $3R$. Typical values for the entropy of an insulating solid are developed in Problem 1.3. [These results for insulating solids will be interpreted in terms of statistical mechanics in Chapter 10.]

1.11 THE VARIATION OF ENTROPY WITH INTERNAL ENERGY

In this Section we shall see how far we can go using plausibility arguments based on the second law of thermodynamics and the definition of absolute temperature by equation (1.19), namely

$$T = (\partial U/\partial S)_V . \tag{1.19}$$

It is generally agreed in the context of classical thermodynamics that there is an absolute zero of temperature. The present discussion is restricted to positive temperatures. Using equation (1.19), equation (1.28) can be rewritten as

$$C_V = T(\partial S/\partial T)_V = \frac{T}{(\partial^2 U/\partial S^2)_V} = \frac{(\partial U/\partial S)_V}{(\partial^2 U/\partial S^2)_V} . \tag{1.29}$$

Stability considerations require C_V to be positive. (Reference: ter Haar and Wergeland [8].) As an example, consider two systems in thermal equilibrium when they are in thermal contact. If, due to fluctuations, a small quantity of heat were transferred from system 1 to system 2, and if C_V were negative, according to equation (1.28), when U_1 was decreased T_1 would increase and T_2 would decrease. According to the Clausius statement of the second law, heat would continue to flow from system 1 to system 2 and thermal equilibrium would be unstable. According to equation (1.29), for C_V to be greater than zero when $T > 0$, we must have

$$(\partial^2 U/\partial S^2)_V > 0.$$

If $\partial^2 U/\partial S^2$ is always positive, the curve relating the internal energy U to the entropy S must always curve upwards with ever increasing slope as shown in Figure 1.5. The point (S_0, U_0) where $\partial U/\partial S$ is zero corresponds to the absolute zero of temperature.

On the basis of the second law of thermodynamics and the definition of absolute temperature, we have predicted a variation of the internal energy U with the entropy S of the type illustrated in Figure 1.5. According to the Planck statement of the third law of thermodynamics, S_0 should be zero for pure solids in internal thermodynamic equilibrium at the absolute zero. It

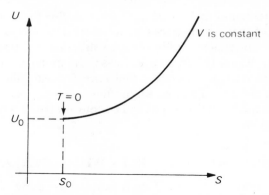

Figure 1.5—The form of the variation of the internal energy U of a system with its entropy S, predicted by classical equilibrium thermodynamics.

is more conventional to plot S as ordinate with U as abscissa, as shown in Figure 1.6. The term U_0 is the internal energy of the system at the absolute zero. Figure 1.6 summarises our predictions of the qualitative form of the variation of S with U based on the laws of classical thermodynamics. We shall find that the quantitative predictions of statistical mechanics, based on quantum mechanics, confirm the qualitative predictions presented in Figure 1.6. (See for example Figure 2.7(b) of Section 2.5.1 of Chapter 2.) The reverse argument can be used to show that a variation of S with U of the type shown in Figures 1.5 and 1.6 implies a positive value for C_V, since $\partial U/\partial S$ and $\partial^2 U/\partial S^2$ are both positive in Figures 1.5 and 1.6.

One interesting point the reader might like to speculate about at this stage is the value to be expected for C_V at $T = 0$ on the basis of the second

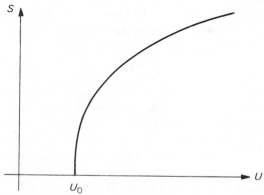

Figure 1.6—The form of the variation of the entropy S of a system with its internal energy U predicted by classical equilibrium thermodynamics.

law of thermodynamics. (Reference: Problem 1.4). Generally one would expect $\partial^2 U/\partial S^2$ to be greater than zero at $T = 0$, so that the curve of U plotted against S shown in Figure 1.5 can start curving upwards to give a finite positive thermodynamic temperature for internal energies $U > U_0$. According to equation (1.29), if $\partial^2 U/\partial S^2$ is finite at $T = 0$, then C_V must be zero at $T = 0$. There is no need to invoke the third law of thermodynamics in the above plausibility argument, though the conclusion that C_V is zero at the absolute zero does follow from the third law of thermodynamics.

REFERENCES

[1] Callen, H. B., *Thermodynamics*, 1960, Wiley.
[2] Adkins, C. J., *Equilibrium Thermodynamics* 2nd edition, 1975, McGraw Hill.
[3] Pippard, A. B., *The Elements of Classical Thermodynamics*, 1957, Cambridge University Press.
[4] Sears, F. W. and Salinger, G. L., *Thermodynamics, Kinetic Theory and Statistical Thermodynamics*, Third Edition, 1975, Addison-Wesley.
[5] Zemansky, M. W., *Heat and Thermodynamics*, Fifth Edition, 1968, McGraw-Hill.
[6] Reif, F., *Fundamentals of Statistical and Thermal Physics*, 1965, McGraw-Hill.
[7] Wilks, J. *The Third Law of Thermodynamics*, 1961, Oxford University Press.
[8] ter Haar, D. and Wergeland, H., *Elements of Thermodynamics*, 1966, page 48, Addison-Wesley.

PROBLEMS

Problem 1.1
A mole of ball bearings is stacked uniformly all over the United Kingdom. Each ball bearing needs a volume of 1 cm^3. Calculate the height of the stack of ball bearings.

Avogadro's constant $= 6.02 \times 10^{23}$ mol^{-1}.
Area of the U.K. $= 10^6$ km^2.

Problem 1.2
One kilogram of water at 273K is heated to a temperature of 373K by placing it in thermal contact with a large heat reservoir, which is at a temperature of 373K. What is the change in the total entropy of the Universe? How could the temperature of the water be raised from 273K to 373K without increasing the total entropy of the Universe?

Specific heat capacity of water $= 4.18 \times 10^3$ J K^{-1} kg^{-1} .

Problem 1.3
According to Debye's theory of heat capacities, at very low temperatures,

the molar heat capacity of an insulating solid is given by

$$C_V = 12R\pi^4 T^3/5\theta_D{}^3$$

where $R = 8.314$ J mol^{-1} K^{-1} is the molar gas constant. For a typical solid, $\theta_D = 300$K. Show that the molar entropy of such a solid at 0.01K, 0.1K, 1K, 10K and 30K is 2.4×10^{-11}, 2.4×10^{-8}, 2.4×10^{-5}, 2.4×10^{-2} and 0.65 J K^{-1} respectively.

Problem 1.4
Assume that when the volume of a system is kept constant, the variation of its internal energy U with its entropy S, shown in Figure 1.5, is given by

$$(U - U_0) = \alpha(S - S_0)^n$$

where α is a constant for a fixed value of volume. Show that

$$T = (\partial U/\partial S)_V = \alpha n(S - S_0)^{n-1}$$
$$(\partial^2 U/\partial S^2)_V = \alpha n(n - 1)(S - S_0)^{n-2}$$
$$C_V = (S - S_0)/(n - 1) = [1/(n - 1)](T/\alpha n)^{1/(n-1)} .$$

(i) Show that for $n = \frac{4}{3}$, $C_V = 81T^3/64\alpha^3$. This corresponds to the Debye T^3 law for the heat capacity of an insulating solid at temperatures $\simeq 1$K.

(ii) Show that for $n = 2$, $C_V = T/2\alpha$. This corresponds to the contribution of the free electrons to the heat capacity of a metal at temperatures $\simeq 1$K.

(iii) Show that for $n > 2$, even though $(\partial^2 U/\partial S^2)_V$ is zero when $S = S_0$ at $T = 0$, C_V is still zero at $T = 0$.

(iv) Sketch the variation of entropy with temperature for the case when $n = \frac{4}{3}$.

Chapter 2

Introduction to Statistical Mechanics

In this Chapter simple numerical examples are used to introduce two of the main postulates of statistical mechanics, namely the existence of microstates and the principle of equal *a priori* probabilities. The main formulae needed in Chapter 3 are also developed. Introductions are given to the order of magnitude of the fluctuations to be expected with macroscopic thermodynamic systems, and to the number of microstates accessible to a typical macroscopic thermodynamic system.

2.1 EXISTENCE OF MICROSTATES

2.1.1 Introduction

Quantum theory will be used from the outset. To quote Feynman[1]: '... **although most problems are more difficult in quantum mechanics than classical mechanics, problems in statistical mechanics are much easier in quantum theory**'. For the benefit of readers not yet fully familiar with quantum mechanics, we shall now summarise the main results necessary for our development of statistical mechanics. One important example is solved in detail in Appendix 3.

Consider an N particle system contained in a volume V. According to quantum mechanics, a measurement of the total energy of the N particle system will show that the N particle system is in one of a set of N particle quantum eigenstates, which can be determined by solving Schrödinger's time independent equation using the appropriate boundary conditions. According to Schrödinger's time independent equation

$$H\psi_i = \varepsilon_i\psi_i \tag{2.1}$$

where ψ_i is the wave function of the N-particle system, when it is in the ith N particle quantum state (microstate), ε_i is the energy eigenvalue corresponding to the wave function ψ_i and H is the Hamiltonian operator. In classical mechanics, the Hamiltonian is the expression for the sum of the total kinetic energy and the total potential energy of the N particles expressed in terms of the momenta and positions of the N particles. The Hamiltonian operator is obtained from the Hamiltonian by replacing the position and momentum variables by the appropriate operators. For given boundary conditions, equation (2.1) can only be solved for certain values of the energy ε_i. These are the **energy eigenvalues**.

It is sometimes possible to have more than one independent solution of equation (2.1) (that is different wave functions representing different quantum states), for the same energy eigenvalue. The **energy level** is then said to be **degenerate**. The degeneracy of the energy level is equal to the number of independent solutions of equation (2.1) for the given value of energy. We shall call each independent N particle quantum state a microstate of the N particle system. The microstate is specified by the appropriate wave function. [In practice the microstate can often be specified in terms of a set of quantum numbers.]

2.1.2 The hydrogen atom and the periodic table

Most readers will probably have met this example in an introductory course on atomic physics. In the first order theory of the hydrogen atom, the electron must be in one of a set of energy levels which have energies equal to $(-13.6/n^2)$ eV, where n is the principal quantum number, which can have the integral values 1, 2, 3 .... . The electron is normally in the lowest energy level $(n = 1)$, which is generally referred to as the ground state. The Frank-Hertz experiment and the existence of line spectra are experimental evidence that the electrons in atoms are in discrete energy levels. The energy levels of the hydrogen atom are degenerate. The different solutions of Schrödinger's equation can be specified by the quantum numbers n, l, m and s, where n is the principal quantum number, l is the orbital angular momentum quantum number, m is the magnetic quantum number and s is the spin angular momentum quantum number. In general, for a given value of n, l can have any integral value between 0 and $(n - 1)$. For a given value of l, m can have any integral value between $-l$ and $+l$. For each value of n, l and m the spin quantum number s can have the values $\pm\frac{1}{2}$. The total degeneracy of an energy level having principal quantum number n is $2n^2$.

Atoms of atomic number greater than hydrogen have more than one electron. According to the **Pauli exclusion principle**, only one of the electrons in an atom can have a given set of quantum numbers. Different electrons must differ in at least one of their quantum numbers. In the case of helium, in the ground state the two electrons have the quantum numbers $n = 1$, $l = 0$, $m = 0$, $s = \pm\frac{1}{2}$. These two electrons fill the K shell. For the elements lithium to neon, the L shell is progressively filled up. The degeneracy of the L shell, for which $n = 2$, is $2n^2 = 8$. For full details of the interpretation of the periodic table, the interested reader is referred to a textbook on atomic physics. This brief discussion was included, primarily, as an example of the Pauli exclusion principle.

2.1.3 The harmonic oscillator

In classical mechanics, the equation of motion of a particle of mass m moving with simple harmonic motion in one dimension is

$$m(d^2x/dt^2) = -kx$$

where kx is the restoring force on the particle, when its displacement from its equilbrium position is x. The angular frequency of the simple harmonic motion is given by

$$\omega = (k/m)^{1/2} .\tag{2.2}$$

The potential energy of the particle, when its displacement from its equilibrium position is x, is $\frac{1}{2}(kx^2)$. With this expression for potential energy, Schrödinger's time independent equation is

$$(\hbar^2/2m)(d^2\psi/dx^2) + (\varepsilon - \tfrac{1}{2}kx^2)\psi = 0\tag{2.3}$$

where $\hbar$ is Planck's constant h divided by 2π and ε is the total energy, which is the sum of the kinetic energy and the potential energy of the particle. It is shown in text books on quantum mechanics, e.g. Pauling and Wilson [2], that equation (2.3) only has regular solutions when ε has one of the values

$$\varepsilon_n = (n + \tfrac{1}{2})\hbar\omega = (n + \tfrac{1}{2})hv\tag{2.4}$$

where ω is given by equation (2.2), and the frequency v is equal to $\omega/2\pi$. Equation (2.4) gives the energy eigenvalues of the linear harmonic oscillator. The quantum number n can have the positive integer values 0, 1, 2, ... There is no degeneracy in the case of the one dimensional simple harmonic oscillator. The energy levels are equally spaced. The lowest energy level corresponding to $n = 0$ has energy $\frac{1}{2}\hbar\omega$. This is the zero point energy. There is no upper limit to the value of the quantum number n, so that there is no upper bound to the energy the one dimensional harmonic oscillator can have.

2.1.4 Particle in a box

An example which will be used extensively throughout the text is the case of a free particle inside a cubical box. To illustrate the methods of quantum mechanics, this important example is solved in detail in Appendix 3. According to equation (A3.13) of Appendix 3, the energy eigenvalues for a particle of mass m, in a cubical box of side L and volume V, are given by

$$\varepsilon = (\hbar^2\pi^2/2mV^{2/3})(n_x^2 + n_y^2 + n_z^2) = C(n_x^2 + n_y^2 + n_z^2)\tag{2.5}$$

where

$$C = \hbar^2\pi^2/2mV^{2/3} . \tag{2.6}$$

In equations (2.5) and (2.6), $\hbar$ is equal to $h/2\pi$ where Planck's constant h is equal to 6.626×10^{-34} J s. The quantum numbers n_x, n_y, and n_z can have the positive integer values $1, 2, 3, \ldots$. According to quantum mechanics, a measurement of the energy of the particle will give one of the energy eigenvalues given by equation (2.5).

The lowest energy eigenvalue, which is the ground state, has the quantum numbers $n_x = n_y = n_z = 1$. The energy of this state is

$$\varepsilon_{1,1,1} = (1^2 + 1^2 + 1^2)C = 3C . \tag{2.7}$$

As a numerical example, consider a single monatomic helium gas atom in a cubical box of side 0.282 m and volume 0.0224 m³. (At STP, which is one atmosphere pressure and 273K temperature, such a volume would contain a mole of helium atoms.) The mass of a helium atom is 6.65×10^{-27} kg. Substituting in equation (2.6), we find that

$$C = 1.04 \times 10^{-40} . \tag{2.8}$$

Equation (2.7) then gives

$$\varepsilon_{1,1,1} = 3C = 3.12 \times 10^{-40} \text{ J} . \tag{2.9}$$

One electron volt (1 eV) is equal to 1.602×10^{-19} J. Hence

$$\varepsilon_{1,1,1} = 1.95 \times 10^{-21} \text{ eV} . \tag{2.10}$$

The lowest energy level, specified by $n_x = n_y = n_z = 1$ is not degenerate. The next set of quantum states have quantum numbers n_x, n_y, and n_z equal to 2,1,1; 1,2,1 and 1,1,2 respectively corresponding to three independent solutions of Schrödinger's equation having the same energy eigenvalue. This energy level is threefold degenerate. According to equation (2.5), this energy level has energy $(2^2 + 1^2 + 1^2)C = 6C$, which is double the value of $\varepsilon_{1,1,1}$ given by equation (2.7). The energy level with quantum numbers 2,2,1 is also threefold degenerate, the energy eigenvalues being equal to $9C$.

According to the kinetic theory of ideal gases, the mean energy of a helium atom in a helium gas at an absolute temperature T is $\frac{3}{2}kT$, where k is Boltzmann's constant. (This result will be derived in Section 4.9 of Chapter 4.) At 273K, the mean energy of a helium gas atom is

$$\bar{\varepsilon} = \tfrac{3}{2}kT = \tfrac{3}{2} \times 1.38 \times 10^{-23} \times 273 = 5.65 \times 10^{-21} \text{ J} . \tag{2.11}$$

This is 1.81×10^{19} times larger than the value of the energy of the ground state given by equation (2.9). Putting the energy given by equation (2.5) equal to $\frac{3}{2}kT = 5.65 \times 10^{-21}$ J, to represent a typical helium gas atom at 273K, and substituting for C from equation (2.8) into equation (2.5), we have

$$n_x{}^2 + n_y{}^2 + n_z{}^2 = 3kT/2C = (5.65 \times 10^{-21})/(1.04 \times 10^{-40}) \tag{2.12}$$
$$= 5.43 \times 10^{19} \; .$$

In order to get a rough order of magnitude for the quantum numbers n_x, n_y, and n_z for a helium gas atom of energy $\frac{3}{2}kT$ at 273K, consider the special case when $n_x = n_y = n_z$. Equation (2.12) then gives

$$n_x{}^2 = 1.81 \times 10^{19} \tag{2.13}$$

so that $$n_x \simeq 4 \times 10^9 \; . \tag{2.14}$$

This shows that a helium gas atom in a cubical box of volume 0.0224 m^3 kept at 273K is likely to be in a single particle quantum state having quantum numbers in the range 10^9 to 10^{10}.

According to equation (2.5), the values of the energy eigenvalues of the single particle states for a particle in a cubical box are all inversely proportional to $V^{2/3}$, where V is the volume of the box. It is illustrated in Appendix 3 that this dependence on V comes in via the boundary conditions used to solve Schrödinger's equation. If the volume of the box is increased, according to equation (2.5), for fixed values of n_x, n_y and n_z, the values of all the energy eigenvalues decrease as illustrated in Figure 2.1. For example, if the volume of the box is increased from V to $8V$, the single particle energy eigenvalues go down to a quarter of their values at volume V. If the volume V is decreased, the energy eigenvalues are increased, as shown in Figure 2.1.

There is no upper limit to the integral values of n_x, n_y, and n_z. Hence there is no upper bound to the energy of a particle in a box. If the particle has spin, then in order to specify the state of the particle, the spin quantum number must also be specified in addition to n_x, n_y, n_z.

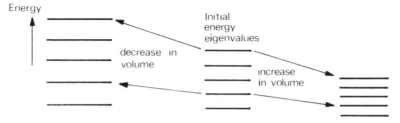

Figure 2.1—The effect of changing the volume of the box on the energy eigenvalues of a particle in a box. If the volume is increased, the energy eigenvalues are all decreased, whereas if the volume is decreased, the energy eigenvalues are all increased and are further apart.

2.1.5 N-particle systems

As an example, consider a system of N particles in a cubical box. In this case the wave function ψ_i, which is the solution of equation (2.1), is a function of the coordinates of all the N particles in the box. The energy eigenvalue ε_i of the ith state of the N particle system, is a function of both V, the volume of the box, and N the number of particles in the box. The energy eigenvalue ε_i is the sum of the total potential energy and the total kinetic energy of all the N particles in the system.

In some special cases, the interactions between the particles are small enough so that, as a first approximation, the particles can be treated as independent particles. This would be true of an ideal gas. (In practice, with real gases there would be interactions between the molecules. For example, in the van der Waals approximation, the term a/V^2 is added to the pressure to allow for the attractions between the gas molecules.) If there is no interaction between the particles, the total energy of the N-particle system is given by

$$\varepsilon_i(V, N) = E_1 + E_2 + \ldots + E_N = \sum_{j=1}^{N} E_j$$

where E_j is the energy of the jth independent particle.

Following the development of quantum mechanics in the 1920's, it was concluded that one could not distinguish between non-localised identical particles, such as the atoms of an ideal monatomic gas inside a box. A reader interested in a full discussion of the implications of the indistinguishability of particles in quantum mechanics is referred to a textbook on quantum mechanics such as Eisberg [3], or French and Taylor [4]. Some of the consequences of these results will be summarised in this Section and illustrated by a simple numerical example in Section 2.2.

Particles of half integral spin ($\frac{1}{2}$, $\frac{3}{2}$, ...) obey Fermi–Dirac statistics and are called fermions. The wave function of an N-particle system of identical fermions is antisymmetric, in the sense that the wave function changes sign if the coordinates of any two fermions are interchanged. This leads to the result that only one of the fermions in the N-particle system can be in any particular single particle quantum state, with a given set of single particle quantum numbers, but one cannot say which of the fermions is in which particular single particle quantum state. This leads to the **Pauli exclusion principle**, which was mentioned in Section 2.1.2.

Particles of integral spin (0, 1, ...) obey Bose–Einstein statistics and are called bosons. The wave function of a system of N identical bosons is symmetric, in the sense that, if the coordinates of any two bosons are interchanged, the wave function of the N-particle system is unchanged. In **Bose–Einstein statistics**, there is no limit to the number of bosons that can be in a particular single particle quantum state with a given set of single particle quantum numbers, but one cannot say which of the bosons is in which particular single particle state.

2.2 IDEALISED EXAMPLE ON THE COUNTING OF ACCESSIBLE MICROSTATES

2.2.1 Distinguishable particles

Consider an idealised system in which the single particle energy levels are non degenerate, are equally spaced and have the energy eigenvalues 0, 1, 2, 3 . . . energy units. (This would be true of a one dimensional harmonic oscillator in which the zero point energy $\frac{1}{2}\hbar\omega$ in equation (2.4) is ignored and in which $\hbar\omega$ is put equal to one energy unit.) It will be assumed, initially, that the system consists of three *distinguishable* particles of spin zero, which can be distinguished by their colours, which are red, white and blue. These are denoted R, W and B respectively in Figure 2.2. We shall consider the various ways in which three energy units can be distributed between the three distinguishable particles.

A division of energy between the three distinguishable particles, in which it is not specified which of the three distinguishable particles has which amount of energy, will be called a **configuration**. One possible configuration is one in which one of the particles has three energy units and the other two have zero energy. Either the red, the white or the blue particle could have the three units of energy, as illustrated in Figures 2.2(a), 2.2(b) and 2.2(c) respectively. Thus there are three microstates in the 3;0;0 configuration. These microstates will be denoted 3,0,0; 0,3,0 and 0,0,3 respectively, where the first, second and third numbers refer to the energies of the red, white and blue particles respectively.

The other configurations, consistent with the condition that the total energy is three units, are 2;1;0 and 1;1;1. The various microstates associated with these two configurations are summarised in Table 2.1 and illustrated in Figures 2.2(d) to 2.2(j) inclusive.

For three distinguishable particles of spin zero, the total number of configurations is equal to three, and the total number of microstates is equal to ten. In the context of statistical mechanics, one would say that each one of the microstates listed in Table 2.1 and illustrated in Figures 2.2(a) to 2.2(j) is an **accessible microstate** of the three particle system, consistent with the condition that the total energy of the system is three energy units. The system has other possible microstates such as 4,2,3 and 0,1,1 but these are not consistent with the condition that the total energy of the system is equal to three energy units and they are therefore not accessible microstates of the system.

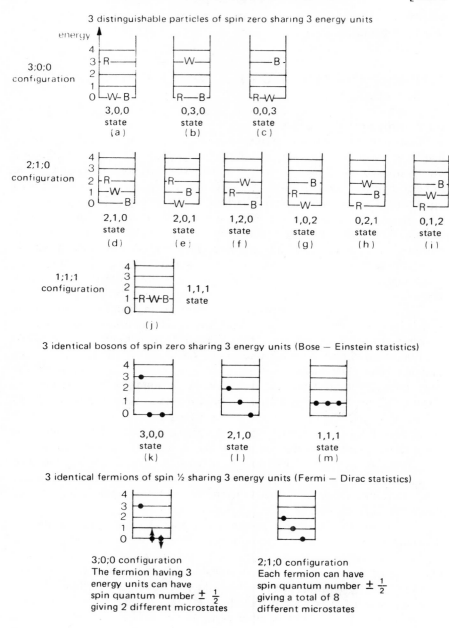

Figure 2.2—Numerical example on the counting of microstates.

Table 2.1

Configuration	Possible microstates		
3;0;0	3,0,0	0,3,0	0,0,3
2;1;0	2,1,0	2,0,1	1,2,0
	1,0,2	0,2,1	0,1,2
1;1;1		1,1,1	

According to equation (A1.4) which is derived in Section A1.3 of Appendix 1, the total number of microstates in a configuration of a system of N *distinguishable* particles of spin zero in which there are n_i particles in the ith single particle quantum state is equal to $N!/(n_0! \, n_1! \ldots)$. (The reader should remember that 0! is equal to unity.) For example, for the 3;0;0 configuration in Table 2.1, $N = 3$, $n_0 = 2$, $n_1 = 0$, $n_2 = 0$ and $n_3 = 1$. The number of microstates in this configuration should be $3!/2!$, which is equal to 3.

According to equation (A1.5), which is derived in Section A1.3 of Appendix 1, if U units of energy are distributed between N *distinguishable* particles of spin zero, the total number of accessible microstates g_T is given by

$$g_T = (N + U - 1)!/(N - 1)! \, U! \; . \tag{2.15}$$

For the example shown in Figure 2.2, $N = 3$ and $U = 3$ so that $(N + U - 1) = 5$ and $(N - 1) = 2$. Hence, according to equation (2.15), in this example g_T is equal to $5!/(2! \, 3!)$ which is equal to 10. This is in agreement with the results presented in Table 2.1.

The number of accessible microstates is often called the **degeneracy (or statistical weight)**. Even though the single particle energy levels in Figure 2.2 are not degenerate, the energy levels of the 3 particle system are degenerate. In the example shown in Figure 2.2, the total degeneracy is 10, when the total energy of the 3 particle system is 3 energy units.

2.2.2 Indistinguishable particles

If the 3 particles in Figure 2.2 are identical, one must use either Bose–Einstein or Fermi–Dirac statistics depending on the spins of the particles. If the particles have zero or integral spin, they are bosons obeying Bose-Einstein statistics. In this case there is no restriction on the number of particles which can be in any particular single particle quantum state. For bosons of spin zero, one cannot distinguish between the cases 3,0,0; 0,3,0 and 0,0,3 illustrated in Figure 2.1(a), 2.1(b) and

2.1(c) respectively. According to Bose–Einstein statistics they are the same microstate. Thus for three identical bosons of spin zero, all the microstates in a given configuration in Table 2.1 are the same state. The total number of accessible microstates in the case of 3 identical bosons is equal to 3. These are the 3,0,0; 2,1,0 and 1,1,1 states. The order of the numbers no longer has any significance. These microstates are illustrated in Figures 2.2(k), 2.2(l) and 2.2(m) respectively.

If the particles are identical fermions of spin $\frac{1}{2}$, such as electrons, one must use Fermi–Dirac statistics, according to which only one of the fermions can have a given set of quantum numbers. The fermions must differ in at least one of their quantum numbers. In the present example, two quantum numbers are required to specify the single particle quantum state occupied by a fermion. Let the energy of the single particle state be specified by the quantum number n, which is equal to the number of energy units, and let the component of the spin of the fermion be denoted by the spin quantum number s, which for fermions of spin $\frac{1}{2}$ can have the values $\pm\frac{1}{2}$. There can be 2, but only 2, fermions of spin $\frac{1}{2}$ with a given value of n, but these must have opposite spins. The 1;1;1 configuration is not allowed in Fermi–Dirac statistics. The 3;0;0 configuration is allowed provided the two fermions with $n = 0$ have different spin quantum numbers $+\frac{1}{2}$ and $-\frac{1}{2}$. The fermion with $n = 3$ can have either $s = +\frac{1}{2}$ or $s = -\frac{1}{2}$. Hence for fermions of spin $\frac{1}{2}$ there are two accessible microstates in the 3;0;0 configuration. The 2;1;0 configuration is also allowed for fermions of spin $\frac{1}{2}$. In this case the 3 fermions have different n quantum numbers. Each of the fermions can have $s = +\frac{1}{2}$ or $s = -\frac{1}{2}$. The two possibilities for $n = 2$ can be combined with the two possibilities for $n = 1$ to give 4 possibilities, each of which can be combined with the 2 possibilities for $n = 0$. Hence, when one allows for the spin quantum number, there are 8 accessible microstates in the 2;1;0 configuration for fermions of spin $\frac{1}{2}$. The total number of accessible microstates for 3 identical fermions of spin $\frac{1}{2}$ is equal to 10.

Originally in the Nineteenth Century it was assumed that, in principle at least, one could distinguish between the atoms of a monatomic gas. This assumption led to the difficulty of the Gibbs paradox, which will be discussed in Section 5.9.3 of Chapter 5. Our approach from the outset will be to assume that one cannot distinguish between non-localised identical particles, such as the atoms of a monatomic gas in a box. Such particles can be anywhere in the box and obey either Bose–Einstein or Fermi–Dirac statistics.

2.3 THERMAL INTERACTION

2.3.1 Introductory example

A simple example of thermal interaction will be used in this Section to introduce the principle of equal *a priori* probabilities and to develop the formulae which will be used in the general case of thermal interaction in Section 3.2 of Chapter 3.

Consider two subsystems surrounded by rigid, impermeable, adiabatic outer walls and separated by a rigid, impermeable, adiabatic partition, as shown in Figure 2.3(a). Let the single particle energy levels in both subsystems be non-degenerate, equally spaced and have the energy eigenvalues 0, 1, 2, 3 . . . energy units as shown in Figure 2.3(b). Subsystem 1 consists of two distinguishable particles of spin zero, which are coloured red and green. These are denoted R and G respectively in Figure 2.3(b). Subsystem 2 consists of two distinguishable particles of spin zero which are coloured white and blue. These are denoted by W and B respectively in Figure 2.3(b). Initially, the total energy of subsystem 1 is

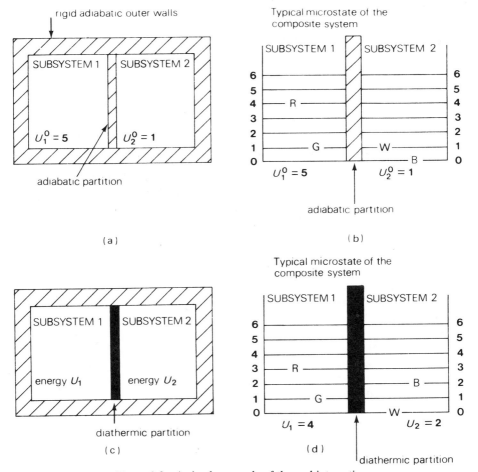

Figure 2.3—A simple example of thermal interaction.

$U_1^0 = 5$ energy units and the total energy of subsystem 2 is $U_2^0 = 1$ energy units. The total energy U^* of subsystem 1 plus subsystem 2 is

$$U^* = U_1^0 + U_2^0 = 6 \ . \tag{2.16}$$

The 5 units of energy which subsystem 1 has initially, can be distributed between the red and the green particles in the following six different ways: 5,0; 0,5; 4,1; 1,4; 3,2 and 2,3 where the first number is the energy of the red particle and the second number is the energy of the green particle. (The microstate 4,1 is illustrated in Figure 2.3(b).) Initially subsystem 1 has $g_1^0 = 6$ accessible microstates consistent with $U_1^0 = 5$ energy units. Initially, subsystem 2 has $g_2^0 = 2$ accessible microstates consistent with $U_2^0 = 1$ energy units. These microstates are 1,0 and 0,1, where the first and second numbers refer to the energies of the white and the blue particles respectively. (The 1,0 microstate is illustrated in Figure 2.3(b).)

Subsystems 1 and 2 in Figure 2.3(a) will be treated as a composite *closed* system. The adiabatic partition in Figure 2.3(a) is an internal constraint which prevents the flow of heat from one subsystem to the other, and keeps the total energies of subsystems 1 and 2 constant and equal to 5 and 1 energy units respectively. At any instant, subsystem 1 could be in any one of its 6 accessible microstates and subsystem 2 could be in any one of its 2 accessible microstates. The total number of microstates g_T^0 accessible to the *composite* system is

$$g_T^0 = g_2^0 \times g_1^0 = 2 \times 6 = 12 \ . \tag{2.17}$$

The 12 accessible microstates of the composite system are: 5,0,1,0; 5,0,0,1; 0,5,1,0; 0,5,0,1; 4,1,1,0; 4,1,0,1; 1,4,1,0; 1,4,0,1; 3,2,1,0; 3,2,0,1; 2,3,1,0 and 2,3,0,1. The first and second numbers are the energies of the red and green particles of subsystem 1 respectively and the third and fourth numbers are the energies of the white and blue particles of subsystem 2 respectively.

It will now be assumed that the adiabatic partition separating subsystems 1 and 2 is changed to a fixed, rigid, *diathermic* partition, as shown in Figure 2.3(c). The diathermic partition keeps the volumes V_1 and V_2 of subsystems 1 and 2 constant. (This means that the energy eigenvalues of the single particle quantum states of subsystems 1 and 2 are unchanged.) It will also be assumed that the particles cannot penetrate the partition and go from one subsystem to the other. Energy, in the form of heat, can flow across the diathermic partition, subject to the condition that U^*, the total energy of subsystems 1 and 2, is constant, that is

$$U^* = U_1 + U_2 = 6 \ . \tag{2.18}$$

In equation (2.18), U_1 and U_2 are the instantaneous values of the energies of subsystems 1 and 2 respectively.

Table 2.2

U_1	Microstates accessible to subsystem 1		$g_1(U_1)$	U_2	Microstates accessible to subsystem 2		$g_2(U_2)$
6	6,0. 5,1. 4,2. 3,3.	0,6. 1,5. 2,4.	7	0	0,0.		1
5	5,0. 4,1. 3,2.	0,5. 1,4. 2,3.	6	1	1,0.	0,1.	2
4	4,0. 3,1. 2,2.	0,4. 1,3.	5	2	2,0. 1,1.	0,2	3
3	3,0. 2,1.	0,3. 1,2.	4	3	3,0. 2,1.	0,3. 1,2.	4
2	2,0. 1,1.	0,2.	3	4	4,0. 3,1. 2,2.	0,4. 1,3.	5
1	1,0.	0,1.	2	5	5,0. 4,1. 3,2.	0,5. 1,4. 2,3.	6
0	0,0.		1	6	6,0. 5,1. 4,2. 3,3.	0,6. 1,5. 2,4.	7

The possible microstates of subsystems 1 and 2 are listed in Table 2.2 for various values of U_1 and U_2. All the possible microstates of the composite system shown in Figure 2.3(c) are listed in Table 2.3. Let $g_1(U_1)$ denote the number of microstates accessible to subsystem 1, when it has energy U_1 and let $g_2(U_2)$ denote the number of microstates accessible to subsystem 2, when it has energy U_2. By analogy with equation (2.17), the total number of microstates $g_1(U_1, U_2)$ accessible to the composite system, when subsystems 1 and 2 have energies U_1 and U_2 respectively, is

$$g(U_1, U_2) = g_2(U_2)g_1(U_1) \ . \tag{2.19}$$

Table 2.3

Microstates accessible to the closed composite system shown in Figure 2.3(c). The first, second, third and fourth numbers are the energies of the red, green, white and blue particles respectively.

$U_1 = 6$;	$U_2 = 0$					
6,0,0,0	0,6,0,0					
5,1,0,0	1,5,0,0					
4,2,0,0	2,4,0,0					
3,3,0,0						
$U_1 = 5$;	$U_2 = 1$					
5,0,1,0	5,0,0,1	0,5,1,0	0,5,0,1			
4,1,1,0	4,1,0,1	1,4,1,0	1,4,0,1			
3,2,1,0	3,2,0,1	2,3,1,0	2,3,0,1			
$U_1 = 4$;	$U_2 = 2$					
4,0,2,0	4,0,0,2	4,0,1,1				
0,4,2,0	0,4,0,2	0,4,1,1				
3,1,2,0	3,1,0,2	3,1,1,1				
1,3,2,0	1,3,0,2	1,3,1,1				
2,2,2,0	2,2,0,2	2,2,1,1				
$U_1 = 3$;	$U_2 = 3$					
3,0,3,0	3,0,0,3	3,0,2,1	3,0,1,2			
0,3,3,0	0,3,0,3	0,3,2,1	0,3,1,2			
2,1,3,0	2,1,0,3	2,1,2,1	2,1,1,2			
1,2,3,0	1,2,0,3	1,2,2,1	1,2,1,2			
$U_1 = 2$;	$U_2 = 4$					
2,0,4,0	2,0,0,4	2,0,3,1	2,0,1,3	2,0,2,2		
0,2,4,0	0,2,0,4	0,2,3,1	0,2,1,3	0,2,2,2		
1,1,4,0,	1,1,0,4	1,1,3,1	1,1,1,3	1,1,2,2		
$U_1 = 1$;	$U_2 = 5$					
1,0,5,0,	1,0,0,5	1,0,4,1	1,0,1,4	1,0,3,2	1,0,2,3	
0,1,5,0	0,1,0,5	0,1,4,1	0,1,1,4	0,1,3,2	0,1,2,3	
$U_1 = 0$;	$U_2 = 6$					
0,0,6,0	0,0,0,6	0,0,5,1	0,0,1,5	0,0,4,2	0,0,2,4	0,0,3,3

Table 2.4

U_2	$g_2(U_2)$	$g_1(6 - U_2)$	g_1g_2	$P_2(U_2) = g_2g_1/g_T$
0	1	7	7	7/84
1	2	6	12	12/84
2	3	5	15	15/84
3	4	4	16	16/84
4	5	3	15	15/84
5	6	2	12	12/84
6	7	1	7	7/84

For example, for a distribution of energy in which $U_1 = 4$ and $U_2 = 2$, subsystem 1 can be in any one of the 5 microstates: 4,0; 0.4; 3,1; 1,3 and 2,2, where the first number is the energy of the red particle and the second number is the energy of the green particle. When $U_2 = 2$, subsystem 2 can be in any one of the microstates: 2,0; 0,2 and 1,1, where in this case the first and second numbers refer to the energies of the white and the blue particles respectively. These two sets of microstates of subsystems 1 and 2 can be combined to give $5 \times 3 = 15$ different microstates of the composite system. These 15 microstates of the composite system are listed under $U_1 = 4$; $U_2 = 2$ in Table 2.3.

The values of the product $g_2(U_2)g_1(U_1)$ for various values of U_1 and U_2, subject to the constraint that $(U_1 + U_2) = 6$, are given in Table 2.4. The total number of microstates g_T accessible to the *composite* system shown in Figure 2.3(c) is obtained by summing $g_2(U_2)g_1(U_1)$ over all possible divisions of energy between subsystem 1 and 2 consistent with $(U_1 + U_2) = 6$. Expressed in terms of $g_1(U_1)$ and $g_2(U_2)$, g_T is given by

$$g_T = \sum_{U_2} g_2(U_2)g_1(U_1) = \sum_{U_2} g_2(U_2)g_1(U^* - U_2) \ . \tag{2.20}$$

Alternatively, $g_T = \sum_{U_1} g_1(U_1)g_2(U^* - U_1) \ .$

Summing all the values of $g_2(U_2)g_1(U_1)$ in Table 2.4 for the case shown in Figure 2.3(c), we find that

$$g_T = 84 \ .$$

The reader can check this result by counting all the microstates accessible to the composite system listed in Table 2.3. Thus when subsystem 2 is in thermal contact with subsystem 1 in Figure 2.3(c) the *composite* system has

84 accessible microstates. Notice that, when an internal constraint is removed, that is when the adiabatic partition is changed to a diathermic partition so that heat can flow from one subsystem to the other in Figure 2.3(c), the number of accessible microstates increases from 12 to 84.

One of our basic postulates of statistical mechanics will be the principle of equal *a priori* probabilities, according to which, when a *closed* system is in internal thermodynamic equilibrium, a measurement on the system is equally likely to find the closed system in any one of the microstates accessible to the closed system. A closed system is an *isolated* system of constant total energy, constant total volume and constant total number of particles. Since a closed isolated system is not under any external influences whatsoever, and its total energy, total volume and total number of particles are all constant, there is no reason whatsoever to believe that, at thermodynamic equilibrium, any particular one of the microstates accessible to the closed isolated system is any more probable than any of the other accessible microstates. Hence it seems reasonable, at least as a starting point, to assume that when a *closed* isolated system has had time to reach internal thermodynamic equilibrium, we are equally likely to find the closed isolated system in any of the microstates accessible to the closed isolated system. The principle of equal *a priori* probabilities will be checked *a posteriori*, that is it will be shown, later in the book, that predictions made on the assumption that the principle of equal *a priori* probabilities is correct are in agreement with the experimental results.

Since the composite system shown in Figure 2.3(c) is an isolated system of constant total energy $(U_1 + U_2)$, constant total volume $(V_1 + V_2)$ and constant total number of particles $(N_1 + N_2)$, it is a *closed* system. According to the principle of equal *a priori* probabilities, when the closed composite system shown in Figure 2.3(c) has had time to reach internal thermal equilibrium, the composite closed system is equally likely to be found in any one of the 84 accessible microstates of the composite closed system listed in Table 2.3. At thermal equilibrium, the probability of finding the composite system in any one of the 84 accessible microstates of the composite system is 1/84. The probability that at thermal equilibrium the composite system is in any one of the 15 microstates accessible to the composite system in which $U_2 = 2$ and $U_1 = 4$ is therefore 15/84. Thus the probability that at thermal equilibrium subsystem 2 has energy U_2 equal to 2 energy units is 15/84. This result will now be expressed in terms of $g_1(U_1)$ and $g_2(U_2)$.

Let $P_2(U_2)$ be the probability that at thermal equilibrium subsystem 2 has energy U_2, in which case $U_1 = (U^* - U_2)$. Subsystem 2 has energy U_2 in $g_2(U_2)g_1(U^* - U_2)$ of the g_T microstates accessible to the composite system. If, at thermal equilibrium, we are equally likely to find the composite closed system in any one of its accessible microstates, the probability of finding the composite system in any one of the $g_2(U_2)g_1(U^* - U_2)$ accessible microstates of the composite system in which subsystem 2 has energy U_2 is

$$P_2(U_2) = g_2(U_2)g_1(U^* - U_2)/g_T \qquad (2.21)$$

where g_T is given by (2.20). Similarly $P_1(U_1)$, the probability that at thermal equilibrium subsystem 1 has energy U_1, is

$$P_1(U_1) = g_1(U_1)g_2(U^* - U_1)/g_T \quad . \qquad (2.22)$$

The probability $P_2(U_2)$ that subsystem 2 has energy U_2 is equal to the probability $P_1(U_1)$ that subsystem 1 has energy $U_1 = (U^* - U_2)$.

It can be seen from Table 2.3 that, for the example shown in Figure 2.3(c), the number of microstates of the composite system for which $U_1 = 6$ is equal to 7. In the initial state shown in Figure 2.3(a), $U_1^0 = 5$ and $U_2^0 = 1$ energy units. Thus in 7 of the 84 microstates accessible to the composite system in Figure 2.3(c), subsystem 1 has gained energy and subsystem 2 has lost energy. In 12 cases the energies of subsystems 1 and 2 remain the same. It can be seen from Table 2.3 that in 65 cases subsystem 2 gains energy. We have not yet introduced the concept of absolute temperature. In classical thermodynamics, the temperature of a system increases when the internal energy of the system is increased, corresponding to a positive heat capacity. Thus it is reasonable to assume that, since the single particle quantum states of subsystems 1 and 2 have identical energy eigenvalues and the subsystems have the same number of distinguishable particles, since subsystem 1 starts with more energy than subsystem 2, initially subsystem 1 is 'hotter' than subsystem 2. In the example illustrated in Figure 2.3(c), in 65 cases heat has gone from the 'hot' subsystem 1 to the 'cold' subsystem 2. However, in 7 of the 84 possible cases heat has gone from the 'cold' subsystem 2 to the 'hot' subsystem 1. This shows that, according to statistical mechanics, it is possible for heat to go from the 'cold' to the 'hot' subsystem, though it is more probable for heat to go the other way.

It can be seen from Table 2.3 that out of the 84 microstates accessible to the composite system shown in Figure 2.3(c), U_2 is equal to 0, 1, 2, 3, 4, 5 and 6 energy units in 7, 12, 15, 16, 15, 12 and 7 cases respectively. Hence the average energy $\bar{U}_2$ of subsystem 2, averaged over the 84 equally probable accessible microstates of the composite closed system, shown in Figure 2.3(c), is

$$\bar{U}_2 = \frac{(7 \times 0 + 12 \times 1 + 15 \times 2 + 16 \times 3 + 15 \times 4 + 12 \times 5 + 7 \times 6)}{84}$$

$$(2.23)$$

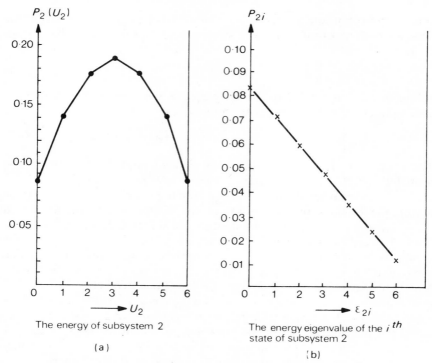

Figure 2.4—(a) The variation with U_2 of the probability $P_2(U_2)$ that at thermal equilibrium, subsystem 2 in Figure 2.3(c) has an energy equal to U_2. (b) The variation with ε_{2i} of the probability P_{2i} that, at thermal equilibrium, subsystem 2 in Figure 2.3(c) is in one particular one of the microstates of subsystem 2, namely the ith microstate which has energy eigenvalue ε_{2i}. Notice that P_{2i} decreases as ε_{2i} increases.

which is
$$\bar{U}_2 = 3 \ .$$

Similarly, $\bar{U}_1 = 3$.† Since U_2 goes up from an initial value of 1 to a mean value of 3 energy units, whereas U_1 goes down from an initial value of 5 to a mean value of 3 energy units, *on average* heat flows from the 'hot' subsystem to the 'cold' subsystem. The variation of $P_2(U_2)$ with U_2, calculated using equation (2.21), is shown in Figure 2.4(a). The probability distribution is very broad in this example. It is left as an exercise for the reader to use equation (A1.8) of Appendix 1 to check that the standard deviation σ_2 is equal to 1.73, so that in this example $\sigma_2/\bar{U}_2$ is equal to 0.58.

---

†In the general case, the mean energies $\bar{U}_1$ and $\bar{U}_2$ are not equal at thermal equilibrium. (See problems 2.5 and 2.9.) The mean energies were equal in the above special case, since the energy eigenvalues of the single particle states and the number of distinguishable particles were the same in both subsystems.

[An introduction to the standard deviation is given in Section A1.4 of Appendix 1.]

It will be shown in Section 3.2 and in Appendix 4 that, for two macroscopic systems, which are in thermal equilibrium when they are in thermal contact, a rough order of magnitude for $\sigma_2/\bar{U}_2$ is $1/N_2^{1/2}$ or less, where N_2 is the number of particles in the smaller of the two macroscopic subsystems. (We shall generally adopt the convention of labelling the smaller of the two subsystems in Figure 2.3(c) subsystem 2.) It will be shown in Chapter 3 that, though the probability that heat goes from the 'cold' to the 'hot' subsystem is finite, the overwhelming probability for macroscopic systems is for heat to go from the hotter to the colder subsystem when they are placed in thermal contact.

2.3.2 Ensemble averages

In statistical mechanics, the averages of physical quantities could be determined in two ways. The system could be followed over a very long period of time during which time the system should sample a large number of its accessible microstates. The average of the physical quantity over this long time period would give the time averaged value of the physical quantity. Alternatively, one could consider a large ensemble (collection) of identical closed systems prepared under identical conditions and average the physical quantity over all these systems at one instant of time to determine the *ensemble average*. In practice when estimating an ensemble average, it is convenient to restrict the ensemble to g_T systems containing one and only one of each of the g_T microstates accessible to the closed system. This makes the probability of finding each of the accessible microstates of the closed system exactly $1/g_T$. Though the reader probably did not realise it at the time, this is precisely what we did when we determined the mean energy $\bar{U}_2$ of subsystem 2 in Figure 2.3(c) by averaging U_2 over all the 84 microstates accessible to the composite system using equation (2.23). We took an ensemble of 84 identical composite systems and assumed that, in the ensemble, there was one of each of the 84 microstates accessible to the composite system.

Equation (2.23) can be rewritten in the form

$$\bar{U}_2 = (0 \times 7/84) + (1 \times 12/84) + (2 \times 15/84) + (3 \times 16/84)$$
$$+ (4 \times 15/84) + (5 \times 12/84) + (6 \times 7/84) \ . \tag{2.24}$$

It can be seen from Table 2.4 that $7/84$ is the probability $P_2(0)$ that subsystem 2 has energy $U_2 = 0$ etc. Hence the expression for the ensemble average energy of subsystem 2 in Figure 2.3(c) can be written in the form

$$\bar{U}_2 = \sum_{U_2} U_2 P_2(U_2) \tag{2.25}$$

In equation (2.25) the summation is over all the *energy levels* of subsystem 2, some of which may be degenerate. Equation (2.25) is the same as equation (A1.7) of Appendix 1. Sometimes it is more convenient to sum over all the microstates of subsystem 2. This approach will now be developed.

2.3.3 Summation over microstates

The reader can omit this Section at this stage, and return to read it when the reader reaches Sections 3.4, 3.9* and Chapter 4.

It can be seen from Table 2.2 that, when $U_2 = 2$, subsystem 2 in Figure 2.3(c) can be in any one of the three microstates 2,0; 0,2 or 1,1. From time to time we shall use the probability that, at thermal equilibrium, subsystem 2 is in one *particular* microstate of subsystem 2. For example, consider the 1,1 microstate of subsystem 2. When subsystem 2 is in this particular microstate having 2 energy units, $U_1 = 4$ and subsystem 1 can be in any one of the $g_1(U_1) = 5$ microstates of subsystem 1 for which $U_1 = 4$. (These microstates of subsystem 1 are 4,0; 0,4; 3,1; 1,3 and 2,2.) Thus there are $g_1(U_1) = 5$ microstates of the composite system in which subsystem 2 is in the 1,1 microstate of subsystem 2. They are 4,0, 1,1; 0,4, 1,1; 3,1, 1,1; 1,3, 1,1 and 2,2, 1,1. According to the principle of equal *a priori* probabilities, the probability that, at thermal equilibrium, the composite closed system shown in Figure 2.3(c) is in any one of its accessible microstates is 1/84. Hence, the probability that at thermal equilibrium subsystem 2 is in the 1,1 microstate of subsystem 2 is 5/84.

Generalising, the probability $P_{2i}(\varepsilon_{2i})$ that at thermal equilibrium subsystem 2 is in the ith microstate of subsystem 2, which has an energy eigenvalue ε_{2i}, is

$$P_{2i}(\varepsilon_{2i}) = g_1(U^* - \varepsilon_{2i})/g_T \ . \qquad (2.26)$$

In equation (2.26), $g_1(U^* - \varepsilon_{2i})$ is the number of microstates accessible to subsystem 1 when subsystem 1 has energy U_1 equal to $(U^* - \varepsilon_{2i})$.

If the energy levels of subsystem 2 are degenerate, such that the ith microstate of subsystem 2 is one of the $g_2(U_2)$ microstates of subsystem 2 which have energies U_2 equal to ε_{2i}, the probability that at thermal equilibrium subsystem 2 is in *any* one of these $g_2(U_2)$ independent microstates of subsystem 2, is

$$P_2(U_2) = g_2(U_2)P_{2i}(\varepsilon_{2i}) \ .$$

This is the same as equation (2.21), if we substitute for $P_{21}(\varepsilon_{2i})$ from equation (2.26).

The probability $P_{2i}(\varepsilon_{2i})$ that, at thermal equilibrium, subsystem 2 in Figure 2.3(c) is in its ith microstate is tabulated in Table 2.5 and plotted in Figure 2.4(b). It can be seen from Figure 2.4(b) that subsystem 2 is more

Table 2.5

Energy eigenvalue ε_{2i}	$P_{2i}(\varepsilon_{2i}) = g_1(U^* - \varepsilon_{2i})/g_T$
0	7/84
1	6/84
2	5/84
3	4/84
4	3/84
5	2/84
6	1/84

The number of microstates of subsystem 2 which have energy eigenvalues 0, 1, 2, 3, 4, 5 and 6 energy units are 1, 2, 3, 4, 5, 6 and 7 respectively.

likely to be found in one of the microstates of subsystem 2 for which the energy of subsystem 2 is zero than in any other microstate. This is due to the fact that, in equation (2.26), $g_1(U_1)$ increases as U_1 increases and has its maximum value when U_1 is a maximum, which is when ε_{2i} has its minimum value, which is zero in the present example.

The reader should be absolutely clear that according to the principle of equal *a priori* probabilities it is the 84 accessible microstates of the composite *closed* isolated system listed in Table 2.3 which are equally probable at thermal equilibrium. The principle of equal *a priori* probabilities only applies to closed isolated systems. Subsystems 1 and 2 in Figure 2.3(c) are not closed isolated systems. The energies of subsystems 1 and 2 can both vary by the transfer of heat across the diathermic partition in Figure 2.3(c). The probability that subsystem 2 is in any particular microstate of subsystem 2 depends on the energy eigenvalue of that microstate of subsystem 2 in the way given by equation (2.26) and illustrated in Figure 2.4(b).

To determine the ensemble average energy of subsystem 2, when the composite closed system shown in Figure 2.3(c) has reached internal thermal equilibrium, put $x_i = \varepsilon_{2i}$ in equation (A1.7) of Appendix 1. We have

$$\bar{U}_2 = \sum_{(states)} \varepsilon_{2i} P_{2i}(\varepsilon_{2i}) \qquad (2.27)$$

In equation (2.27), the probability P_{2i} that subsystem 2 is in its ith microstate, having energy eigenvalue ε_{2i}, is given by equation (2.26). The summation in equation (2.27) is over all the microstates of subsystem 2.

Let the variable α_2 of subsystem 2 have the value α_{2i} when subsystem 2 is in the ith microstate of subsystem 2. Putting $x_i = \alpha_{2i}$ in equation (A1.7) of Appendix 1, it follows that when subsystem 2 is in thermal equilibrium with subsystem 1 in Figure 2.3(c), the ensemble average of α_2 is

$$\bar{\alpha}_2 = \sum_{\text{(states)}} \alpha_{2i} P_{2i} \tag{2.28}$$

where P_{2i}, the probability that, at thermal equilibrium, subsystem 2 is in its ith microstate, is given by equation (2.26). The summation in equation (2.28) is over all the microstates of subsystem 2. It will be shown in Chapter 3 that, for macroscopic systems, the ensemble average $\bar{\alpha}_2$ corresponds to the equilibrium value of α_2 in classical equilibrium thermodynamics.

2.3.4 Definition of entropy

It will be shown in Section 2.5 that for macroscopic systems the number of accessible microstates g_T is such a fantastically large number that one invariably works with the natural logarithm of g_T. In order to fit in with the entropy introduced into classical thermodynamics, it is conventional to multiply $\log g_T$ by Boltzmann's constant k and to *define* the entropy S in statistical mechanics by the equation†

$$S = k \log g_T \tag{2.29}$$

where $\log g_T$ is the natural logarithm of g_T to the base e. In the example illustrated in Figures 2.3(a) and 2.3(c), the total number of microstates accessible to the composite system increases from 12 to 84, when the adiabatic partition is changed to a diathermic partition which allows heat to flow from one subsystem to the other. The entropy of the composite system increases from $k \log 12$ in Figure 2.3(a) to $k \log 84$ in Figure 2.3(c).

When an internal constraint of a system is removed, for example, when the adiabatic partition in Figure 2.3(a) is changed into a diathermic partition to allow heat to flow from one subsystem to the other, all the original accessible microstates of the composite system are still accessible. Hence the total number of accessible microstates of the composite closed system must either increase (or stay the same), when an internal constraint is removed. Consequently the total entropy of the composite closed system must either increase (or stay the same) when an internal constraint is removed, and the composite closed system has had time to reach internal thermodynamic equilibrium, subject to the remaining internal constraints.

2.4 FLUCTUATIONS IN MACROSCOPIC SYSTEMS

With small systems of the type discussed in Section 2.3, the fluctuations in energy, defined as the ratio of the standard deviation to mean value, are large, as can be seen in the distribution plotted in Figure 2.4(a). Many of

---

†Strictly, the natural logarithm of x to the base e should be written as $\log_e x$. Since we shall only use logarithms to the base e it will simplify the expressions just to refer to it as $\log x$.

the distributions we shall meet in statistical mechanics are very sharp Gaussians. To illustrate the orders of magnitude of the fluctuations in macroscopic systems, we shall consider an idealised example which can be solved using the binomial distribution. For large numbers of particles the binomial distribution goes over to a Gaussian distribution.

As an idealised example, consider N *spatially separated* particles as illustrated in Figure 2.5. It will be assumed that each particle can be in one or other of two degenerate states, which we shall call the up state and the down state respectively, as shown in Figure 2.5. It will be assumed in this idealised example that the probability that any particular particle is in the up state or in the down state is exactly $\frac{1}{2}$. For example, we could toss N spatially separated coins. The probability that any particular coin has heads upwards (the up state) is $\frac{1}{2}$. Another imaginary system would be a system of N spatially separated atomic magnetic dipoles which could point either parallel to the magnetic field (the up state) or antiparallel to the magnetic field (the down state) with equal probability. [The real case of atomic magnetic moments in an applied magnetic field will be discussed in Sections 4.4 and 5.8.]

Since the particles in Figure 2.5 are spatially separated, they can be 'distinguished' by virtue of their different positions. As an example, consider a system of N spatially separated coins. The first coin can give either a heads (H) or a tails (T), so that for one coin there are two equally probable possibilities, namely H or T. For two spatially separated coins, there are four, that is 2^2 possibilities, namely HH, HT, TH and TT. For three such coins, there are 8 that is 2^3 possibilities, namely HHH, HHT, HTH, THH, HTT, THT, TTH and TTT. The reader should now be able to recognise the pattern and realise that, for N spatially separated coins, there are 2^N possibilities. Similarly the first choice between the up and the down state of the first particle in the N particle system shown in Figure 2.5 can be made in two ways. Both of these can be combined with the two possibilities for the state of the second particle etc., so that the total number of possibilities for the N spatially separated particles is equal to 2^N. Each of these is a possible

N spatially separated particles which can be in either the up state or the down state.

Direction of the up state   Direction of the down state

Figure 2.5—The N spatially separated independent particles can be in either the up state or the down state.

microstate of the N particle system. Hence the total number of microstates g_T is given by

$$g_T = 2^N \qquad (2.30)$$

The probability $P(n)$ that, out of the total of N particles in Figures 2.5, n are in the up state, is given by the binomial distribution. [An account of the binomial distribution is given in Section A1.5 of Appendix 1.] According to equation (A1.10) of Appendix 1, we have

$$P(n) = \frac{N!}{n!\,(N-n)!}p^n(1-p)^{(N-n)} \qquad (2.31)$$

where p is the probability that any particular one of the particles is in the up state. It is being assumed that p is *exactly* equal to $\frac{1}{2}$, so that $(1-p)$ is also $\frac{1}{2}$. Since there are a total of 2^N microstates, it follows from equations (2.30) and (2.31) that the number of microstates $g(n)$, in which n particles are in the up state is

$$g(n) = g_T P(n) = 2^N \frac{N!}{n!\,(N-n)!}\,(\tfrac{1}{2})^n(\tfrac{1}{2})^{(N-n)}= \frac{N!}{n!\,(N-n)!} \;. \qquad (2.32)$$

This result follows directly from equation (A1.3) of Appendix 1. According to equation (A1.13) of Appendix 1, for a binomial distribution the mean number of particles in the up state is

$$\bar{n} = Np = N/2 \;. \qquad (2.33)$$

According to equation (A1.17) of Appendix 1, the standard deviation σ, which is defined by equation (A1.8) of Appendix 1, is given for a binomial distribution by

$$\sigma = [Np(1-p)]^{1/2} = N^{1/2}/2 \;. \qquad (2.34)$$

The ratio of the standard deviation to the mean value $\bar{n}$ is

$$\sigma/\bar{n} = (N^{1/2}/2) \div (N/2) = 1/N^{1/2} \;. \qquad (2.35)$$

For a system of only 3 spatially separated particles, the number of microstates in which $n = 0, 1, 2$ and 3 are 1, 3, 3 and 1 respectively. In this example, according to equations (2.33), (2.34) and (2.35), $\bar{n} = 1.5$, $\sigma = 0.866$ and $\sigma/\bar{n} = 0.58$. In this case, the predicted fluctuations, measured by the ratio of standard deviation to mean value, are large. However, consider a macroscopic system of exactly $N = 6.4 \times 10^{23}$ spatially separated particles. This is just over a mole. (The choice of 6.4×10^{23} gives a simple square root.) For this system, according to equation (2.33), the mean number of particles in the up state is

$$\bar{n} = Np = 3.2 \times 10^{23} \;.$$

According to equations (2.34) and (2.35)

$$\sigma = N^{1/2}/2 = 4 \times 10^{11}$$

$$\sigma/\bar{n} = 1/N^{1/2} = 1.25 \times 10^{-12} \ .$$

These results are sketched in Figure 2.6. It is impossible to show in Figure 2.6 how extremely sharp the distribution is. If we take $2\sigma = 8 \times 10^{11}$ to represent the order of magnitude of the width of the curve, $\bar{n}/2\sigma = 4 \times 10^{11}$. This means that the distance along the abscissa from the origin to the mean value $\bar{n}$ is 4×10^{11} times the width of the curve, defined as 2σ. If the numbers are written out in full, we have

$$\bar{n} = 320\ 000\ 000\ 000\ 000\ 000\ 000\ 000 \qquad (2.36)$$

$$\sigma = \qquad\qquad\qquad 400\ 000\ 000\ 000 \ . \qquad (2.37)$$

For large values of N, the binomial distribution can be approximated by a Gaussian distribution. For a Gaussian distribution the mean value is equal to the most probable value. For a Gaussian distribution, 68% of the values of n should be within $\pm\sigma$ of the mean value $\bar{n}$, and 99.7% should be within $\pm3\sigma$ of the mean value. It is shown in Problem 2.11 that, for a Gaussian distribution, the probability of getting a value of n deviating from the mean value $\bar{n}$ by more than $\pm100\sigma$ is 2.69×10^{-2174}. On average one would have

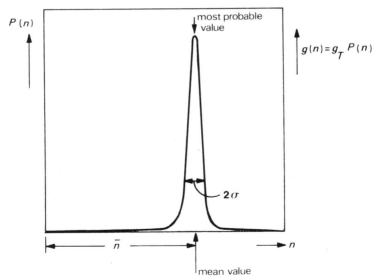

Figure 2.6—The variation with n of the probability $P(n)$ that n of the N spatially separated independent particles in Figure 2.5 are in the up state. For large N, the mean value $\bar{n}$ is very much bigger than the standard deviation σ. By changing the scale on the ordinate, the same curve can be used to represent the variation of $g(n)$ with n.

to sample a total of 3.7×10^{2173} microstates to find one whose value of n deviated from $\bar{n}$ by more than 100σ. Even a deviation of 100σ from the mean would only move the number 4 two places to the left in equation (2.37), and n would only differ from $\bar{n}$ in the eleventh significant figure. For purposes of discussion it will be assumed that each particle changes its state every 10^{-12} s, that is 10^{12} times per second. (The relaxation time of an atomic magnetic dipole in a weak magnetic field is generally longer than 10^{-12} s.) The reader can see from Figure 2.5 that each change of state of *one* of the N particles in the system changes the microstate of the N-particle system. For a system of 6.4×10^{23} particles there would be $6.4 \times 10^{23} \times 10^{12} = 6.4 \times 10^{35}$ changes of microstate per second. If we take 4.5×10^9 years as the age of the Earth, there are $(4.5 \times 10^9) \times 365 \times 24 \times 3600 = 1.42 \times 10^{17}$ seconds in a time equal to the age of the Earth. Hence, in a time equal to the age of the Earth, there would be approximately 10^{53} changes of microstate of the N particle system. On average, to get a variation from the mean exceeding 100σ one would have to wait approximately $3.7 \times 10^{2173} \div 10^{53}$, that is 3.7×10^{2120} ages of the Earth. The chances of a deviation from the mean exceeding 100σ is so completely and utterly remote that, even though the mathematical probability of such an event is finite, one would be safe in concluding that it would 'never' be observed in practice. It is shown in Problem 2.11 that the probability of obtaining a deviation from the mean exceeding 10σ is 1.54×10^{-23}. Hence the overwhelming probability is that the N particle system in Figure 2.5 will be found to be in a microstate having a value of n within $\pm 10\sigma$ of the mean value $\bar{n}$.

2.5 THE NUMBER OF MICROSTATES ACCESSIBLE TO A MACROSCOPIC SYSTEM

2.5.1 Introductory examples
To illustrate how rapidly the number of accessible microstates, (sometimes called the degeneracy or statistical weight) increases as the energy of a macroscopic system is increased, we shall return to the example illustrated in Figure 2.2. In this example, an idealised system was considered, in which the single particle energy levels were non degenerate and equally spaced. The system consisted of three *distinguishable* particles of spin zero, which were assumed to be coloured red, white and blue. The number of accessible microstates is given by equation (2.15), which for $N = 3$ and total energy U energy units becomes

$$g_T = \frac{(3 + U - 1)!}{(3 - 1)!\, U!} = (U + 2)(U + 1)/2 \quad . \qquad (2.38)$$

The values of g_T for $N = 3$ and for selected values of U in the range 0 to 20 energy units are given in Table 2.6 and plotted in Figure 2.7(a). It can be seen from Figure 2.7(a) that, as the energy U increases, the number of accessible microstates increases rapidly.

By writing out all the various possibilities, the reader can check the values of g_T given in Table 2.6 for the case of three identical bosons of spin zero (Bose–Einstein statistics). The results for bosons of spin zero are also plotted in Figure 2.7(a). The shape of the variation of g_T with U is similar to the result for 3 distinguishable particles.

Returning to the case of *three* distinguishable particles of spin zero, using equation (2.38) it can be shown that, for U equal to 10^2, 10^3, 10^4, 10^5 and 10^6 energy units, the values of g_T, the total number of accessible microstates are 5.15×10^3, 5.02×10^5, 5.00×10^7, 5.00×10^9 and 5.00×10^{11} respectively. For values of N of the order of 10^{23} distinguishable particles, with an average of about 10^6 energy units per particle, the values of g_T given by equation (2.38) would be truly enormous. These enormous degeneracies of the N particle system were calculated for an idealised system in which the single particle energy levels occupied by individual particles were not degenerate. If the single particle energy levels were degenerate, and if the particles had spin greater than zero the values of g_T would be even larger.

Due to the enormous values of g_T for a macroscopic system, one almost invariably works with the natural logarithm of g_T rather than with g_T. The

Table 2.6

Total energy U	3 distinguishable particles		3 identical bosons	
	g_T	$\log g_T$	g_T	$\log g_T$
0	1	0	1	0
1	3	1.10	1	0
3	10	2.30	3	1.10
5	21	3.04	5	1.61
8	45	3.81	10	2.30
10	66	4.19	14	2.64
12	91	4.51	19	2.94
15	136	4.91	27	3.30
18	190	5.25	37	3.61
20	231	5.44	44	3.78

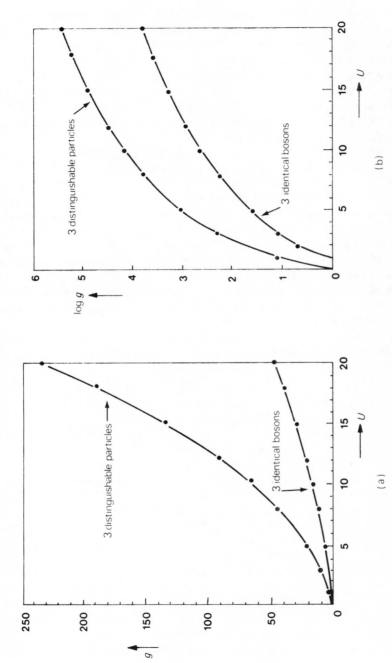

Figure 2.7—(a) The variation with total energy U of the number of microstates $g(U)$ accessible to a system whose non-degenerate energy eigenvalues are one energy unit apart. The system consists of either 3 distinguishable particles of spin zero or 3 identical bosons of spin zero. (b) The variation of log g with U for both 3 distinguishable particles of spin zero and 3 identical bosons of spin zero.

variation of $\log g_T$ with U for values of U up to 20 energy units is shown in Figure 2.7(b), for the case of both three distinguishable particles of spin zero and for three identical bosons of spin zero.

It is conventional to denote the slope of the curve relating $\log g_T$ to U by β, so that

$$\beta = \partial(\log g_T)/\partial U \ . \qquad (2.39)$$

It can be seen from Figures 2.7(b) that the slope β tends to infinity as U, the energy of the system tends to zero. The slope β decreases with increasing U. This is true of all normal thermodynamic systems.[†] The relation of these results to absolute temperature will be developed in Section 3.3 of Chapter 3.

In the example illustrated in Figure 2.7(b), the zero point energy of the lowest single particle energy level was ignored. Let the energy eigenvalue of the lowest single particle quantum state be ε_0. In the case of N distinguishable particles and in the case of N identical bosons of spin zero, all of which can be in the lowest energy single particle state, the total energy of the lowest energy microstate of the N particle system is $U_0 = N\varepsilon_0$. To allow for this zero point energy, U_0 should be added to U in Figure 2.7(b). According to equation (2.29) the entropy is related to $\log g_T$ by the relation

$$\log g_T = S/k \qquad (2.29)$$

where k is Boltzmann's constant. When the zero point energy U_0 is included, the variation of $\log g_T$ with U in Figure 2.7(b) has the same general form as the variation of S/k with U, predicted in Section 1.11 on the basis of classical equilibrium thermodynamics, and illustrated in Figure 1.6.

2.5.2 Macroscopic systems

To illustrate how large the number of accessible microstates g_T is for a macroscopic system, consider again the system of N spatially separated particles illustrated in Figure 2.5. According to equation (2.30), for one mole, that is for $N = 6.02 \times 10^{23}$,

$$g_T = 2^N = 2^{6.02 \times 10^{23}} = 10^{1.8 \times 10^{23}} \ . \qquad (2.40)$$

[†]There is a very special class of system, in which there is an upper limit to the energy one can add to the system. For example, one cannot add more energy to a system of atomic magnetic dipoles in a magnetic field, when all the magnetic dipoles are pointing in a direction opposite to the direction of the applied magnetic field. This type of system will be discussed in Chapter 8, when the concept of negative temperatures will be introduced. Until then, we shall only consider normal thermodynamic systems, whose (internal) energy can increase indefinitely.

To illustrate how really big the number ten to the power 1.8×10^{23} is, we shall work out how long the number would be, if it were typed out on a typical typewriter. To type 10^4 one would start with a 1 followed by four zeros. To type out ten to the power 1.8×10^{23}, one would start with a 1 followed by 1.8×10^{23} zeros. A typical typewriter types about 5 digits per centimetre, which is 500 digits per metre. Hence the number 2^N would be $1.8 \times 10^{23} \div 500$ that is 3.6×10^{20} m long. One light year is equal to the velocity of light times the number of seconds in a year, which is $(3 \times 10^8) \times 365 \times 24 \times 3600$ which is 9.5×10^{15} m. Hence, if the number given by equation (2.40) were typed out, it would be about 3.8×10^4 light years long. This is approximately equal to the radius of our galaxy. Thus one needs astronomical distances to type out the number of accessible microstates (statistical weight) of a typical macrostate of a macroscopic system. Hence, one generally works with $\log g_T$. For the N particle system shown in Figure 2.5, with $N = 6.02 \times 10^{23}$, from equation (2.40) we have

$$\log g_T = \log 2^N = N \log 2 = 4.17 \times 10^{23} . \qquad (2.41)$$

It only needs 24 digits to type $\log g_T$ out in full. In order to fit in with the definition of entropy introduced into classical thermodynamics, it is conventional to multiply $\log g_T$ by Boltzmann's constant k and to define the entropy S in statistical mechanics by the relation

$$S = k \log g_T . \qquad (2.29)$$

It will be shown in Section 5.9.5 of Chapter 5 that the experimental value of k is 1.3807×10^{-23} J K^{-1}. If $N = 6.02 \times 10^{23}$, the entropy of the system of N spatially separated particles shown in Figure 2.5 is

$$S = k \log 2^N = 1.38 \times 10^{-23} \times 6.02 \times 10^{23} \log 2 = 5.76 \text{ J K}^{-1} .$$

Thus the number g_T, which requires astronomical distances to write out in full, is reduced to a number less than 10 by using equation (2.29).

A rough order of magnitude for the entropy of most substances, measured in joule per kelvin, is

$$S = k \log g_T = \alpha Nk = \alpha nR \qquad (2.42)$$

where N is the number of particles in the system, n is the number of moles in the system, k is Boltzmann's constant, R is the molar gas constant and α generally has a value in the range 0.5 to 100. For example, for the N particle system shown in Figure 2.5, according to equations (2.41) and (2.29)

$$S = kN \log 2 = 0.693Nk = 0.693nR . \qquad (2.43)$$

Another example of equation (2.42) is the Sackur-Tetrode equation for the entropy of an ideal monatomic gas (Reference: Section 5.9.6). In the case of neon just above its boiling point, α has a value of 11.6.

It follows from equation (2.29) that the number of accessible microstates is

$$g_T = e^{S/k} \qquad (2.44)$$

Using equation (2.42) we find that a very rough guide for the order of magnitude of the number of accessible microstates is $\exp(\alpha N)$, where N is the number of particles in the system and α is generally in the range 0.5 to 100.

The equivalence of the entropy defined by equation (2.29) and the entropy introduced into classical equilibrium thermodynamics will be developed in Chapter 3. The results of classical equilibrium thermodynamics can be used to illustrate the types of changes in the number of microstates accessible to a macroscopic system, to be expected when the (internal) energy of the system is changed. According to equation (1.13), if a quantity of heat dQ is added to a system, which is at an absolute temperature T, the increase in the entropy of the system is

$$dS = dQ/T \ . \qquad (1.13)$$

[Equation (1.13) will be developed from statistical mechanics in Section 3.6.4.] Consider a macroscopic system such as 1 kg of water at 300 K. If one joule of heat is added to the water, which has a heat capacity of 4.2×10^3 J K^{-1}, the temperature of the water goes up by only $1/4.2 \times 10^3$ that is 2.38×10^{-4}K. According to equation (1.13), assuming T is constant, the increase in entropy

$$\Delta S = 1/T = 1/300 \ . \qquad (2.45)$$

According to equation (2.44) the number of accessible microstates is now given by

$$g_T + \Delta g_T = \exp[(S + \Delta S)/k] \ .$$

Dividing by equation (2.44), substituting for ΔS from equation (2.45) and putting k equal to 1.38×10^{-23} J K^{-1}, we find that the ratio of the number of accessible microstates after adding the one joule of energy to the number of accessible microstates before adding the joule of energy is

$$(g_T + \Delta g_T)/g_T = \exp(\Delta S/k) = \exp(2.4 \times 10^{20}) \ .$$

Thus the addition of just one joule of energy to a macroscopic system at 300K increases the number of accessible microstates by a factor of $\exp(2.4 \times 10^{20})$, even though the increase in temperature is only 2.38×10^{-4}K.

2.5.3 Approximations that can be made when calculating the logarithm of a very large number

It is possible to make approximations when working with the logarithm of a number as large as the number of accessible microstates of a macroscopic system, which may appear very strange at first sight. As an illustrative example we shall return to the system of N spatially separated particles illustrated in Figure 2.5. Taking the natural logarithm of both sides of equation (2.30), we have

$$\log g_T = \log 2^N = N \log 2 \ .$$

Consider a number which is 10^{20} times greater than 2^N. Taking the natural logarithm of this number, for $N = 6.02 \times 10^{23}$, we have

$$\log(2^N \times 10^{20}) = N \log 2 + 20 \log 10$$
$$= 4.17 \times 10^{23} + 46 \ .$$

The second number on the right hand side is negligible compared with the first number. Hence, for $N = 6.02 \times 10^{23}$, multiplying 2^N by 10^{20} makes no significant difference to the logarithm of the number. Similarly if 2^N were divided by 10^{20}, the only effect on the natural logarithm of the number would be to reduce $N \log 2 = 4.17 \times 10^{23}$ by 46, which is again insignificant.

It is shown in Section A1.10 of Appendix 1, that, for large values of N, the binomial distribution given by equation (2.31) goes over to the Gaussian distribution:

$$g(n) = g_T P(n) = [g_T/(2\pi\sigma^2)^{1/2}] \exp[-(n - \bar{n})^2/2\sigma^2]$$

where, for $p = \frac{1}{2}$, the mean value $\bar{n}$ is given by

$$\bar{n} = N/2 \ .$$

The standard deviation σ is given by

$$\sigma = (Npq)^{1/2} = \tfrac{1}{2}N^{1/2} \ .$$

The maximum value of $g(n)$ is when $n = \bar{n} = N/2$, when $g(n)$ has the value

$$g_{max} = g_T/(2\pi\sigma^2)^{1/2} = g_T/(\pi N/2)^{1/2} \ .$$

Taking the natural logarithm of both sides, we have

$$\log g_{max} = \log g_T - \tfrac{1}{2} \log(\pi N/2) \ .$$

For the example shown in Figure 2.5, $g_T = 2^N$, $\log g_T = N \log 2$. Hence,

$$\log g_{max} = N \log 2 - \tfrac{1}{2} \log(\pi N/2) \ . \tag{2.46}$$

For $N = 6.02 \times 10^{23}$, equation (2.46) gives

$$\log g_{max} = 4.17 \times 10^{23} - 27.6 \ .$$

The second term on the right hand side is negligible compared to the first term. Hence to an excellent approximation, for a very sharp Gaussian distribution,

$$\log g_T = \log g_{max} \ . \tag{2.47}$$

Many of the distributions we shall meet in statistical mechanics are very sharp Gaussian distributions, and the approximation given by equation (2.47) can be used. For such distributions, to calculate $\log g_T$ we can use the natural logarithm of the maximum value of $g(n)$.

2.5.4 Dependence of the number of accessible microstates and the entropy on U, V and N

It was illustrated in Figure 2.7(a), for the special case of a system of 3 particles, that the number of accessible microstates g_T increases rapidly as U, the energy of the system, is increased. In this example, the number of particles N and the values of the energy eigenvalues were constant. If the volume of a system is changed, the values of the energy eigenvalues change. For example, according to equation (2.5), for a particle in a cubical box the energy eigenvalues are proportional to $V^{-2/3}$. As the volume is increased the energy eigenvalues are decreased as illustrated in Figure 2.1. If, in the idealised example shown in Figure 2.2, the volume of the system is changed, (or the restoring force per unit displacement is changed in the case of a harmonic oscillator), such that all the energy eigenvalues go down to half their original values, with $U = 3$ energy units and $N = 3$ distinguishable particles of spin zero, the total number of accessible microstates is equal to the number of ways six half units of energy can be distributed between the 3 distinguishable particles. According to equation (2.15)

$$g_T = (N + U - 1)!/(N - 1)! \ U! = 8!/2! \ 6! = 28 \ .$$

The total number of accessible states increases from 10 to 28, when the volume is changed such that all the energy eigenvalues are reduced to a half of their original values.

To illustrate that g_T depends on N, assume that in the example illustrated in Figure 2.2, $U = 3$, but N is increased from 3 to 4 *distinguishable* particles of spin zero. In this case, according to equation (2.15)

$$g_T = (N + U - 1)!/(N - 1)! \ U! = 6!/3! \ 3! = 20 \ .$$

Thus increasing N from 3 to 4 increases g_T from 10 to 20.

The above idealised examples illustrate that, in general, the number of accessible microstates (the degeneracy or statistical weight of the macrostate) is a function of U, V and N. Since in statistical mechanics the entropy S is defined in terms of $\log g_T$ by equation (2.29), the entropy S can also be expressed as a function of U, V and N.

REFERENCES

[1] Feynman, R. P., Leighton, R. D. and Sands, M. *The Feynman Lectures on Physics*, Vol. 1, 1963, page 40–9, Addison-Wesley.
[2] Pauling, L. and Wilson, E. B., *Introduction to Quantum Mechanics*, 1935, page 67, McGraw Hill.
[3] Eisberg, R. M., *Fundamentals of Modern Physics*, 1964, page 360, Wiley.
[4] French, A. P. and Taylor, E. F., *An Introduction to Quantum Physics*, 1979, page 557, Thomas Nelson.

PROBLEMS

Problem 2.1

Consider a particle of mass 6.65×10^{-27} kg in a cubical box of volume 1 m³. Use equation (2.5) to determine the energy eigenvalues of the states up to $n_x = n_y = n_z = 2$. What is the degeneracy of each energy level? Plot the results on an energy level diagram of the type shown in Figure 2.1. What would be the difference in the values of the energy eigenvalues if the volume were (a) 0.125 m³ (b) 8 m³? (Planck's constant $h = 6.626 \times 10^{-34}$ J s.)

Problem 2.2

Consider again the problem illustrated in Figure 2.2. The system has non-degenerate single particle energy levels of energy 0, 1, 2, 3, 4 . . . energy units. The system consists of three distinguishable particles of spin zero. The total energy is 4 units. List the accessible microstates.

Repeat the calculation, if (i) the system consists of three identical bosons of spin zero (ii) the system consists of three identical fermions of spin $\frac{1}{2}$. (Allow for the spin degeneracy.)

Problem 2.3

Consider a system in which the non-degenerate single particle energy levels have energies 0, 1, 2, 3 . . . energy units. Assume that the system has 4 distinguishable particles of spin zero. The total energy is 3 units. List the accessible microstates. Repeat the calculation (i) for 4 identical bosons of spin zero, and (ii) 4 identical fermions of spin $\frac{1}{2}$. (Allow for the spin degeneracy.)

Problem 2.4.

Assume that the single particle energy levels of a system are non-degenerate and have energies 0, 1, 2, 3, 4 . . . energy units. A total of 8 energy units is shared among 6 particles.

(i) Assume that the particles are non-interacting distinguishable particles of spin zero. Show that there are 20 configurations and 1287 accessible microstates.

(ii) Assume that the particles are non-interacting identical bosons of spin zero. Determine the number of accessible microstates.

(iii) Assume that the particles are non-interacting identical fermions of spin $\frac{1}{2}$. Determine the number of accessible microstates (Allow for the spin degeneracy.)

Problem 2.5
Assume that in the problem shown in Figure 2.3, the single particle energy level of subsystem 2 which has an energy of 4 units is two fold degenerate. Show that the total number of microstates accessible to the composite closed system is increased from 84 to 96 and the ensemble average energy of subsystem 2 is increased from 3.0 to 3.21 energy units.

Problem 2.6
Assume that the red particle in subsystem 1 in Figure 2.3(c) goes from subsystem 1 to subsystem 2. Calculate the total number of microstates accessible to the composite closed system and the ensemble average energy of subsystem 2 at thermal equilibrium.

Problem 2.7
Isolated system A has non-degenerate single particle quantum energy levels of energy 0, 2, 4, 6, 8, 10 energy units and consists of one red particle of spin zero having an energy of 6 energy units. Isolated system B has non-degenerate single particle quantum energy levels of energy 0, 1, 2, 3 ... 10 energy units, and consists of two non-interacting particles of spin zero, coloured white and blue respectively. The total energy of the white and the blue particles is 2 energy units. List the microstates accessible to each isolated system. The two systems are placed in thermal contact. What is the number of microstates accessible to the composite system? What is the increase in entropy? Calculate the ensemble averages of the energies of both systems after they are placed in thermal contact.

Repeat the calculations for the case when the particles are identical bosons of spin zero.

Problem 2.8
Consider two idealised systems labelled 1 and 2 which have equally spaced non-degenerate single particle energy levels having energies 0, 1, 2, 3, 4, ... ∞ energy units. Each system consists of three distinguishable particles of spin zero. The two systems are placed in thermal contact. The total energy of the composite system is 20 energy units. Calculate the probability $P_2(U_2)$ that at thermal equilibrium system 2 has energy U_2 for values of U_2 from 0 to 20 energy units. Calculate the ensemble average of U_2 and the standard deviation from this mean value. Plot $\log P_2(U_2)$ against U_2 and show that $\log P_2(U_2)$ is a maximum when $P_2(U_2)$ is a maximum.

Problem 2.9
Consider two idealised systems which have equally spaced non-degenerate energy levels having energies 0, 1, 2, 3, 4 ... energy units. System 1 consists of three distinguishable particles of spin zero, and system 2 consists of 5 distinguishable particles of spin zero. The systems are placed in thermal contact and allowed to reach thermal equilibrium. The total energy is 20 energy units. Plot $P_2(U_2)$, $\log(g_2(U_2))$, $\log(g_1(20 - U_2))$, $\log(g_1 g_2)$ and $\log(P_2(U_2))$ against U_2, where $g_1(U_1)$ and $g_2(U_2)$ are the number of microstates accessible to systems 1 and 2 when they have energies U_1 and U_2 respectively, and $P_2(U_2)$ is the probability that system 2 has energy U_2 at thermal equilibrium. Illustrate how $\log(P_2(U_2))$ has its maximum value when $P_2(U_2)$ has its maximum value.

Problem 2.10
A box containing an ideal gas in internal thermodynamic equilibrium is divided conceptually into two equal parts, A and B. How many gas molecules must the box contain if the relative fluctuations in the number of molecules in part A of the box are to be (a) about 1%? (b) about 1 in 10^6? Calculate in each case the probability that all the molecules are found in part A of the box.

Problem 2.11
Show that if $y \gg 1$,

$$\int_y^\infty e^{-x^2}\, dx \simeq \frac{e^{-y^2}}{2y}$$

(*Hint:* Use $e^{-x^2} = (-1/2x)\, d(e^{-x^2})$ and continue to integrate by parts.)
Use this result to show that the probability of getting a deviation from the mean of more than (a) $\pm 10\sigma$ and (b) $\pm 100\sigma$ for a Gaussian distribution is (a) 1.54×10^{-23} (b) 2.69×10^{-2174}, where σ is the standard deviation.

Problem 2.12
A typist has a typewriter with 26 keys, one for each letter of the English alphabet. The typist types 26 letters completely at random and then starts again doing another complete set in 30 seconds. How long on average would it be between cases when the typist happened to type the 26 letters of the alphabet in their correct sequence?

Problem 2.13
For the purpose of this problem only, assume that the number of molecules in a mole of a gas at STP is $\exp(6 \times 10^{23})$ rather than 6×10^{23}. Assume that the volume of the mole of gas at STP is still $0.0224\ m^3$. Calculate the fractional change in the natural logarithm of the number of particles in the

hypothetical gas, if the volume of the gas system were changed from a sphere of diameter 10^{-10} m (the size of an atom) to a sphere of diameter 10^5 light years (the size of the galaxy). The gas remains at STP.

Problem 2.14
For large N and n, the maximum value of $g(n)$, given by equation (2.32), is

$$g_{max} = N!/n!\,(N-n)! = N!/(N/2)!(N/2)! \ .$$

Use Stirling's approximation, equation (A1.19) of Appendix 1, to show that for large N, to an excellent approximation we have

$$\log g_{max} = N \log 2 \ .$$

Use a calculator to show that for $N = 50$, $\log g_{max} = 32.47$ whereas $N \log 2 = 34.66$.

Problem 2.15
Consider an idealised system in which the single particle energy levels are non-degenerate and have the integer values 0, 1, 2, ... energy units. The system consists of 10 distinguishable particles of spin zero. Calculate the number of accessible microstates g_T, if the total energy is 0, 1, 10, 10^2, 10^3, 10^4 and 10^5 energy units. Plot the variation of $\log g_T$ against U for U in the range 0 to 10^3 energy units.

Problem 2.16
Use equations (1.13) and (2.29) to calculate by what factor the number of accessible microstates increases if 1 J of heat is added to a large system, kept at constant volume if the temperature of the macroscopic system is (a) 0.1K, (b) 1K, (c) 10K, (d) 100K, (e) 1000K. Comment on how the factor varies with the temperature of the macroscopic system. (Boltzmann's constant $k = 1.3807 \times 10^{-23}$ J K^{-1}.)

Chapter 3

The Interpretation of Classical Equilibrium Thermodynamics Using Statistical Mechanics

3.1 LIST OF POSTULATES

In this Chapter the ideas of statistical mechanics, introduced in Chapter 2, will be used to develop the laws of classical equilibrium thermodynamics. Our main postulates in this Chapter will be:

Postulate 1: The existence of microstates

A measurement of an N particle system will show that the system is in one of the quantum eigenstates (microstates) of the N particle system. In principle, these microstates could be determined by solving Schrödinger's time independent equation. According to quantum mechanics, it is only for certain values of the energy that Schrödinger's time independent equation for an N particle system, of fixed total volume, can have solutions which satisfy the boundary conditions. These values of energy are the energy eigenvalues. It is sometimes possible to have more than one independent solution of Schrödinger's equation for a given value of energy. The number of independent quantum eigenstates (microstates) of the N particle system associated with a given value of energy is the degeneracy (or the statistical weight) of that energy level. In the text, we shall generally refer to the degeneracy (statistical weight), when the N particle system has a specified value of total energy, as the **number of accessible microstates**. It was shown in Section 2.5.4 that the number of accessible microstates $g(U, V, N)$ is a function of the energy U, the volume V and the number of particles N in the system. It was shown in Section 2.5.2 that, for macroscopic system, g is an extremely large number. The variation of $\log g$, the natural logarithm of g to the base e, with energy for a normal thermodynamic system is shown in

Figure 2.7(b).† In Sections 3.2–3.7 we shall use a simplified model in which it is assumed that the N particle system has discrete energy levels each of which has a very large degeneracy g. In practice, due to the interactions between the particles in the system, these large degeneracies are generally split, so that it is more realistic to use the density of states D in the way outlined later in Section 3.8*.

Postulate 2: The principle of equal *a priori* probabilities

According to the principle of equal *a priori* probabilities, if a measurement is made on a **closed system** which is in internal thermodynamic equilibrium, we are equally likely to find the closed system in any one of the microstates accessible to the closed system. A closed system is an *isolated* system of constant total energy, constant total volume and constant total number of particles. Since a closed isolated system is not under any external influences whatsoever, and its total energy, total volume and total number of particles are all constant, there is no reason whatsoever to believe that, at thermodynamic equilibrium, any particular one of the microstates accessible to the closed isolated system is any more probable than any of the other accessible microstates. Hence it seems reasonable, at least as a starting point, to assume that when a *closed* isolated system has had time to reach internal thermodynamic equilibrium, we are equally likely to find the closed isolated system in any one of the microstates accessible to the closed isolated system. The principle of equal *a priori* probabilities will be checked *a postereori*, that is it will be shown that predictions made on the assumption that the principle of equal *a priori* probabilities is correct are in agreement with the experimental results.

It will be shown in Section 3.10 that, as an alternative postulate, one could assume that, when a closed system is in internal thermodynamic equilibrium, the information we have about the system is a minimum. (Alternatively, one could assume that the uncertainty is a maximum.) It will be shown in Section 3.10 that for a closed system this postulate leads to the principle of equal *a priori* probabilities. It is a matter of taste which form of postulate 2 the reader prefers to adopt from Section 3.10 onwards.

Postulate 3: The law of conservation of energy

Since the law of conservation of energy is being taken as axiomatic, we are in effect assuming the main part of the first law of classical thermodynamics. From time to time we shall bear in mind the uncertainty principle, according to which, if the lifetime of a state is Δt, the indeterminancy ΔE in the energy of the state is of the order of $\hbar/\Delta t$, where $\hbar = 1.055 \times 10^{-34}$ J s.

---

†The natural logarithm of x to the base e is generally denoted $\log_e x$ or $\ln x$. As we shall only use logarithms to the base e in the text, it will simplify the expressions to denote the logarithm of x to the base e by $\log x$.

Postulate 4: The equilibrium values of the macroscopic variables of classical equilibrium thermodynamics are given by the ensemble averages of the corresponding quantities in our microscopic approach

This postulate relates our microscopic theory to the macroscopic variables of classical equilibrium thermodynamics. It is not necessary to raise the above statement to the status of a postulate. It will be shown during our development of the interpretation of classical equilibrium thermodynamics in this Chapter, that due to the sharpness of the probability distributions for macroscopic systems, it is sensible to assume that the equilibrium values of classical equilibrium thermodynamics are given by the ensemble averages of the corresponding microscopic quantities. It will also be shown that, for macroscopic systems, the ensemble averages are given to an excellent approximation by the most probable values. In practice it is often simpler to estimate the most probable value than the ensemble average.

3.2 THERMAL INTERACTION

Consider two subsystems, labelled 1 and 2, surrounded by fixed, rigid, impermeable, adiabatic walls as shown in Figure 3.1(a). The two subsys-

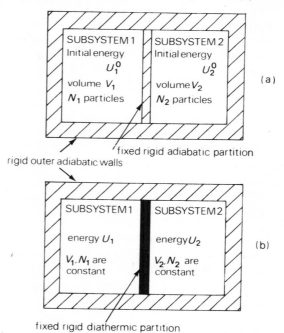

Figure 3.1—(a) Idealised example of two subsystems inside rigid, outer adiabatic walls. The two subsystems are separated by a fixed, rigid, adiabatic partition. (b) The adiabatic partition is replaced by a fixed, rigid, diathermic partition, which allows heat energy to flow from one subsystem to the other.

tems make up a composite *closed* isolated system of fixed total energy, fixed total volume and fixed total number of particles. In Figure 3.1(a), the two subsystems are separated by a fixed, rigid, impermeable, *adiabatic* partition. This partition is an internal constraint of the composite system, which keeps the number of particles N_1 and N_2 in subsystems 1 and 2 and the volumes V_1 and V_2 of subsystems 1 and 2 constant. The adiabatic partition prevents heat from flowing from one subsystem to the other, so that the energies U_1^0 and U_2^0 of subsystems 1 and 2 respectively are constant in Figure 3.1(a). The total energy U^* of the composite system is

$$U^* = U_1^0 + U_2^0 \ . \tag{3.1}$$

A simplified model will be used initially, in which it will be assumed that each subsystem has discrete energy levels. It will be assumed that the degeneracy of each energy level is extremely large, being typically of the order of e^N, where N is the number of particles in the subsystem. (Reference: Section 2.5.2.) The more general case when the degeneracies of the levels may be split due to the interactions between the particles will be considered later in Section 3.8*.

Let the number of microstates accessible to subsystem 1, when it has energy U_1^0, be denoted by $g_1(U_1^0)$, and let the number of microstates accessible to subsystem 2 when it has energy U_2^0 be denoted by $g_2(U_2^0)$. Since subsystem 2 can be in any one of its $g_2(U_2^0)$ accessible microstates, when subsystem 1 is in any one of its $g_1(U_1^0)$ accessible microstates, the total number of microstates accessible to the composite system, for the conditions shown in Figure 3.1(a), is

$$g_T^0 = g_2(U_2^0)g_1(U_1^0) \ . \tag{3.2}$$

This is the same as equation (2.17), which was developed in Section 2.3 using a numerical example.

Let the partition in Figure 3.1(a) be changed to a fixed, rigid, impermeable, *diathermic* partition, as shown in Figure 3.1(b). This removes one of the internal constraints of the composite system, and heat can now flow from one subsystem to the other through the diathermic partition. The volumes V_1 and V_2 of the subsystems and the numbers of particles N_1 and N_2 in the subsystems are still constant. If U_1 and U_2 are the instantaneous values of the energies of subsystems 1 and 2 respectively, and if energy is conserved (Postulate 3), then

$$U_1 + U_2 = U^* \ . \tag{3.3}$$

Since the composite system of subsystem 1 in thermal equilibrium with subsystem 2 has fixed total energy, fixed total volume and consists of a fixed total number of particles, it is a *closed* system. [It is being assumed that the thermal capacity of the diathermic partition in Figure 3.1(b) is negligible.]

Let the number of microstates accessible to subsystem 2, when it has energy U_2, be denoted by $g_2(U_2)$, and let the number of microstates accessible to subsystem 1, when it has energy U_1, be denoted by $g_1(U_1)$. Since every one of the $g_2(U_2)$ microstates accessible to subsystem 2, when subsystem 2 has energy U_2, can be combined with every one of the $g_1(U_1)$ microstates accessible to subsystem 1, when subsystem 1 has energy U_1 equal to $(U^* - U_2)$, to form a different accessible microstate of the closed composite system shown in Figure 3.1(b), the total number of microstates accessible to the composite system, when subsystem 2 has energy U_2 and subsystem 1 has energy $(U^* - U_2)$, is $[g_2(U_2)g_1(U^* - U_2)]$. The total number of microstates g_T accessible to the composite system shown in Figure 3.1(b) is obtained by summing over all possible values of U_2. We have

$$g_T = \sum_{U_2} g_2(U_2)g_1(U^* - U_2) \ . \tag{3.4}$$

This equation is the same as equation (2.20).

According to the *principle of equal a priori probabilities* (Postulate 2), if the *closed* isolated composite system, made up of subsystem 1 in thermal contact with subsystem 2, is left long enough for subsystems 1 and 2 to reach thermal equilibrium, the *closed* composite system is equally likely to be found experimentally to be in any one of the g_T microstates accessible to the closed composite system. The probability $P_2(U_2)$ that, at thermal equilibrium, subsystem 2 has energy U_2 is given by the ratio of the number of microstates $[g_2(U_2)g_1(U^* - U_2)]$ accessible to the closed composite system in which subsystem 2 has energy U_2 to the total number g_T of microstates accessible to the closed composite system. Hence at thermal equilibrium, according to the principle of equal *a priori* probabilities, we should have

$$P_2(U_2) = g_2(U_2)g_1(U^* - U_2)/g_T = Cg_2(U_2)g_1(U^* - U_2) \ . \tag{3.5}$$

where $C = 1/g_T$ is a constant independent of U_2. [Equation (3.5) is the same as equation (2.21), which was developed in Section 2.3 using the simple numerical example illustrated in Figure 2.3(c).] In the case illustrated in Figure 3.1(b), V_1, V_2, N_1 and N_2 are all constant. Since U_1 is equal to $(U^* - U_2)$, $P_2(U_2)$ can be expressed as a function only of U_2. (The probability $P_1(U_1)$ that at thermal equilibrium subsystem 1 has energy U_1 equal to $(U^* - U_2)$ is equal to $P_2(U_2)$.)

For a macroscopic system, $g_2(U_2)$, the number of microstates accessible to subsystem 2 when subsystem 2 has energy U_2, increases very rapidly as U_2 increases, as shown in Figure 3.2(a). However, as U_2 increases, U_1 the energy of subsystem 1 decreases so that $g_1(U^* - U_2)$, the number of microstates accessible to subsystem 1 when it has energy U_1 equal to $(U^* - U_2)$, decreases rapidly as U_2 increases, as shown in Figure 3.2(a).

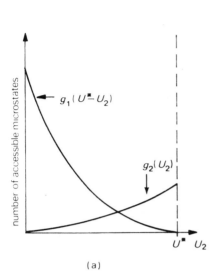

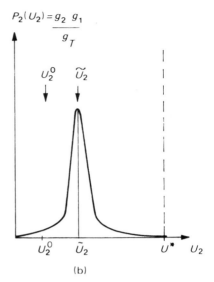

$$P_2(U_2) = g_2 \, g_1$$
$$\frac{}{g_T}$$

(a)   (b)

Figure 3.2—(a) The variation with U_2 of $g_2(U_2)$, the number of micro-states accessible to subsystem 2 in Figure 3.1(b), when it has energy U_2; $g_2(U_2)$ increases with U_2. The variation with U_2 of $g_1(U^* - U_2)$, the number of microstates accessible to subsystem 1 in Figure 3.1(b) when it has energy U_1 equal to $(U^* - U_2)$, is also shown. When U_2 increases, U_1 decreases so that $g_1(U^* - U_2)$ decreases. (b) The variation with U_2 of the probability $P_2(U_2) = g_1 g_2/g_T$ that, at thermal equilibrium, subsystem 2 in Figure 3.1(b) has energy U_2. The probability $P_2(U_2)$ is proportional to the product of the ordinates $g_2(U_2)$ and $g_1(U^* - U_2)$ at the value of U_2 in Figure 3.2(a). For macroscopic subsystems the probability distribution is extremely sharp.

The product $g_2(U_2)g_1(U^* - U_2)$ has an extremely sharp maximum at a particular value of U_2, denoted by $\tilde{U}_2$, as shown in Figure 3.2(b). According to equation (A4.5) of Appendix 4*, in the vicinity of $\tilde{U}_2$, the probability distribution $P_2(U_2)$ can be approximated by the Gaussian distribution

$$P_2(U_2) = g_1 g_2/g_T = \left[\frac{(g_1 g_2)_{\text{max}}}{g_T} \right] \exp(-x^2/2\sigma^2) \, , \qquad (3.6)$$

where x is equal to $(U_2 - \tilde{U}_2)$ and the standard deviation σ, defined by equation (A1.8) of Appendix 1, is given by equation (A4.7) of Appendix 4*. [An introduction to the standard deviation is given in Sections A1.4 and A1.5 of Appendix 1.] For a Gaussian distribution, the mean value $\bar{U}_2$ is equal to the most probable value $\tilde{U}_2$, which is the value of U_2 at which $P_2(U_2)$ is a maximum. We shall adopt the convention of labelling the smaller of the two subsystems in Figure 3.1(b) subsystem 2. It is shown in Appendix 4* that the ratio of the standard deviation σ_2 in the energy of the smaller subsystem 2 at thermal equilibrium to the mean value $\bar{U}_2$ of the energy of subsystem 2 at thermal equilibrium is of order $1/N_2^{1/2}$ or less. For example, if $N_2 = 10^{18}$, corresponding to a cube of matter of side $\sim 10^{-4}$ m and volume $\sim 10^{-12}$ m^3, $\sigma_2/\bar{U}_2$ is of the order of 10^{-9}. If the width of the Gaussian distribution in Figure 3.2(b) is represented by $2\sigma_2$, the distance from the origin in Figure 3.2(b) to the value of U_2 which is equal to $\bar{U}_2 = \tilde{U}_2$ is 5×10^8 times the width of the Gaussian distribution represented by $2\sigma_2$. This is an extremely sharp distribution. It is shown in Problem 2.11 that for a Gaussian distribution, the probability of a fluctuation from the mean exceeding $100\sigma_2$ is 2.69×10^{-2174}. It was shown in Section 2.4 that it is safe to conclude that a fluctuation as large as $100\sigma_2$ will not occur in a time equal to the age of the Earth. For a cube of matter of volume 10^{-12} m^3, $100\sigma_2/\bar{U}_2$ would be about 10^{-7}. For larger macroscopic systems $100\sigma_2/\bar{U}_2$ would be even smaller.

Since, according to the principle of equal *a priori* probabilities, at thermal equilibrium one is equally likely to find the closed composite system shown in Figure 3.1(b) in any one of the g_T microstates accessible to the composite system, the probability of obtaining any value of U_2 between 0 and U^* is finite. However, due to the extremely sharp distribution in Figure 3.2(b), for macroscopic subsystems with $N_2 > 10^{18}$ particles, the overwhelming probability is that every measurement of the energy of subsystem 2 will give a value of U_2 which is equal to $\bar{U}_2 = \tilde{U}_2$ to within at least 1 part in 10^7. Thus it is reasonable to assume that either $\bar{U}_2$, the ensemble average, or $\tilde{U}_2$, the most probable value of U_2, gives the equilibrium value of U_2 to be used in classical equilibrium thermodynamics. This is adopted as Postulate 4. In our development of the theory, we shall find it easier to find the value of $\tilde{U}_2$ at which $P_2(U_2)$ is a maximum, rather than calculate the ensemble average of U_2.

According to equation (3.5), the probability $P_2(U_2)$ that, at thermal equilibrium, subsystem 2 in Figure 3.1(b) has energy U_2 can be expressed as a function only of U_2. The condition that specifies the value of U_2, at which the probability distribution given by equation (3.5) is a maximum, for constant V_1, V_2, N_1 and N_2, is

$$(\partial P_2/\partial U_2) = 0 \ . \tag{3.7}$$

[An introduction to partial differentiation is given in Section A1.2 of Appendix 1.]

Differentiating equation (3.5), for constant V_1, V_2, N_1 and N_2, we have

$$\left(\frac{\partial P_2}{\partial U_2}\right) = C\left[g_1\frac{\partial g_2}{\partial U_2} + g_2\frac{\partial g_1}{\partial U_2}\right] \ . \tag{3.8}$$

Differentiating equation (3.3), for constant V_1, V_2, N_1 and N_2, we have

$$\left(\frac{\partial U_1}{\partial U_2}\right) = -1 \ .$$

Hence, for constant V_1, V_2, N_1 and N_2, we have

$$\left(\frac{\partial g_1}{\partial U_2}\right) = \left(\frac{\partial g_1}{\partial U_1}\right)\left(\frac{\partial U_1}{\partial U_2}\right) = -\left(\frac{\partial g_1}{\partial U_1}\right) \ . \tag{3.9}$$

Substituting in equation (3.8), we have

$$\left(\frac{\partial P_2}{\partial U_2}\right) = C\left[g_1\left(\frac{\partial g_2}{\partial U_2}\right)_{V_2,N_2} - g_2\left(\frac{\partial g_1}{\partial U_1}\right)_{V_1,N_1}\right] \ . \tag{3.10}$$

The value of U_2 at which P_2 is a maximum is when $\partial P_2/\partial U_2$ is zero. This is given by the condition

$$g_1\left(\frac{\partial g_2}{\partial U_2}\right)_{V_2,N_2} = g_2\left(\frac{\partial g_1}{\partial U_1}\right)_{V_1,N_1} \ . \tag{3.11}$$

Dividing by $g_1 g_2$, the condition specifying the value of U_2, at which $P_2(U_2)$ is a maximum, becomes

$$\frac{1}{g_1}\left(\frac{\partial g_1}{\partial U_1}\right)_{V_1,N_1} = \frac{1}{g_2}\left(\frac{\partial g_2}{\partial U_2}\right)_{V_2,N_2} \ . \tag{3.12}$$

The reader can check by differentiating that equation (3.12) can be re-written in the form

$$\left(\frac{\partial \log g_1}{\partial U_1}\right)_{V_1,N_1} = \left(\frac{\partial \log g_2}{\partial U_2}\right)_{V_2,N_2} \ . \tag{3.13}$$

It is conventional to introduce a parameter β, which is defined in the general case for any system which has $g(U, V, N)$ accessible microstates, when the system has energy U, volume V and consists of N particles, by the relation

$$\beta = \left(\frac{\partial \log g}{\partial U}\right)_{V,N} = \frac{1}{g}\left(\frac{\partial g}{\partial U}\right)_{V,N} \ . \tag{3.14}$$

Since the number of accessible microstates g is a number, β has the dimensions of the reciprocal of energy. Using equation (3.14), equation (3.13), which specifies the condition when $U_2 = \tilde{U}_2$, becomes

$$\beta_1 = \beta_2 \ . \tag{3.15}$$

If we start with a division of energy between subsystems 1 and 2 in Figure 3.1(b) such that subsystem 2 has an initial energy U_2^0 significantly less than $\tilde{U}_2$, as shown in Figure 3.2(b), the overwhelming probability is that heat will flow through the diathermic partition in Figure 3.1(b) from subsystem 1 to subsystem 2 until at thermal equilibrium U_2 is very close to $\tilde{U}_2 = \bar{U}_2$, and β_1 is very close to β_2.

3.3 THE ZEROTH LAW OF THERMODYNAMICS, ABSOLUTE TEMPERATURE AND THE DEFINITION OF ENTROPY

If system 1 in Figure 3.1(b) had energy $\tilde{U}_1$ and system 2 had energy $\tilde{U}_2$ before they were placed in thermal contact, the systems would start with the most probable division of energy and, on average, no heat would flow from one system to the other by thermal conduction through the diathermic partition when they were placed in thermal contact. For macroscopic systems there would be small fluctuations in the energies of the two systems at thermal equilibrium, but the energies of the systems would always be extremely close to the most probable values, and, in the context of classical equilibrium thermodynamics, one would say that they were in thermal equilibrium immediately they were placed in thermal contact. (Actually, in classical equilibrium thermodynamics the expression that 'two systems are in thermal equilibrium' is often applied to two isolated systems which are in such thermodynamic states that, if they were connected through a diathermic partition they would be in thermal equilibrium immediately.) It follows from equation (3.15) that, for U_2 to be equal to $\tilde{U}_2$ and for U_1 to be equal to $\tilde{U}_1$ before the two systems were placed in thermal contact, systems 1 and 2 would have to have been in such states before they are placed in thermal contact that,

$$\beta_2 = \beta_1 , \tag{3.15}$$

where β is defined by Equation (3.14). Hence, if systems 1 and 2 had equal values of β before they were placed in thermal contact, they would be in thermal equilibrium immediately after they were placed in thermal contact.

Consider a third macroscopic system, labelled system 3. If, on average, no energy was transferred from one system to the other when systems 1 and 3 were placed in thermal contact, the initial division of energy between systems 1 and 3 was the most probable one and systems 1 and 3 would be in thermal equilibrium immediately, when placed in thermal contact. The condition which specifies the most probable division of energy between systems 1 and 3, when they are in thermal contact and in thermal equilibrium, is

$$\beta_1 = \beta_3 . \tag{3.16}$$

Hence, if systems 1 and 3 were in such thermodynamic states before they were placed in thermal contact that they had the same value of β, they would be in thermal equilibrium immediately after they were placed in thermal contact.

Comparing equations (3.15) and (3.16), we see that

$$\beta_1 = \beta_2 = \beta_3 . \tag{3.17}$$

Hence, if systems 2 and 3 were placed in thermal contact, since they would start with the same value of β, they would start with the most probable division of energy. On average no energy would be transferred from one system to the other, when systems 2 and 3 were placed in thermal contact, and they would be in thermal equilibrium immediately, when placed in thermal contact. It follows from equation (3.17) that, if two systems are in thermal equilibrium with a third system, they are in thermal equilibrium with each other. This is the **zeroth law of thermodynamics**.

The parameter β, defined by equation (3.14), could be used as a measure of whether or not thermodynamic systems would be in thermal equilibrium, if they were placed in thermal contact. It will be shown in Section 3.5 that, on average heat flows from a system having a low value of β to one having a higher value of β. According to the Clausius statement of the second law of classical thermodynamics, heat should flow from the hotter to the colder system. To be in accord with classical thermodynamics, we shall introduce the **absolute temperature** T, defined by the relation

$$\beta = 1/kT; \qquad T = 1/k\beta . \tag{3.18}$$

Notice that the absolute temperature is inversely proportional to β. The positive constant of proportionality k is called Boltzmann's constant. Using equations (3.14) and (3.18), we have

$$T = \frac{g}{k\,(\partial g/\partial U)_{V,N}} = \frac{1}{k}\left(\frac{\partial U}{\partial \log g}\right)_{V,N} . \tag{3.19}$$

The numerical value of the absolute temperature of a body depends on the particular temperature scale chosen. For the Kelvin scale, discussed in Section 1.4, the value of k in equation (3.19) must be adjusted such that, if the temperature of a system in thermal equilibrium with water at its triple point is calculated from equation (3.19) using the 'measured' value of $(\partial \log g/\partial U)_{V,N}$, the value of the absolute temperature will be exactly 273.16. It will be shown in Section 5.9.5 that Boltzmann's constant k is equal to the ratio of the molar gas constant R of an ideal gas to Avogadro's constant N_A. It has the experimental value of

$$k = 1.3807 \times 10^{-23} \text{ J K}^{-1} .$$

The value of temperature assigned to the triple point of water, and the unit of energy could be changed by international agreement. This would change the numerical value of k.

Using equation (3.18), equation (3.15) can be rewritten in the form

$$T_1 = T_2. \tag{3.20}$$

Thus at thermal equilibrium, when on average no energy is being transferred from one subsystem to the other in the example shown in Figure 3.1(b), the two subsystems should have the same value of absolute temperature, estimated using equation (3.19).

In the case of three systems in thermal equilibrium, it follows from equations (3.17) and (3.18) that

$$T_1 = T_2 = T_3 \tag{3.21}$$

corresponding to

$$\frac{1}{k}\left(\frac{\partial U_1}{\partial \log g_1}\right)_{V_1,N_1} = \frac{1}{k}\left(\frac{\partial U_2}{\partial \log g_2}\right)_{V_2,N_2} = \frac{1}{k}\left(\frac{\partial U_3}{\partial \log g_3}\right)_{V_3,N_3}.$$

System 3 could be used as a thermometer to measure the temperatures of systems 1 and 2. If the variation of $\log g_3$ with U_3 were determined, at constant V_3 and constant N_3, then, according to equations (3.21) and (3.19), when systems 1, 2 and 3 were in thermal equilibrium,

$$T_1 = T_2 = T_3 = \frac{1}{k}\left(\frac{\partial U_3}{\partial \log g_3}\right)_{V_3,N_3}.$$

This method would give an **absolute scale**, since *whatever* system 3 were composed of, the equilibrium temperature determined from the variation of $\log g_3$ with U_3 should give the same value for $T_1 = T_3$ as would be determined from the variation of $\log g_1$ with U_1 at thermal equilibrium using the equation

$$T_1 = \frac{1}{k}\left(\frac{\partial U_1}{\partial \log g_1}\right)_{V_1,N_1}.$$

It is not convenient in practice to determine the asbolute temperature of a system by trying to measure the variation of $\log g$ with U. It will be shown in Section 5.9.5 that the absolute temperature, defined by equation (3.19), is identical to the ideal gas scale. It will be shown in Section 7.4* that the thermodynamic scale defined in terms of the heat taken in and given out by the working substance in a Carnot engine is also the same as the absolute temperature scale defined by equation (3.19). Most determinations of absolute temperatures are based either directly on gas thermometers, corrected to the ideal gas scale by extrapolating to zero pressure as outlined in

Section 1.4, or using secondary thermometers which have been calibrated using ideal gas thermometers. At very low temperatures, the thermodynamic temperature is sometimes estimated using a thermodynamic relation which contains the thermodynamic temperature.

In our microscopic interpretation of classical thermodynamics, the important quantity is the total number of accessible microstates g. It was illustrated in Section 2.5.2 that for a macroscopic system, g is such a fantastically large number that it is more convenient to work with $\log g$. To correspond to the quantities used in classical equilibrium thermodynamics, the entropy S will be used as an *abbreviation* for $k \log g$, so that in our approach to statistical mechanics the entropy is *defined* by the equation

$$S = k \log g \ . \qquad (3.22)$$

In our microscopic approach, the entropy is a logarithmic measure of the total number of accessible microstates. It was shown in Section 2.5.4 that g is a function of the internal energy U, the volume V and the number of particles N in the system. Hence the entropy S of a system is also a function of U, V and N. Using equation (3.22), equation (3.19) can be rewritten in the form

$$T = (\partial U/\partial S)_{V,N} \ . \qquad (3.23)$$

This is the same as equation (1.19) developed in Section 1.8 from classical equilibrium thermodynamics.

The parameter β, which is defined by equation (3.14), is generally called the **temperature parameter**. It is an extremely useful abbreviation to use instead of $1/kT$ and it will be used extensively from Chapter 4 onwards. The reader should be thoroughly familiar with the relation $\beta = 1/kT$ and should always realise that, whenever β appears in an equation, it stands for $1/kT$.

The variation of $\log g = S/k$ with energy U for a typical thermodynamic system is shown in Figure 3.3. The precise variation of $\log g$ with U depends on the quantum mechanical properties of the system. A simple case was illustrated previously in Figure 2.7(b) and discussed in Section 2.5.1. A variation of S/k with U of the type shown in Figure 3.3 was predicted in Section 1.11 on the basis of classical equilibrium thermodynamics. When U, the energy of the system, is close to the ground state energy U_0, the slope $\beta = (\partial \log g/\partial U)_{V,N}$ tends to infinity. As the energy U increases, the slope β decreases. Since β is equal to $1/kT$, the absolute temperature tends to zero as U tends to U_0. Since β decreases as U increases, the absolute temperature T increases as U increases, so that $\partial U/\partial T$ is positive. Hence the molar heat capacity at constant volume, defined in Section 1.10 by the equation

$$C_V = (\partial U/\partial T)_V \ . \qquad (1.28)$$

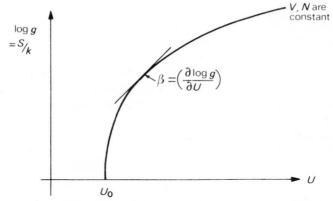

Figure 3.3—Sketch of the variation of $\log g = S/k$ with the (internal) energy U for a normal thermodynamic system. The slope of the curve is equal to the temperature parameter β.

must be positive for a normal thermodynamic system, for which the variation of $\log g$ with U is of the form illustrated in Figure 3.3. [It is shown in Section A4.2 of Appendix 4*, that, if the thermal equilibrium between subsystems 1 and 2 in Figure 3.1(b) is to be stable, C_V must be positive.]

There is a class of system, of which a spin system in a magentic field is one, which has an upper limit to the energy the system can have. The variation of S with U in these cases has a maximum at a particular value of U and for higher energies $(\partial U/\partial S)_{V,N}$ is negative and corresponds to negative absolute temperatures. Negative temperatures will be discussed in Chapter 8. Apart from Chapter 8, we shall only consider normal thermodynamic systems at positive absolute temperatures, for which the variation of $\log g$ with U is as shown in Figure 3.3.

3.4 THE FIRST LAW OF THERMODYNAMICS

In classical equilibrium thermodynamics, in addition to incorporating the law of conservation of energy, the first law of thermodynamics leads to the concept of the internal energy U, which is assumed to have a precise value, and is a function of the thermodynamic macrostate of the system. The law of conservation of energy is being taken as axiomatic in our microscopic approach. (It is our Postulate 3.) When subsystem 2 is in thermal equilibrium with subsystem 1 in Figure 3.1(b), subsystem 2 can be in any one of a very large number of its microstates and these can have different energy eigenvalues. The energy eigenvalue of each microstate of subsystem 2 is equal to the sum of the kinetic energies and the potential energies of all the particles in subsystem 2, when subsystem 2 is in that microstate.

According to equation (2.27) the ensemble average energy of subsystem 2 in Figure 3.1(b), when subsystem 2 is in thermal equilibrium with subsystem 1, is

$$\bar{U}_2 = \sum_{\text{(states)}} P_{2i}\varepsilon_{2i} \tag{3.24}$$

where P_{2i}, the probability that at thermal equilibrium subsystem 2 is in its ith microstate having energy eigenvalue ε_{2i}, is given by equation (2.26).

It was shown in Appendix 4* and Section 3.2 that, when subsystems 1 and 2 are in thermal equilibrium in Figure 3.1(b), the probability $P_2(U_2)$ that subsystem 2 has energy U_2 is a sharp Gaussian distribution of the type illustrated in Figure 3.2(b). It was shown in Section 3.2 that the ratio of the standard deviation σ_2 in the energy of subsystem 2 to its mean energy $\bar{U}_2$ is of order $1/N_2^{1/2}$. It was pointed out in Section 1.3 that, if surface energy effects are to be negligible so that the internal energy U of classical equilibrium thermodynamics can be treated as an extensive variable, then one must take the thermodynamic limit. In the thermodynamic limit, the number of particles N and the volume V both tend to infinity, but the ratio N/V is kept constant. In the thermodynamic limit, N_2 would tend to infinity, so that $\sigma_2/\bar{U}_2$ would tend to zero and the energy per unit volume at thermal equilibrium would tend to a definite value in the thermodynamic limit. For finite macroscopic systems, our microscopic approach based on statistical mechanics gives both the mean value of the energy of subsystem 2 (which is equal to the most probable value), and the magnitude of the fluctuations in the energy of subsystem 2 that would be expected in practice, when subsystem 2 is in thermal equilibrium with subsystem 1 in Figure 3.1(b). For example, for $N_2 = 10^{18}$, $\sigma_2/\bar{U}_2$ is of the order of 10^{-9}.

According to the first law of classical thermodynamics the increase in the internal energy U is equal to the sum of the heat supplied to the system and the work done on the system. We shall return to discuss heat and work in Chapter 7*, and equation (1.6) will be interpreted in Section 7.2.2*.

3.5 THE SECOND LAW OF THERMODYNAMICS: THE DIRECTION OF HEAT FLOW

Let the initial division of energy between subsystems 1 and 2 in Figure 3.1(a) be such that U_2^0, the initial energy of subsystem 2 is significantly less than $\tilde{U}_2$, as shown in Figure 3.2(b). The energy $\tilde{U}_2$ is the most probable value of U_2, when subsystems 1 and 2 are in thermal equilibrium in Figure 3.1(b). It was shown in Section 3.2 that for macroscopic systems (for which $N_2 > 10^{18}$ atoms) the overwhelming probability is that, at thermal equilibrium, U_2, the instantaneous value of the energy of subsystem 2, is equal to $\tilde{U}_2$, at least to within 1 part in 10^7. On average, if $U_2^0 < \tilde{U}_2$, in the transition from the initial state to the final equilibrium state, energy (heat) flows from subsystem 1 to subsystem 2. It was shown in Section 3.3 that for normal thermodynamic systems the heat capacity at constant volume C_V is positive, where according to equation (1.28)

$$C_V = (\partial U/\partial T)_V \ . \tag{1.28}$$

According to equation (1.28), since the heat capacity of subsystem 2 is positive, as the energy of subsystem 2 increases from $U_2{}^0$ to a value close to $\tilde{U}_2$, the temperature of subsystem 2 must increase until at thermal equilibrium, according to equation (3.20), the temperature of subsystem 2 is equal to the final equilibrium temperature of subsystem 1. On the other hand, as subsystem 1 loses energy, its temperature drops until it is equal to the final increased temperature of subsystem 2. Hence during the transition from the initial state, in which $U_2{}^0 < \tilde{U}_2$, to the final state of thermal equilibrium, the temperature of subsystem 1 must be greater than the temperature of subsystem 2. Since in the approach to thermal equilibrium heat flows, on average, from subsystem 1 to subsystem 2, heat flows on average from the subsystem at the higher absolute temperature to the subsystem at the lower absolute temperature. This is the Clausius statement of the second law of thermodynamics. Since the temperature parameter β is equal to $1/kT$, on average heat flows from the subsystem having the lower value of β to the subsystem having the higher value of β.

In classical equilibrium thermodynamics the concept of entropy is developed from the second law of thermodynamics in the way outlined in Chapter 1. In our statistical approach to thermodynamics, it is more convenient to go directly to the entropy approach to thermodynamics than to repeat the arguments of Section 1.7.

3.6 THE SECOND LAW OF THERMODYNAMICS: THE ENTROPY APPROACH

3.6.1 Principle of the increase of entropy

Consider again the example shown in Figure 3.1. Initially, the partition between subsystems 1 and 2 in Figure 3.1(a) is an adiabatic partition. The energies of subsystems 1 and 2 in Figure 3.1(a) are $U_1{}^0$ and $U_2{}^0$ respectively. According to equation (3.2) the number of microstates accessible to the composite system in Figure 3.1(a), when the adiabatic partition is in position, is

$$g_0 = g_2(U_2{}^0)g_1(U_1{}^0) \tag{3.25}$$

where $\qquad U_1{}^0 + U_2{}^0 = U^* \ \ .$

According to equation (3.22), the total entropy of the composite system shown in Figure 3.1(a) is

$$S_i = k \log g_0 = k \log g_1(U_1{}^0) + k \log g_2(U_2{}^0)$$

or
$$S_i = S_1(U_1{}^0) + S_2(U_2{}^0) \qquad (3.26)$$

where $S_1(U_1{}^0)$ and $S_2(U_2{}^0)$ are the entropies of subsystems 1 and 2 respectively, which with the rigid adiabatic partition and the rigid adiabatic outer walls can be treated as two isolated systems having energies $U_1{}^0$ and $U_2{}^0$ respectively.

When the adiabatic partition is replaced by a diathermic partition, as shown in Figure 3.1(b), one of the internal constraints of the composite system is removed and heat can flow from one subsystem to the other. According to the principle of equal *a priori* probabilities (Postulate 2), when the composite closed system shown in Figure 3.1(b) has had time to reach thermal equilibrium, it is equally likely to be found in any one of the g_T microstates accessible to the composite closed system. According to equation (3.4)

$$g_T = \Sigma g_2(U_2) g_1(U_1) . \qquad (3.27)$$

The summation in equation (3.27) is over all possible values of U_1 and U_2 consistent with the law of conservation of energy, namely

$$U_1 + U_2 = U^* .$$

Since the right hand side of equation (3.25) is included in equation (3.27) for one of the possible values of U_2, namely for $U_2 = U_2{}^0$, it is clear that g_T must be greater than g_0. (In the simple example discussed previously in Section 2.3.1. and illustrated in Figure 2.3, $g_0 = 12$ and $g_T = 84$.) When an internal constraint is removed, the number of microstates accessible to the composite system cannot decrease, since all the original possibilities are still there. Hence g_T must always be greater than or equal to g_0, so that

$$k \log g_T \geqslant k \log g_0 .$$

Using equation (3.22), $\qquad S_f \geqslant S_i \qquad (3.28)$

where $\qquad\qquad S_f = k \log g_T \qquad (3.29)$

is the total entropy of the *composite system* in Figure 3.1(b), when subsystems 1 and 2 are in thermal equilibrium.

An excellent approximation for the value of the entropy S_f of the *composite system* shown in Figure 3.1(b), when subsystems 1 and 2 are in thermal equilibrium, is developed in Section A4.3[*] of Appendix 4. (A similar introductory example was given in Section 2.5.3.) According to equation (A4.16) of Appendix 4[*], to an excellent approximation, we have

$$\log g_T = \log g_1(\tilde{U}_1) + \log g_2(\tilde{U}_2) .$$

Using equation (3.29), to an excellent approximation we have

$$S_f = k \log g_1(\tilde{U}_1) + k \log g_2(\tilde{U}_2) \ . \tag{3.30}$$

Using equation (3.22), equation (3.30) can be rewritten in the form

$$S_f = S_1(\tilde{U}_1) + S_2(\tilde{U}_2) \ . \tag{3.31}$$

where
$$S_1(\tilde{U}_1) = k \log g_1(\tilde{U}_1) \tag{3.32}$$

and
$$S_2(\tilde{U}_2) = k \log g_2(\tilde{U}_2) \ . \tag{3.33}$$

are the entropies subsystems 1 and 2 in Figure 3.1(b) would have, *if* they were two separate *isolated* systems having the most probable energies $\tilde{U}_1$ and $\tilde{U}_2$ respectively. Since $\log g_T$ is greater than $\log (g_2 g_1)_{max}$, strictly the total entropy S_f of the composite system, defined as $k \log g_T$ is greater than $[S_1(\tilde{U}_1) + S_2(\tilde{U}_2)]$ but, due to the sharpness of the probability distribution given by equation (3.6), the difference is completely and utterly negligible for a macroscopic system and equation (3.31) is an excellent approximation for S_f. (Reference: Section 2.5.3.)

Since $g_2(\tilde{U}_2)g_1(\tilde{U}_1)$ is the maximum value the product $g_2(U_2)g_1(U_1)$ can have, corresponding to the maximum value of $P_2(U_2) = g_2 g_1/g_T$, we must always have

$$g_2(\tilde{U}_2)g_1(\tilde{U}_1) \geqslant g_2(U_2{}^0)g_1(U_1{}^0) \ . \tag{3.34}$$

Taking the natural logarithm of both sides and multiplying by Boltzmann's constant k, we have

$$k \log[g_2(\tilde{U}_2)g_1(\tilde{U}_1)] \geqslant k \log[g_2(U_2{}^0)g_1(U_1{}^0)] \ .$$

Using equations (3.26) and (3.31), we have

$$S_f = S_1(\tilde{U}_1) + S_2(\tilde{U}_2) \geqslant S_i = S_1(U_1{}^0) + S_2(U_2{}^0) \ . \tag{3.35}$$

Equation (3.35) is a simplified form of equation (3.28). Thus the total entropy of the composite system in Figure 3.1(b), when subsystems 1 and 2 are in thermal equilibrium, is greater than, or equal to, the total entropy of the composite system in the initial state shown in Figure 3.1(a). For the equality sign to hold in equation (3.35), we must have $U_2{}^0 = \tilde{U}_2$ and $U_1{}^0 = \tilde{U}_1$, in which case the two subsystems in Figure 3.1(b) would start with the most probable division of energy and, according to equation (3.20), would start at the same absolute temperature. Any infinitesimal transfer of energy from one subsystem to the other, when they are at the same absolute temperature would be a reversible change, according to the criterion of reversibility introduced into classical equilibrium thermodynamics in Section 1.6. Hence the equality sign in equation (3.35) only holds for a reversible change.

Equation (3.35) is often expressed in the form that in the change from the initial state of the closed composite system shown in Figure 3.1(a) to the final equilibrium state of the closed composite system shown in Figure 3.1(b)

$$\Delta S_{total} \geq 0 \qquad (3.36)$$

where ΔS_{total} is the change in the total entropy of the **closed composite system**. The equality sign in equation (3.36) holds for a reversible change.

3.6.2 Entropy as an extensive variable

If, initially in Figure 3.1(a), the absolute temperatures of subsystems 1 and 2 were the same, the subsystems would start with the most probable division of energy when they were placed in thermal contact in Figure 3.1(b), so that $U_1^0 = \tilde{U}_1$ and $U_2^0 = \tilde{U}_2$. For this case, according to equation (3.35), at thermal equilibrium, to an excellent approximation, the final entropy of the composite system is given by

$$S_f = S_1(\tilde{U}_1) + S_2(\tilde{U}_2) = S_1(U_1^0) + S_2(U_2^0) \ . \qquad (3.37)$$

which is just the sum of the entropies of the two isolated systems. Hence, according to equation (3.37), if two *identical* systems at the same absolute temperature and each having initial entropy S_0 were placed in thermal contact, the entropy of the composite system at thermal equilibrium would be

$$S_f = S_0 + S_0 = 2S_0 \ .$$

Hence, the total entropy of the two systems in thermal contact would be double the initial entropy of each separate system. This shows that, if the mass of a system were doubled, keeping the intensive variables such as mass density and temperature constant, the entropy of the system would be doubled. This shows that entropy is an **extensive variable**. It was illustrated in Section 2.5.4 that the entropy can be expressed as a function of the extensive variables, U, V, N. Summarising, to an excellent approximation, the entropy of a macroscopic system is an extensive variable, which can be expressed as a function of the extensive variables: the (internal) energy U, the volume V and the number of particles N in the system.

3.6.3 Alternative form of the condition of thermal equilibrium

The most probable division of energy between subsystems 1 and 2 in Figure 3.1(b), when they are in thermal equilibrium, was obtained in Section 3.2 by finding the value of U_2 at which $P_2(U_2)$ was a maximum, where according to equation (3.5),

$$P_2(U_2) = g_2(U_2)g_1(U^* - U_2)/g_T \ . \qquad (3.5)$$

This led to the condition that $P_2(U_2)$ is a maximum when $\beta_1 = \beta_2$, that is when $T_1 = T_2$. Taking the natural logarithm of both sides of equation (3.5), we have

$$\log[P_2(U_2)] = \log g_1(U^* - U_2) + \log g_2(U_2) - \log g_T .$$

(3.38)

When $P_2(U_2)$ is a maximum, $\log[P_2(U_2)]$ is also a maximum. (This is illustrated in Problems 2.8 and 2.9.) Hence, the most probable value of U_2 is that value of U_2 for which $\log[P_2(U_2)]$ is a maximum. Using equation (3.38) the condition for U_2 to be equal to $\tilde{U}_2$ becomes

$$[\log g_1(U^* - U_2) + \log g_2(U_2) - \log g_T] \text{ is a maximum.}$$

(3.39)

For the case shown in Figure 3.1(b), g_T, the total number of accessible microstates of the composite system, which is given by equation (3.4), is independent of the instantaneous value of U_2, so that $\log g_T$ is a constant. Multiplying by Boltzmann's constant, equation (3.39), which determines the most probable value of U_2, can be rewritten as

$$k \log g_1 + k \log g_2 \text{ is a maximum.} \tag{3.40}$$

According to equation (3.22), the quantities

$$k \log g_1 = S_1(U_1, V_1, N_1) \tag{3.41}$$

and
$$k \log g_2 = S_2(U_2, V_2, N_2) \tag{3.42}$$

are the entropies subsystems 1 and 2 would have *if* they were **isolated systems** having energies U_1 and U_2, volumes V_1 and V_2 and consisted of N_1 and N_2 particles respectively. Equation (3.40) can be rewritten as

$$S_1(U_1, V_1, N_1) + S_2(U_2, V_2, N_2) \text{ is a maximum.} \tag{3.43}$$

According to equation (3.43) the most probable division of energy between subsystems 1 and 2 in Figure 3.1(b), when the subsystems have reached **thermal equilibrium**, is that division of energy which renders $(S_1 + S_2)$ a maximum, subject to the conditions that

$$U_1 + U_2 = U^* \tag{3.44}$$

and U^*, V_1, V_2, N_1 and N_2 are all constant. Equations (3.43) and (3.44) are a mathematical recipe which enables us to calculate the most probable values of U_1 and U_2 at thermal equilibrium. An alternative way of expressing the mathematical recipe is to say that for any virtual displacement away from thermal equilibrium:

$$dS_1 + dS_2 = 0 . \tag{3.45}$$

For example, let energy dU_2 be transferred from subsystem 1 to subsystem 2 through the diathermic partition in Figure 3.1(b), when the two subsystems are in thermal equilibrium. According to equation (A1.32) of Appendix 1, for constant V_1 and N_1 we have

$$dS_1 = (\partial S_1/\partial U_1)_{V_1,N_1} dU_1 \ .$$

Similarly, if V_2 and N_2 are constant

$$dS_2 = (\partial S_2/\partial U_2)_{V_2,N_2} dU_2 \ .$$

Substituting in equation (3.45), for a virtual displacement away from thermal equilibrium we have

$$(\partial S_1/\partial U_1)_{V_1,N_1} dU_1 + (\partial S_2/\partial U_2)_{V_2,N_2} dU_2 = 0 \ . \tag{3.46}$$

From equation (3.44) $dU_1 = -dU_2$. Using equation (3.19) which defines the absolute temperature, we find that equation (3.46) leads to the condition that $T_1 = T_2$ at thermal equilibrium. This is in agreement with equation (3.20), which was developed in Sections 3.2 and 3.3.

As a practical example to which equation (3.43) could be applied, consider a large ensemble of identical closed composite systems of the type shown in Figure 3.1(b). When each of the closed composite systems in the ensemble is in internal thermal equilibrium, let all the diathermic partitions be changed to rigid adiabatic partitions. Each subsystem 1 and subsystem 2 of each composite system can then be treated as an isolated system, having entropies S_1 and S_2 respectively. The division of energy between the two subsystems will be different for different composite systems in the ensemble, but, corresponding to the maximum of $P_2(U_2)$, there will be more composite systems having the maximum value of $(S_1 + S_2)$ than any other values of $(S_1 + S_2)$ and the values of U_1 and U_2 for the members of the ensemble which have the maximum value of $(S_1 + S_2)$ will be $\tilde{U}_1$ and $\tilde{U}_2$ respectively.

3.6.4 Small heat transfer

Consider a macroscopic system which has energy U, volume V and consists of N particles. Let $g(U, V, N)$ be the number of accessible microstates, when the system is in a macrostate specified by U, V and N. It will simplify the algebra to use the entropy $S(U, V, N)$ as an abbreviation for $k \log g$. (By now the reader should have cultivated the habit of always thinking of the entropy S in the context of statistical mechanics as an abbreviation for $k \log g$.) Let a small quantity of heat dQ be added to the system by placing it in thermal contact with a heat reservoir for a short time. The volume V of the system and the number of particles N in the system are both kept constant. Since S is a function of U, V and N, for constant V and N, using equation (A1.32) of Appendix 1 we have

$$dS = (\partial S/\partial U)_{V,N}\, dU \;.$$

Putting $dU = dQ$ and using equation (3.23) for the absolute temperature T, we have

$$dS = dQ/T \;. \tag{3.47}$$

Equation (3.47) is the same as equation (1.13) of classical equilibrium thermodynamics. Equation (3.47) gives the increase in the entropy of a system, if a small quantity of heat dQ is added to the system, when the volume V and the number of particles N in the system are kept constant. The general case, when V and N can also vary, will be considered in Section 3.7.5.

3.6.5* Fluctuations in entropy at thermal equilibrium

It is generally assumed in classical equilibrium thermodynamics that, when subsystems 1 and 2 in Figure 3.1(b) are in thermal equilibrium, each subsystem is in a definite macrostate which has a definite value of entropy. In this Section, we shall examine how reasonable this assumption is. Consider a large ensemble of identical closed composite systems of the type shown in Figure 3.1(b). When all the composite systems are in thermal equilibrium, the diathermic partitions are all changed to rigid adiabatic partitions. Each subsystem can then be treated as an isolated system with a definite entropy, defined by equation (3.22) in terms of the number of accessible microstates. Since there are more members of the ensemble in which the energy of subsystem 2 is equal to $\tilde{U}_2 = \bar{U}_2$ than any other value of energy, there are more members of the ensemble in which the entropy of subsystem 2 is equal to $S_2(\tilde{U}_2)$ than any other value of S_2.

It was shown in Section 3.2 and in Appendix 4* that, when subsystem 2 is in thermal equilibrium with subsystem 1 in Figure 3.1(b), the probability $P_2(U_2)$ that at thermal equilibrium subsystem 2 has energy U_2 is a sharp Gaussian, the standard deviation σ_2 being equal to $\gamma \bar{U}_2/N_2^{1/2}$, where γ is a constant of order of magnitude unity, whose precise value depends on the nature of subsystems 1 and 2. (Reference: Section A4.1* of Appendix 4*.) For a Gaussian distribution, the probability of obtaining a value of U_2 deviating from $\bar{U}_2 = \tilde{U}_2$ by more than $100\sigma_2$ is 2.69×10^{-2174}. Hence only 1 in 3.7×10^{2173} members of the ensemble have values of U_2 deviating from $\tilde{U}_2$ by more than $100\sigma_2$.

For purposes of discussion, it will be assumed that the energy of one of the subsystems 2 in the ensemble is increased from $\bar{U}_2$ to $(\bar{U}_2 + \Delta U_2)$ where

$$\Delta U_2 = 100\sigma_2 \simeq 100\gamma \bar{U}_2/N_2^{1/2} \tag{3.48}$$

where γ is a constant of order of magnitude unity. For a macroscopic system for which $N_2 > 10^{18}$, $\Delta U_2/\bar{U}_2$ is less than 10^{-7}, so that the temperature of subsystem 2 is virtually unchanged. It will be shown in Chapter 10 that, at room temperatures, the heat capacities of many insulating solids are approximately equal to $3N_2k$, so that the internal energy of such a subsystem 2 is given approximately by

$$\bar{U}_2 = 3N_2kT_2$$

where k is Boltzmann's constant and T_2 is the absolute temperature of subsystem 2. Substituting for $\bar{U}_2$ in equation (3.48), we have

$$\Delta U_2 = 300\gamma N_2^{1/2}kT_2 \ .$$

According to equation (3.47), the increase in the entropy of subsystem 2, associated with an increase ΔU_2 in its energy, is

$$\Delta S_2 = \Delta U_2/T_2 = 300\gamma N_2^{1/2}k \ . \tag{3.49}$$

According to equation (2.42) of Section 2.5.2, a rough order of magnitude for the entropy of subsystem 2 is αN_2k, where α is a constant generally in the range 0.5 to 100. Hence, for a value of ΔS_2 given by equation (3.49), we have

$$\Delta S_2/S_2 = (300\gamma/\alpha)(1/N_2)^{1/2} \ . \tag{3.50}$$

According to equation (3.50), only 1 member in 3.7×10^{2173} of the ensemble should have a value of S_2 deviating from $S_2(\tilde{U}_2)$ by more than $300\gamma S_2(\tilde{U}_2)/\alpha N_2^{1/2}$. Since $\gamma \sim 1$ and α is in the range of 0.5 to 100, and since $N_2 > 10^{18}$ for a macroscopic system, the spread in the values of S_2 for various members of the ensemble is very small compared with $S_2(\tilde{U}_2)$. If one takes the thermodynamic limit by letting N_1 and N_2 tend to infinity keeping N_1/V_1 and N_2/V_2 constant, then in the thermodynamic limit $\Delta S_2/S_2$ tends to zero. Thus in the context of classical equilibrium thermodynamics, particulary in the thermodynamic limit, it is just as reasonable to assume that, when subsystems 1 and 2 in Figure 3.1(b) are in thermal equilibrium, they have definite values of entropy as it is to assume that subsystems 1 and 2 have definite values of internal energy.

3.6.6* The approach to thermal equilibrium*

In general we shall only consider equilibrium states, but in this Section we shall make a few qualitative remarks about the approach to thermal equilibrium. (A reader interested in a discussion of Boltzmann's H theorem is referred to Reif [1].)

When the adiabatic partition in Figure 3.1(a) is changed to a diathermic partition, it takes a finite time after they are in thermal contact for the two subsystems to reach thermal equilibrium. If U_2^0, the initial energy of subsystem 2, is significantly less than $\tilde{U}_2$, which is the most probable value of U_2 at thermal equilibrium, as shown in Figure 3.2(b), the overwhelming probability is that heat flows from subsystem 1 to subsystem 2, until at thermal equilibrium U_2 is very, very close to $\tilde{U}_2$. In order to discuss the progress towards thermal equilibrium, it will be assumed that a very short time after the adiabatic partition in Figure 3.1(a) is changed to a diathermic partition, it is changed back to an adiabatic partition, so that each subsystem can be treated as an isolated system again, with a definite value of energy and entropy. This procedure is repeated until the two subsystems reach thermal equilibrium. The values of energy and entropy quoted all refer to times when the adiabatic partitions are in place. In the approach to thermal equilibrium, if $U_2^0 < \tilde{U}_2$, the energy U_2 of subsystem 2 increases *on average*, until it reaches a value close to $\tilde{U}_2$. There may be times when U_2 decreases momentarily. It will be assumed that U_2 is smoothed out, for example, by taking a running mean.† The energy U_2 should then increase 'continuously' to reach a value close to $\tilde{U}_2$ at thermal equilibrium. We shall now consider the change in $S_1(U_1) + S_2(U_2)$, where S_1 and S_2 are the entropies subsystems 1 and 2 would have if they were isolated systems having energies U_1 and U_2 respectively. Using equation (3.22), we have

$$S_1(U_1) + S_2(U_2) = k[\log g_1(U_1) + \log g_2(U_2)] = k \log[g_1 g_2] .$$

(3.51)

The form of the variation of $g_1 g_2$ with U_2 can be determined from another context. According to equation (3.5), the probability $P_2(U_2)$ that *at thermal equilibrium* subsystem 2 in Figure 3.1(b) has energy U_2 is $P_2(U_2) = g_1 g_2/g_T$. For two normal thermodynamic systems, for which the variation of $\log g$ with U is as shown in Figure 3.3, the variation of the function $P_2(U_2)$ with U_2 is as shown in Figure 3.2(b). When $U_2 < \tilde{U}_2$, the function $P_2(U_2)$ increases with U_2 reaching a maximum value when $U_2 = \tilde{U}_2$. Since $g_1 g_2$ is equal to $g_T P_2(U_2)$, where g_T is a constant, if $U_2^0 < \tilde{U}_2$, the function $g_1 g_2$, and hence the function $\log(g_1 g_2)$ increases as U_2 increases, reaching a maximum value when $U_2 = \tilde{U}_2$. Since k is a constant, $[S_1(U_1) + S_2(U_2)]$, which according to equation (3.51) is equal to $k \log(g_1 g_2)$, increases as U_2 increases from U_2^0 to $\tilde{U}_2$ reaching a maximum value when $U_2 = \tilde{U}_2$. [It is left as an exercise for the reader to consider the case when $U_2^0 > \tilde{U}_2$.]

As an alternative approach the reader can generalise some of the results of Problem 2.9. Assume that the variations of $\log g_1$ and $\log g_2$ with U_1 and U_2 respectively are of the form shown in Figure 3.3. Plot $\log g_2(U_2)$ and $\log g_1(U^* - U_2)$ against U_2. Add the ordinates to obtain $\log(g_1 g_2)$. The results will show that $\log(g_1 g_2)$ increases to reach a maximum, when equation (3.13) is valid and $U_2 = \tilde{U}_2$.

† If successive values of a variable are $x_1, x_2, \ldots$, to use a 5 point running mean one would replace x_i by $(x_{i-2} + x_{i-1} + x_i + x_{i+1} + x_{i+2})/5$.

Returning to our example of the composite system approaching thermal equilibrium, if it is smoothed to remove fluctuations, the function $[S_1(U_1) + S_2(U_2)]$ increases as U_2 increases, reaching a maximum value of $[S_1(\tilde{U}_1) + S_2(\tilde{U}_2)]$ at thermal equilibrium. In the approach to thermal equilibrium, the *smoothed* behaviour of $(S_1 + S_2)$ is given by

$$d(S_1 + S_2)/dt > 0 \ . \tag{3.52}$$

3.6.7 Direction of heat flow: the entropy approach to the second law of thermodynamics

Consider again the case of subsystem 1 in thermal contact with subsystem 2, as shown in Figure 3.1(b). It follows from equation (3.35) that, in the approach to thermal equilibrium, the overwhelming probability is that

$$dS_1(U_1) + dS_2(U_2) > 0 \tag{3.53}$$

where $S_1(U_1)$ and $S_2(U_2)$ are the entropies subsystems 1 and 2 would have, if they were isolated systems having the instantaneous values of energy U_1 and U_2 respectively. (Equation (3.53) also follows from equation (3.52).)

Since V_1, V_2, N_1 and N_2 are all constant in Figure 3.1(b), using equation (A1.32) of Appendix 1, the total differentials dS_1 and dS_2 can be expressed in the forms

$$dS_1 = (\partial S_1/\partial U_1)_{V_1,N_1}\, dU_1; \qquad dS_2 = (\partial S_2/\partial U_2)_{V_2,N_2} dU_2 \ .$$

Using equations (3.23) and (3.44) we have

$$dS_1 = dU_1/T_1 = -dU_2/T_1 \tag{3.54}$$

and $$dS_2 = dU_2/T_2 \ . \tag{3.55}$$

Equations (3.54) and (3.55) are consistent with equation (3.47). Substituting for dS_1 and dS_2 into equation (3.53), in the approach to thermal equilibrium we have

$$-(dU_2/T_1) + (dU_2/T_2) > 0 \ .$$

Rearranging, $$[(T_1 - T_2)/T_1 T_2]\, dU_2 > 0 \ .$$

For dU_2 to be positive, that is for heat to flow *on average* from subsystem 1 to subsystem 2 in Figure 3.1(b) during the approach to thermal equilibrium, we must have $T_1 > T_2$. This is the Clausius statement of the second law of classical thermodynamics. (Reference: Section 1.7 of Chapter 1.)

3.6.8 Poincaré's theorem

According to the principle of equal *a priori* probabilities (our Postulate 2), when the closed composite system shown in Figure 3.1(b) is in internal thermal equilibrium, we are equally likely to find the closed composite system in any one of the microstates accessible to the composite system. Since the initial microstate, in which $U_1 = U_1{}^0$ and $U_2 = U_2{}^0$, is one of the accessible microstates of the composite system, there is a finite mathematical probability that, at some time in the future, one would find the composite system in Figure 3.1(b) back in its initial state. This is Poincaré's theorem. There is even a finite probability that, even with $T_1 \gg T_2$ initially, one would sometimes find the composite system in a microstate in which U_2 was less than $U_2{}^0$, in which case, if $T_2 \ll T_1$ initially, heat would have flowed from the colder system to the hotter system. It was shown in Section 3.2 that, at thermal equilibrium, the probability of fluctuations from the mean value $\bar{U}_2$ greater than $100\sigma_2$ can be neglected for times of the order of the age of the Earth, where for a macroscopic system $\sigma_2/\bar{U}_2$ is of the order of $1/N_2{}^{1/2}$. For $N_2 > 10^{18}$, we can ignore the possibility of obtaining values of U_2 which differ from $\bar{U}_2$ by more than 1 part in 10^7 at thermal equilibrium. Thus with macroscopic systems (unless the initial energy $U_2{}^0$ of subsystem 2 in Figure 3.1(a) were *extremely* close to $\tilde{U}_2$), when subsystems 1 and 2 were in thermal equilibrium in Figure 3.1(b), one would 'never' find subsystem 2 back in its initial state with $U_2 = U_2{}^0$. Also if $U_2{}^0 < \tilde{U}_2$ (unless $U_2{}^0$ were extremely close to $\tilde{U}_2$), one would 'never' find a value of U_2 less than $U_2{}^0$, when subsystems 1 and 2 were in thermal equilibrium in Figure 3.1(b). For example, if a piece of metal were dropped into a beaker of water, the probability that the piece of metal would get red hot and the water freeze is finite, but the probability is so completely and utterly negligible that one would be safe in concluding that it would 'never' be observed in practice.

3.7 GENERAL THERMODYNAMIC INTERACTION

3.7.1 Introduction

So far in this Chapter, it was assumed that the volume of each subsystem and the number of particles in each subsystem in Figure 3.1(b), remained constant, so that the systems could only exchange heat energy. The more general cases, illustrated in Figures 3.4(a) and 3.4(b) will now be considered. In Figure 3.4(a) the diathermic partition is free to move without friction so that the volumes of the subsystems can change. In the case shown in Figure 3.4(b), not only can the diathermic partition move without friction but the partition has holes in it so that particles can go from one subsystem to the other. The two subsystems in Figure 3.4(b) are composed of the same type of particles. A practical example of the case shown in Figure 3.4(b) would be a liquid in equilibrium with its vapour in a fixed volume adiabatic enclosure.

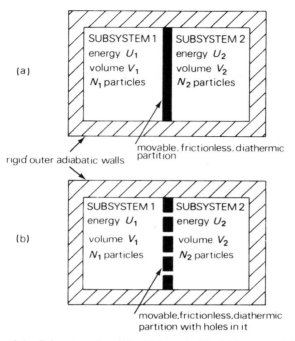

Figure 3.4—Subsystems 1 and 2, which are inside rigid, outer adiabatic walls, form a closed composite system of fixed total energy, fixed total volume and fixed total number of particles. In (a) the two subsystems are separated by a movable, rigid, diathermic partition. Heat can flow across the diathermic partition, and one subsystem can do mechanical work on the other by moving the partition. (b) In this case there are holes in the movable, diathermic partition so that particles can also go from one subsystem to the other until the chemical potentials are equal at thermodynamic equilibrium.

3.7.2 Thermodynamic equilibrium

Consider the composite system shown in Figure 3.4(b). The two subsystems are allowed to reach thermodynamic equilibrium. The two subsystems can exchange heat through the diathermic partition. According to the law of conservation of energy, if U_1 and U_2 are the instantaneous values of the energies of subsystems 1 and 2 respectively,

$$U_1 + U_2 = U^* \tag{3.56}$$

where U^*, the total energy, is a constant. One subsystem can do mechanical work on the other by moving the partition, which moves without friction. The volumes V_1 and V_2 of subsystems 1 and 2 can vary, but the total volume V^* is kept constant by the rigid, outer, adiabatic walls, so that

$$V_1 + V_2 = V^* . \tag{3.57}$$

The holes in the movable partition in Figure 3.4(b) allow particles to go from one subsystem to the other. In this case the numbers of particles N_1 and N_2 in subsystems 1 and 2 respectively can vary subject to the constraint that

$$N_1 + N_2 = N^* \tag{3.58}$$

where N^*, the total number of particles, is a constant.

Let $g_1(U_1, V_1, N_1)$ be the number of microstates accessible to subsystem 1 when it has energy U_1, volume V_1 and consists of N_1 particles, and let $g_2(U_2, V_2, N_2)$ be the number of microstates accessible to subsystem 2 when it has energy U_2, volume V_2 and consists of N_2 particles. Since the composite system shown in Figure 3.4(b) has fixed total energy U^*, fixed total volume V^* and fixed total number of particles N^*, it is a *closed* system. According to the **principle of equal *a priori* probabilities** (Postulate 2), when the closed isolated composite system in Figure 3.4(b) is in thermodynamic equilibrium, one is equally likely to find the closed composite system in any one of the microstates accessible to the closed composite system. Generalising equation (3.5), which was developed in Section 3.2 for the case of thermal interaction, we conclude that the probability $P_2(U_2, V_2, N_2)$ that at thermodynamic equilibrium subsystem 2 has energy U_2, volume V_2 and consists of N_2 particles is

$$P_2(U_2, V_2, N_2) = Cg_2(U_2, V_2, N_2)g_1(U_1, V_1, N_1)$$
$$\tag{3.59}$$

where C is a constant, which is equal to the reciprocal of the total number of accessible microstates. Taking the natural logarithm of equation (3.59), we have

$$\log[P_2(U_2, V_2, N_2)] = \log[g_2(U_2, V_2, N_2)] + \log[g_1(U_1, V_1, N_1)] + \log C .$$
$$\tag{3.60}$$

It can be shown that, for macroscopic systems, the probability distribution $P_2(U_2, V_2, N_2)$ has an extremely sharp maximum. By analogy with the case of thermal interaction only, it is reasonable to assume that the most probable values of U_2, V_2 and N_2, namely the values at which $P_2(U_2, V_2, N_2)$ is a maximum, are equal to the ensemble averages of the quantities and correspond to the equilibrium values of these quantities in classical equilibrium thermodynamics. When $P_2(U_2, V_2, N_2)$ is a maximum $\log[P_2(U_2, V_2, N_2)]$ is also a maximum. Since C is a constant in equation (3.60), the condition that $\log P_2$ is a maximum can be rewritten in the form

$$\log[g_2(U_2, V_2, N_2)] + \log[g_1(U_1, V_1, N_1)] \text{ is a maximum} \tag{3.61}$$

Multiplying by Boltzmann's constant k and using the abbreviations

$$S_1(U_1, V_1, N_1) = k \log[g_1(U_1, V_1, N_1)] \qquad (3.62)$$

and $\qquad S_2(U_2, V_2, N_2) = k \log[g_2(U_2, V_2, N_2)] \qquad (3.63)$

equation (3.61), which specifies when $P_2(U_2, V_2, N_2)$ is a maximum, becomes

$$S_1(U_1, V_1, N_1) + S_2(U_2, V_2, N_2) \text{ is a maximum} \qquad (3.64)$$

subject to the constraints given by equations (3.56), (3.57) and (3.58), namely

$$(U_1 + U_2) = U^* \qquad (3.56)$$

$$(V_1 + V_2) = V^* \qquad (3.57)$$

$$(N_1 + N_2) = N^* \qquad (3.58)$$

In equation (3.64), $S_1(U_1, V_1, N_1)$ is the entropy subsystem 1 would have, if it were an isolated system having energy U_1, volume V_1 and consisted of N_1 particles. Similarly, $S_2(U_2, V_2, N_2)$ is the entropy subsystem 2 would have, if it were an isolated system having energy U_2, volume V_2 and consisted of N_2 particles. Equations (3.64), (3.56), (3.57) and (3.58) are a mathematical recipe that enables us to calculate the most probable values of U_2, V_2 and N_2 at thermodynamic equilibrium. Equations (3.56), (3.57) and (3.58) enable us to express $(S_1 + S_2)$ as a function only of U_2, V_2 and N_2.

For a function $f(x, y, z)$ of three independent variables x, y and z, the condition for a maximum of f is that the following equations hold simultaneously:

$$\partial f/\partial x = 0; \qquad \partial f/\partial y = 0; \qquad \partial f/\partial z = 0 .$$

Applying these relations to find the maximum of $(S_1 + S_2)$, we must have

$$\partial(S_1 + S_2)/\partial U_2 = 0 \qquad (3.65)$$

$$\partial(S_1 + S_2)/\partial V_2 = 0 \qquad (3.66)$$

$$\partial(S_1 + S_2)/\partial N_2 = 0 . \qquad (3.67)$$

3.7.3 Thermal and mechanical interaction

Consider the case shown in Figure 3.4(a), in which the diathermic partition is free to move without friction. In this case U_1, U_2, V_1 and V_2 vary subject to the constraints given by equations (3.56) and (3.57), but both N_1 and N_2 are constant.

It follows from equation (3.56) that $dU_2 = -dU_1$. Hence equation (3.65) reduces to

$$(\partial S_2/\partial U_2)_{V_2, N_2} = (\partial S_1/\partial U_1)_{V_1, N_1} .$$

Using equation (3.23) we have

$$T_1 = T_2 \tag{3.68}$$

where T_1 and T_2 are the absolute temperatures of subsystems 1 and 2 respectively. This was the condition developed earlier for the case of thermal interaction only in Sections 3.2 and 3.3.

It follows from equation (3.57) that $dV_2 = -dV_1$. Hence equation (3.66) reduces to

$$(\partial S_1/\partial V_1)_{U_1,N_1} = (\partial S_2/\partial V_2)_{U_2,N_2} . \tag{3.69}$$

According to equation (A5.8) of Appendix 5,

$$(\partial S/\partial V)_{U,N} = p/T \tag{3.70}$$

where p is the ensemble average pressure. Hence equation (3.69) can be rewritten as

$$p_1/T_1 = p_2/T_2 . \tag{3.71}$$

According to equation (3.68), when $(S_1 + S_2)$ is a maximum, $T_1 = T_2$. Hence, when $(S_1 + S_2)$ is a maximum equation (3.71) reduces to

$$p_1 = p_2 . \tag{3.72}$$

Thus for the case shown in Figure 3.4(a), in which the numbers of particles N_1 and N_2 are kept constant, the probability distribution given by equation (3.59) has a very sharp maximum when the two subsystems have the same absolute temperature and the same pressure. On average, heat will be conducted through the partition in Figure 3.4(a) until $T_1 = T_2$ and one subsystem will move the partition doing mechanical work on the other subsystem until $p_1 = p_2$.

According to equation (3.53), in going from the initial state to the final equilibrium state, unless the two subsystems are already in thermodynamic equilibrium in their initial states, the average behaviour is described by

$$dS_1(U_1, V_1) + dS_2(U_2, V_2) > 0 .$$

For N_1 and N_2 constant, expanding dS_1 and dS_2 using equation (A1.32) of Appendix 1 and substituting in the above inequality, we have

$$\left(\frac{\partial S_1}{\partial U_1}\right)_{V_1,N_1} dU_1 + \left(\frac{\partial S_1}{\partial V_1}\right)_{U_1,N_1} dV_1 + \left(\frac{\partial S_2}{\partial U_2}\right)_{V_2,N_2} dU_2 + \left(\frac{\partial S_2}{\partial V_2}\right)_{U_2,N_2} dV_2 > 0 .$$

$$\tag{3.73}$$

It follows from equations (3.56) and (3.57) that $dU_2 = -dU_1$ and $dV_2 = -dV_1$. Using equations (3.23) and (3.70), equation (3.73) becomes

$$\left(\frac{1}{T_1} - \frac{1}{T_2}\right) dU_1 + \left(\frac{p_1}{T_1} - \frac{p_2}{T_2}\right) dV_1 > 0 \ . \tag{3.74}$$

If $T_2 = T_1$, equation (3.74) reduces to

$$(p_1 - p_2) \, dV_1 > 0 \ . \tag{3.75}$$

For dV_1 to be positive, p_1 must be greater than p_2. Hence, if the pressure of subsystem 1 is greater than the pressure of subsystem 2, subsystem 1 expands in volume at the expense of subsystem 2.

3.7.4 Thermal and mechanical interaction with exchange of particles

It will now be assumed that the movable, frictionless, diathermic partition has holes in it, as shown in Figure 3.4(b). Particles can now move from one subsystem to the other, subject to the condition that the total number of particles is conserved, namely

$$N_1 + N_2 = N^* \ . \tag{3.58}$$

It is being assumed that subsystem 1 and 2 consist of the same type of particle. In this example, for $[S_1(U_1, V_1, N_1) + S_2(U_2, V_2, N_2)]$ to be a maximum, equation (3.67) gives the extra condition

$$(\partial S_1/\partial N_2)_{U_2,V_2} + (\partial S_2/\partial N_2)_{U_2,V_2} = 0 \ . \tag{3.76}$$

Since $(N_1 + N_2)$ is constant, $dN_2 = -dN_1$. Hence at thermodynamic equilibrium equation (3.76) becomes

$$(\partial S_1/\partial N_1)_{U_1,V_1} = (\partial S_2/\partial N_2)_{U_2,V_2} \ . \tag{3.77}$$

It is convenient to introduce a quantity μ, called the **chemical potential**, which is defined in the general case when a system has entropy S, (internal) energy U and absolute temperature T, by the equation

$$\mu = -T(\partial S/\partial N)_{U,V} = -kT(\partial \log g/\partial N)_{U,V} \ . \tag{3.78}$$

Alternative expressions for μ are given by equations (6.11) and (6.43) of Chapter 6.

Using equation (3.78), equation (3.77) becomes

$$\mu_1/T_1 = \mu_2/T_2 \tag{3.79}$$

Since $T_1 = T_2$ when $(S_1 + S_2)$ is a maximum, equation (3.79) reduces to

$$\mu_1 = \mu_2 \ . \tag{3.80}$$

When the exchange of particles can take place, one must add the extra term $[(\partial S_1/\partial N_1)_{U_1,V_1} \, dN_1 + (\partial S_2/\partial N_2)_{U_2,V_2} \, dN_2]$ to the left hand side of equation (3.73). Since $dN_1 = -dN_2$, using equation (3.78) the extra term on the left hand side of equation (3.74) is $[(\mu_1/T_1 - \mu_2/T_2)] \, dN_2$. If $T_1 = T_2$

and $p_1 = p_2$, the modified form of equation (3.74) gives

$$(\mu_1 - \mu_2)\, \mathrm{d}N_2 > 0 \ .$$

For $\mathrm{d}N_2$ to be positive, that is for particles to go, on average from subsystem 1 to subsystem 2 we must have $\mu_1 > \mu_2$, which means that on average particles will go from a region of high chemical potential to a region of low chemical potential until $\mu_1 = \mu_2$. Just as absolute temperature is an indicator, which shows the average direction of heat flow during the approach to thermodynamic equilibrium, the chemical potential is an indicator of the average direction of particle flow, during the approach to thermodynamic equilibrium.

The chemical potential is used extensively in chemistry. When defined by equation (3.78), for a pure substance μ is equal to the Gibbs Free Energy per particle. It is sometimes defined as

$$\mu = -T(\partial S/\partial n)_{U,V} \ . \tag{3.81}$$

where n is the number of moles in the system. For a pure substance the chemical potential is then equal to the Gibbs Free Energy per mole. A brief account of the role of the chemical potential in chemical reactions will be given in Chapter 6. The chemical potential will be used extensively in our discussion of Fermi-Dirac and Bose-Einstein statistics in Chapters 11 and 12. In the free electron theory of solids, the chemical potential μ is sometimes called the Fermi-level. A reader interested in an account of statistical mechanics in which the chemical potential is used extensively, is referred to the text book by Kittel and Kroemer [2].

The overwhelming probability is that, during the approach to thermodynamic equilibrium, subsystems 1 and 2 in Figure 3.4(b) will exchange energy in the form of heat until $T_1 = T_2$, do mechanical work on each other by moving the partition until $p_1 = p_2$ and exchange particles until $\mu_1 = \mu_2$. The equations $T_1 = T_2; p_1 = p_2; \mu_1 = \mu_2$ are the conditions for maximising $(S_1 + S_2)$, which is equivalent to finding the maximum of $P_2(U_2, V_2, N_2)$, the probability that at thermodynamic equilibrium subsystem 2 in Figure 3.4(b) has energy U_2, volume V_2 and consists of N_2 particles. For macroscopic subsystems, the probability distribution $P_2(U_2, V_2, N_2)$ is extremely sharp and the values $\tilde{U}_2$, $\tilde{V}_2$ and $\tilde{N}_2$ at which P_2 is a maximum are equal to the mean (ensemble average) values $\bar{U}_2$, $\bar{V}_2$ and $\bar{N}_2$ and correspond to the equilibrium values of classical equilibrium thermodynamics. (This is Postulate 4 of Section 3.1 of Chapter 3.)

3.7.5 The thermodynamic identity

Consider a system which is in internal thermodynamic equilibrium. The number of accessible microstates g is a function of the (internal) energy U, the volume V of the system and the number N of particles in the system.

(Reference: Section 2.5.4.) To simplify the algebraic expressions, we shall use the abbreviation:

$$S(U, V, N) = k \log g(U, V, N) \ . \tag{3.82}$$

where $S(U, V, N)$ is the entropy of the system. The energy of the system is increased by dU, the volume by dV and the number of particles by dN, by placing the system in contact with suitable reservoirs. The system is allowed to reach a new state of internal thermodynamic equilibrium. Since S is a function of U, V and N, using equation (A1.32) of Appendix 1, we have

$$dS = (\partial S/\partial U)_{V,N} \, dU + (\partial S/\partial V)_{U,N} \, dV + (\partial S/\partial N)_{U,V} \, dN \ .$$

$$\tag{3.83}$$

From equations (3.23), (3.70) and (3.78)

$$(\partial S/\partial U)_{V,N} = 1/T; \qquad (\partial S/\partial V)_{U,N} = p/T;$$
$$(\partial S/\partial N)_{U,V} = -\mu/T \ .$$

Substituting in equation (3.83), we have

$$dS = (dU/T) + (p/T) \, dV - (\mu/T) \, dN \ . \tag{3.84}$$

Rearranging, we have

$$dU = T \, dS - p \, dV + \mu \, dN \ . \tag{3.85}$$

If there is more than one chemical component in the system, the entropy S can be expressed as a function of U, V, N_1, N_2 .... Equation (3.85) can then be generalised to

$$dU = T \, dS - p \, dV + \Sigma \mu_i \, dN_i \tag{3.86}$$

where N_i is the number of particles of the ith chemical component and $\mu_i = -T(\partial S/\partial N_i)_{U,V,N_1,...,N_n(\text{except } N_i)}$ is the chemical potential of the ith chemical component in the system.

Equation (3.85) is the **thermodynamic identity**. It is the same as equation (1.22) of classical equilibrium thermodynamics.

3.8* USE OF THE DENSITY OF STATES*

3.8.1* Introduction*

The exposition in Sections 3.2 to 3.7 was simplified by making it as close as possible to the simple numerical example discussed in Section 2.3.1. It was assumed that subsystems 1 and 2 in Figure 3.1(b) had discrete energy levels and that each energy level had an extremely large degeneracy, of the order of e^N, where N was the number of particles in the subsystem. The entropy of a system, when it had energy U, was defined as

$$S = k \log g(U) \tag{3.22}$$

where $g(U)$ was the degeneracy of the energy level of energy U. This approximation would be valid for an idealised example, similar to the one illustrated in Figure 2.5, in which the N spatially separated particles were atomic magnetic dipoles, whose magnetic dipole moment μ could point either parallel to or antiparallel to the direction of an applied magnetic field **B**, provided there was no interaction between the N magentic moments themselves. It will be shown in Section 4.4.2, that if μ is parallel to the magnetic field, the potential energy of the dipole is $-\mu B$ whereas if μ is pointing in the direction opposite to **B** its potential energy is $+\mu B$. If n of the magnetic dipoles are pointing parallel to the magnetic field, according to equation (8.2) of Chapter 8, the total potential energy of the N particle system is

$$U = (N - 2n)\mu B \ . \tag{3.87}$$

According to equation (2.32) the degeneracy of the energy level in which n magnetic dipoles point in the direction of **B** is

$$g(n) = N!/n! \, (N - n)! \ . \tag{3.88}$$

For a macroscopic system, for which $N > 10^{18}$, $g(n)$ is a very large number. In practice there would be some magnetic interaction between the spatially separated magnetic dipoles. Different magnetic dipoles point parallel to the magnetic field in each of the $g(n)$ microstates with a given value of n. Due to slight variations in the separations of the dipoles due to deviations from a perfect array of dipoles and surface effects, in practice the contribution to the potential energy due to magnetic interactions between the dipoles would probably be different in all the $g(n)$ microstates associated with a given value of n. Hence the energies of each of the $g(n)$ microstates associated with a given value of n would probably be slightly different in practice. This would lead to the splitting of the degeneracy $g(n)$ associated with a given value of n.

To allow for the splitting of the degeneracies of energy levels, it is better in the general case to use the **density of states** rather than the degeneracies of individual energy levels. If an N particle system has a total of X independent microstates having energy eigenvalues in the range of energy U to $(U + \Delta U)$ where $\Delta U \ll U$, but ΔU is substantially bigger than energy separation of neighbouring microstates, the density of states is given by

$$D = X/\Delta U \ . \tag{3.89}$$

The density of states is the number of microstates, that is the number of independent quantum eigenstates of the N particle system, per unit energy range. In a macroscopic theory the density of states can be treated as a smooth continuous function of the (internal) energy of the system.

3.8.2* The energy of a thermodynamic system*

It is not possible in practice to specify precisely the energy of a thermodynamic system. For example, if the internal energy U of a thermodynamic system is measured experimentally there is always an experimental uncertainty in the measured value of U. When subsystem 2 in Figure 3.1(b) is in thermal equilibrium with subsystem 1, according to equation (A4.10) of Appendix 4, at thermal equilibrium, there is a spread in the energy U_2 of subsystem 2 of the order of $U_2/N_2^{1/2}$. For example, consider a large ensemble of identical composite systems of the type shown in Figure 3.1(b). If in each case, subsystem 2 were removed when it was in thermal equilibrium with subsystem 1, the energy of subsystem 2 would be different for different members of the ensemble. This example shows that, if we try to reproduce the thermodynamic macrostate of a system using identical methods of preparation, the energy of the thermodynamic system is not always exactly the same. The role of the uncertainty principle will be discussed in Section 3.8.3. For the above reasons, in a microscopic theory one cannot attribute a precise value to the energy of a thermodynamic system. Hence the total number of accessible microstates (degeneracy of the macrostate) cannot be known exactly, and the entropy defined by equation (3.22) cannot be specified exactly, but should in principle depend on the spread ΔU in the value of the energy. However, the number of accessible microstates is such a fantastically large number that the precise value of ΔU, the indeterminacy in the energy, is not important in practice when the entropy is calculated using equation (3.22).

For example, in the idealised model used in Sections 3.2–3.7, it was assumed that each energy level had a large degeneracy. For purposes of discussion, assume that there are x such energy levels, each of degeneracy g, in the energy range ΔU. The total number of microstates in the energy range ΔU is xg. Consider the expression

$$k \log xg = k(\log g + \log x) \ . \tag{3.90}$$

For the example shown in Figure 2.5, $\log g$ is of order N, whereas even if x were 10^{20}, $\log x$ would only be equal to 46 and $\log x$ could be ignored in comparison with $\log g$. Comparing equations (3.90) and (3.22), it can be seen that the entropy S can be put equal to $k \log xg$. In general, due to the interactions between the N particles in the system, the xg microstates are spread over the energy range ΔU. If D is the density of states, the quantity xg is equal to $D \Delta U$, and the entropy is given by

$$S = k \log xg = k \log(D \Delta U) = k[\log D + \log \Delta U] \ .$$

According to equation (2.42) of Section 2.5.2, a rough order of magnitude for the entropy S is αNk, where N is the number of particles in the system, k is Boltzmann's constant and α is a number which is generally in the range 0.5 to 100. Even if the energy range ΔU were equal to the rest mass energy of the Earth ($m_0 c^2 = 5.98 \times 10^{24} c^2 = 5.4 \times 10^{41}$ J), $k \log(\Delta U)$ would only be equal to $96k$, which is negligible compared with αNk, if $N > 10^{18}$. Hence in the general case of a macroscopic system, to an excellent approximation we have

$$S(U) = k \log D(U) \tag{3.91}$$

where $D(U)$ is the density states when the system has mean energy U.

Substituting αNK for S in equation (3.91) and taking exponentials, we find that a very rough order of magnitude for the density states is

$$D \sim \exp(\alpha N) \tag{3.92}$$

where α is generally in the range 0.5 to 100. Since for a macroscopic system $N > 10^{18}$, D is a fantastically large number. A very rough order of magnitude for the mean energy of separation of neighbouring microstates is $e^{-\alpha N}$.

The imprecision ΔU in the energy of a thermodynamic system does not affect the value of the absolute temperature of a system. Putting the total number of accessible microstates equal to $D\Delta U$ in equation (3.19), we have

$$1/kT = \partial \log(D\Delta U)/\partial U = \partial[\log D + \log \Delta U]/\partial U \tag{3.93}$$

Since $\log \Delta U \ll \log D$, the absolute temperature T can be defined by the equation

$$T = (1/k)(\partial U/\partial \log D)_{V,N} . \tag{3.94}$$

The temperature parameter β is given by

$$\beta = 1/kT = (\partial \log D/\partial U)_{V,N} . \tag{3.95}$$

3.8.3* Thermal equilibrium re-examined*

In the simple numerical example illustrated in Figure 2.3 and discussed in Section 2.3.1, the separation of the energy levels was exactly one energy unit in both subsystems 1 and 2. If the energy level separation in subsystem 1 were still exactly one energy unit, but in subsystem 2 it were exactly 0.723714632 . . . energy units, one could not transfer one energy unit from subsystem 1 to subsystem 2 and conserve the total energy. When discussing the example shown in Figure 3.1(b) in Sections 3.2–3.7, it was tacitly assumed that the positions of the energy eigenvalues were always the correct values to conserve energy. At first sight it might appear that, in real cases, one would never be able to transfer energy from one subsystem to the other in Figure 3.1(b) and conserve the total energy. In this Section, it will be illustrated how the natural line widths of energy eigenvalues, associated with the uncertainty principle, is more than sufficient to overcome this difficulty. (For a more rigorous discussion based on the density matrix leading up to the **microcanonical ensemble**, the advanced reader is referred to Landau and Lifshitz [3].)

If a system interacts with another system, for example with the rest of the universe, according to the time dependent perturbation theory of quantum mechanics, the system does not remain permanently in one particular microstate. According to the uncertainty principle,

$$\Delta E \, \Delta t \sim \hbar \tag{3.96}$$

where Δt is the lifetime of the state and ΔE is the spread in the energy of the state. The latter is called the natural line width. In equation (3.96), $\hbar = 1.05 \times 10^{-34}$ J s. If Δt were as long as one second, ΔE would be $\sim 10^{-34}$ J. For $\Delta t = 10^{-12}$ second, ΔE would be $\sim 10^{-22}$ J. According to equation (3.92), the number of microstates having mean energy eigenvalues in an energy range $\Delta E \simeq 10^{-22}$ would be approximately $10^{-22} \exp(\alpha N)$. This is an extremely large number for a macroscopic system. This shows that the natural line width of a particular microstate of subsystem 2 in Figure 3.1(b), which has a mean energy eigenvalue U_2, is sufficient for the eigenvalues of a large number of the microstates of subsystem 1 to satisfy the condition that $(U_1 + U_2)$ is constant. (The energy eigenvalues of subsystem 1 also have natural line widths.)

In practice, there is likely to be some interaction between the 'closed' system, of subsystems 1 and 2 in thermal equilibrium in Figure 3.1(b), and the rest of the universe. This can lead to fluctuations in the total energy U^* of the composite system. Since the lifetime of the microstates of the composite system are finite, according to the uncertainty principle, U^* has a natural line width. The spread in the values of U^* due to these effects plus the natural linewidths of subsystems 1 and 2 are sufficient to enable one particular microstate of subsystem 2 to be associated with an extremely large number of microstates of subsystem 1 to form an accessible microstate of the composite system.

In the **microcanonical ensemble** it is assumed that the total energy of a closed system is in the energy range U^* to $(U^* + \Delta U^*)$ where $\Delta U^* \ll U^*$. It is assumed that at thermal equilibrium all the microstates of the *closed* system, which have energies in the range U^* to $(U^* + \Delta U^*)$, are equally probable, but the probability of finding the closed system in any microstate having energy outside this range is zero. The above modified form of the postulate of equal *a priori* probabilities can be applied to the composite system shown in Figure 3.1(b). The analysis given in Sections 3.2–3.7 need only be modified to the extent that in Section 3.2, g_1 should stand for $D_1 \Delta U_1$ and g_2 should stand for $D_2 \Delta U_2$. The spreads in energy ΔU_1 and ΔU_2 are adjusted such that $(U_1 + U_2)$ is always within the range U^* to $(U^* + \Delta U^*)$.

For the case of thermal interaction between subsystems 1 and 2 in Figure 3.1(b), equation (3.5) becomes

$$P_2(U_2) \, \Delta U_2 = D_1 \, \Delta U_1 D_2 \, \Delta U_2 / g_T \tag{3.97}$$

where $P_2(U_2) \, \Delta U_2$ is the probability that subsystem 2 has energy in the range U_2 to $(U_2 + \Delta U_2)$ at thermal equilibrium, and g_T is the total number of accessible microstates. This probability distribution has a very sharp maximum at one particular value of U_2 specified by $\partial P_2 / \partial U_2 = 0$. This leads to the condition $D_1(\partial D_2 / \partial U_2) + D_2(\partial D_1 / \partial U_2) = 0$ for the maximum of P_2. Using equations (3.3) and (3.94), we find that P_2 is a maximum when $T_1 = T_2$. It is straightforward to extend the method of using the density of states to the rest of the analysis in Sections 3.2 to 3.7. The use of the density of states is probably the better approach for the advanced reader to adopt from now on.

3.9* THE THIRD LAW OF THERMODYNAMICS*

When the third law of thermodynamics was introduced in Section 1.10 of Chapter 1, two features were stressed. Firstly, according to the Planck formulation of the third law, **the entropy of a pure condensed substance in internal thermodynamic equilibrium should be zero at the absolute zero.** Secondly, the Nernst postulate that, **near the absolute zero, the entropy of a system is independent of the external parameters, such as external pressure and applied magnetic field,** is used to make predictions about the asymptotic behaviour of materials at temperatures near the absolute zero. In some cases these predictions are valid at temperatures up to 1K and above.

In our microscopic approach, the entropy of a system was defined initially in Section 3.3 as

$$S = k \log g \tag{3.22}$$

where g was the total number of accessible microstates (degeneracy) consistent with the macrostate of the system, which can be specified, by the values of the energy U, the volume V and the number of particles N in the system. Later, it was shown in Section 3.8* that, to allow for the splitting of the degeneracies of the energy levels, equation (3.22) should be replaced by the expression

$$S = k \log D \tag{3.91}$$

where D is the density of states, defined by equation (3.89). The variation of $\log g$ with U for a normal thermodynamic system was illustrated in Figure 3.3. As U is reduced towards U_0, the lowest energy eigenvalue of the N particle system, the slope β of the variation of $\log g$ with U tends to infinity, so that the absolute temperature $T = 1/\beta k$ tends to zero. It is reasonable to assume that the system would be in the ground state, if it were at the absolute zero of temperature.

If the ground state of the system were nondegenerate,

$$S = k \log 1 = 0 \ .$$

Thus, if the system were in a non-degenerate ground state at the absolute zero, its entropy would be exactly zero. It has been suggested that the ground is non-degenerate, though this conjecture is not yet definitely established. Since at room temperatures the entropy S given by equation (3.22) is of the order of Nk, for the entropy to be *negligible* at the absolute zero one must have $\log g_0 \ll N$, where g_0 is the degeneracy of the lowest energy level (ground state). Hence for the third law to be valid for a macroscopic system at $T = 0$, we must have $g_0 \ll e^N$.

In practice there may be several contributions to the entropy which are more or less independent. For example, the nuclear spins to not interact very strongly with the crystal lattice. It is found experimentally that the contribution of the nuclear spins to the total entropy of a system does not start decreasing significantly, for an applied magnetic field of 1 tesla, until the temperature is down to about 10^{-3}K. On the other hand, the contribution of the lattice vibrations to the total entropy has decreased very significantly at a temperature of 1K. (Reference: Problem 1.3.) To allow for such almost independent contributions to the total entropy of a system, **the third law** is now generally expressed in a form due to Simon [4]:

> **The contribution to the entropy of a system by each aspect which is internal thermodynamic equilibrium tends to zero as the absolute temperature tends to zero.**

The third law of thermodynamics should not be applied to substances, such as glasses, which are not in internal thermodynamic equilibrium. Many substances can have significant residual entropies at low temperatures due to the entropy of mixing, for example the mixing of the various isotopes making up the substance. The statement due to Simon enables us to apply the third law to other aspects of such systems, for example to the paramagnetic behaviour of the ions at low temperatures. It will be shown in later chapters that, when statistical mechanical methods based on quantum mechanics are applied, the contributions to the total entropy due to the different aspects of the system, which are in internal thermodynamic equilibrium, do tend to zero near the absolute zero, though the precise variation of entropy with temperature can be very different in different systems.

If 10^{-7} J of energy were added to a mole of copper, which is at the absolute zero, the temperature of the copper would rise to about 0.0168K. Assuming that the entropy of the copper was zero at the absolute zero, the entropy of the copper at 0.0168K would be about 1.18×10^{-5} J K^{-1}. (Reference: Problem 3.1.) According to equation (3.22)

$$g = \exp(S/k) \ . \tag{3.98}$$

Using $k = 1.38 \times 10^{-23}$ J K^{-1}, it follows from equation (3.98) that, at 0.0168K, the mole of copper would have a total of $\exp(8.6 \times 10^{17})$, that is ten to the power 3.7×10^{17}, accessible microstates. This is a very large number. By no stretch of the imagination could one say that at 0.0168K the properties of the copper were dominated by the properties of the ground state and the first excited state of the N particle system. This example illustrates how, for macroscopic systems, it is not the properties of the ground state which normally determine those properties of materials which are in agreement with the predictions of the Nernst form of the third law at temperatures as high as 1K. The entropy S is given by

$$S = k \log D \ . \tag{3.91}$$

where D is the density of states. It is the way that the density of states D varies with temperatures which leads to this asymptotic behaviour of materials at temperatures of 1K. Thus it is the properties of the quantum systems which leads to the particular form of the variation of the density of states D with energy which determines the variation of entropy with temperature at low temperatures. (References: Griffiths [5], ter Haar [6] and Wilks [7].)

One form of the third law often used in classical equilibrium thermodynamics is the law of the unattainability of the absolute zero, according to which it is impossible to reach the absolute zero of temperature in a finite number of operations. It is shown in Section 2.3.3 that the probability $P_{2i}(\varepsilon_{2i})$ that, when subsystem 2 is in thermal equilibrium with subsystem 1 in Figures 2.3(c) and 3.1(b), subsystem 2 is in its ith microstate having energy eigenvalue ε_{2i} is given by equation (2.26), which is

$$P_{2i}(\varepsilon_{2i}) = g_1(U_1)/g_T = g_1(U^* - \varepsilon_{2i})/g_T \ . \tag{3.99}$$

For normal thermodynamic systems, g_1 increases with U_1. Hence $g_1(U^* - \varepsilon_{2i})$ decreases as ε_{2i} is increased so that, according to equation (3.99), P_{2i} has its maximum value when ε_{2i} has its minimum value, which is when subsystem 2 is in its ground state. This is illustrated in Figure 2.4(b). It is an amusing result of statistical mechanics that, when subsystems 1 and 2 are in thermal equilibrium in Figure 3.1(b), subsystem 2 is more likely to be in its ground state, which means that it would be at the absolute zero, than in any other **microstate**. This is true whatever the relative sizes of subsystems 1 and 2. For example, subsystem 1 could be a stone thrown into a lake the size of the Atlantic Ocean (subsystem 2). At thermal equilibrium, the lake is more likely to be found in its ground state, corresponding to $T = 0$, than in any other **microstate**. Due to the enormous number of microstates accessible to subsystem 2 at thermal equilibrium, for macroscopic systems the probability of finding subsystem 2 in one particular microstate (the ground state) is so exceedingly small that it can be ignored in practice.

To obtain the probability $P_2(U_2)$ that at thermal equilibrium subsystem 2 has energy U_2 equal to ε_{2i}, we must multiply $P_{2i}(\varepsilon_{2i})$ by $g_2(U_2)$, the number of microstates accessible to subsystem 2 when it has energy U_2, to give equation (2.21). It was shown in Section 3.2 and Appendix 4 that this product has an extremely sharp maximum and that, for macroscopic systems ($N > 10^{18}$), the overwhelming probability is that at thermal equilibrium one would find subsystem 2 in a microstate having an energy eigenvalue within 1 part in 10^7 of the value $\tilde{U}_2$ at which $P_2(U_2)$ is a maximum. Hence in practice, with macroscopic systems, one can rule out the possibility of finding subsystem 2 in its ground state in a time equal to the age of the Earth.

Whilst the finite probability of finding subsystem 2 in Figure 3.1(b) in its ground state can be ignored for **macroscopic systems**, this probability can be very significant if subsystem 2 is a **microscopic system**. Assume that subsystem 1 is so large that, for values of ε_{2i} in the range of interest, the absolute temperature T_1 of subsystem 1 can be assumed to be constant. According to equation (3.19), we have

$$\partial \log g_1/\partial U_1 = 1/kT_1 \, .$$

Integrating with respect to U_1, at constant V_1 and N_1, between the limits $U_1 = U^*$ and $U_1 = (U^* - \varepsilon_{2i})$, assuming that T_1 is constant, we have

$$\log g_1(U^* - \varepsilon_{2i}) - \log g_1(U^*) = -\varepsilon_{2i}/kT_1$$

Taking exponentials, we have

$$g_1(U^* - \varepsilon_{2i}) = g_1(U^*)\exp[-\varepsilon_{2i}/kT_1] \, . \tag{3.100}$$

Substituting in equation (3.99), since $g_1(U^*)$ is a constant, we have

$$P_{2i} \propto \exp[-\varepsilon_{2i}/kT_1] \, . \tag{3.101}$$

This is the **Boltzmann distribution**, which will be discussed in detail in Chapter 4. The exponential decrease of P_{2i} with increasing ε_{2i} is only valid when subsystem 1 is large enough to be treated as a heat reservoir. In other cases, P_{2i} still decreases with increasing ε_{2i}, as illustrated in Figure 2.4(b), but due to the finite change in the temperature of subsystem 1, the decrease is not exponential.

3.10 USE OF INFORMATION THEORY

To illustrate the relevant aspects of information theory, consider a loaded die. Let the probabilities P_i of throwing the numbers 1, 2, 3, 4, 5 and 6 be 0.1, 0.1, 0.1, 0.1, 0.1 and 0.5 respectively. Knowledge of these probabilities gives us some information about the system. For example, we know that if we throw a die we are more likely to get a six than any other number, The uncertainity H is defined as

$$H = -\langle \log P_i \rangle = -\Sigma P_i \log P_i \ . \tag{3.102}$$

In our example, there are five states for which $P_i = 0.1$ and one state for which $P_i = 0.5$. In this example, the uncertainty is

$$H = -[5(0.1 \log 0.1) + 0.5 \log 0.5]$$
$$= -[0.5(-2.303) + 0.5(-0.693)] = +1.50 \ .$$

If the six possibilities were equally likely, all the P_i would be equal to 1/6, and

$$H = -\Sigma(1/6) \log(1/6) = -6 \times (1/6)(-1.79) = 1.79 \ .$$

By using other numerical values for P_i, subject to the restriction that $\Sigma P_i = 1$, the reader can check that the uncertainty H is a maximum when all the six possibilities are equally probable. This is to be expected since, if all six numbers are equally probable, one cannot make any prediction that, when a die is thrown, one is more likely to get one particular number than any of the others.

Instead of using the uncertainty H defined by equation (3.102), some writers prefer to use the information I defined by the relation

$$I = \langle \log P_i \rangle = \Sigma P_i \log P_i = -H \ . \tag{3.103}$$

When the uncertainty H is a maximum the information I is a minimum. In the case of a closed thermodynamic system, if all the accessible microstates of the closed system are equally probable, the uncertainty is a maximum and the information is a minimum. Conversely, if the uncertainty is a maximum, corresponding to minimum information, all the accessible microstates of the closed system should be equally probable. This is the principle of equal *a priori* probabilities (Postulate 2). Instead of the postulate of equal *a priori* probabilities, one could have assumed that, when a closed system is in thermodynamic equilibrium, either the uncertainty is a maximum, or the information is a minimum, subject to the constraints of the system. (Reference: Turner and Betts [8].) If this alternative form of Postulate 2 were adopted, one approach would be to show that the postulate based on information theory leads to the principle of equal *a priori* probabilities in the case of a closed system. The principle of equal *a priori* probabilities could then be applied in the way described earlier in this Chapter. The principle of equal *a priori* probabilities was chosen as our starting point, so as to avoid giving the impression that information theory is a prerequisite for statistical mechanics. In fact information theory is an alternative way of presenting the axiomatics of the subject. In both approaches, the validity of the form of Postulate 2 chosen depends on the fact that predictions based on the postulate are in agreement with the experimental results. The choice of which form of Postulate 2 the reader wishes to adopt from now on, is a matter of taste.

Putting $P_i = 1/g$ in equation (3.102) gives

$$H = -\Sigma (1/g)\log(1/g) = \Sigma (1/g)\log g .$$

Since there is a total of g terms in the summation, all of which are equal to $(1/g)\log g$, we have $H = \log g$. Comparison with equation (3.22) suggests that the entropy can be rewritten in the forms

$$S = kH = -kI . \tag{3.104}$$

Equation (3.104) illustrates the close connection between the uncertainty H and the entropy S, and the close connection between the information I and the negentropy, which is equal to $-S$. Many approaches to statistical mechanics are based on Boltzmann's definition of entropy as

$$S = -k \Sigma P_i \log P_i = -k \langle \log P_i \rangle \tag{3.105}$$

where P_i is the probability that the system is in its ith quantum state. Equation (3.105) is a combination of equations (3.104) and (3.102). According to the principle of equal *a priori* probabilities, a closed system is equally likely to be found in any of its g accessible microstates. Equation (3.105) becomes

$$S = -k \Sigma (1/g)\log(1/g) = k \log g .$$

This is the same as equation (3.22). [The Boltzmann definition of entropy will be developed more fully in Section 7.1.2* of Chapter 7*.]

3.11 DISCUSSION

In this Chapter, the laws of classical equilibrium thermodynamics were interpreted using statistical mechanics. Our approach was based on the postulates given in Section 3.1. The important new quantity introduced was the number of microstates g accessible to the system, when it was in a particular macrostate. For convenience, we introduced the entropy S

defined in terms of g by the equation

$$S = k \log g .\qquad(3.22)$$

In Section 3.8* it was pointed out that, due to the splitting of the degeneracies of the energy levels, it is better in the general case to use the density of states D and to define the entropy using equation (3.91). It is of interest to note that, in our interpretation of classical equilibrium thermodynamics, we did not at any time need to specify the precise value of the number of accessible microstates. All that was assumed was that g was an extremely large number and that the variation of $\log g$ with U was of the general form given in Figure 3.3. This parallels the fact that classical equilibrium thermodynamics only gives relations between thermodynamic state variables, allowing one to calculate the value of one thermodynamic state variable from the values of other thermodynamic state variables. To calculate the value of individual thermodynamic quantities from the appropriate atomic models, one must use statistical mechanics. This approach will be developed from Chapter 5 onwards.

The consideration of thermal equilibrium in Section 3.2 led up to the idea of the zeroth law of thermodynamics in Section 3.3 and to the definition of absolute temperature by the equation

$$T = (\partial U/\partial S)_{V,N} .\qquad(3.23)$$

The condition for thermal equilibrium, for the example illustrated in Figure 3.1(b), reduced to $T_1 = T_2$.

The most convenient way of expressing the second law of thermodynamics was in the form of the principle of the increase of entropy, according to which, in the approach to thermal equilibrium, the entropy of a system plus surroundings, increases except in the case of a reversible change, when the total entropy remains the same. It was shown in Section 3.7 that for the case of general thermodynamic equilibrium illustrated in Figure 3.4(b), when the two subsystems are in thermodynamic equilibrium.

$$S_1(U_1, V_1, N_1) + S_2(U_2, V_2, N_2) \text{ is a maximum}\qquad(3.64)$$

subject to the constraints

$$U_1 + U_2 = U^*$$
$$V_1 + V_2 = V^*$$
$$N_1 + N_2 = N^*$$

where S_1 and S_2 are the entropies subsystems 1 and 2 in Figure 3.4(b) would have if they were isolated systems having energies U_1 and U_2, volumes V_1 and V_2 and consisted of N_1 and N_2 particles respectively. This led to the condition that, at thermodynamic equilibrium, $T_1 = T_2, p_1 = p_2$ and $\mu_1 = \mu_2$, where the absolute temperature T is given by equation (3.23) and the pressure p and chemical potential μ are given by

$$p = T(\partial S/\partial V)_{U,N} \qquad (3.70)$$

$$\mu = -T(\partial S/\partial N)_{U,V} \qquad (3.78)$$

Hence at thermodynamic equilibrium, the *intensive* variables p, μ and T are the same in both subsystems.

The entropy $S(U, V, N)$ is an extensive variable which can be expressed in terms of the extensive variables U, V and N. If the fundamental relation relating S to U, V and N is known from experiment or from statistical mechanics, the intensive variables T, p and μ can be calculated using equations (3.23), (3.70) and (3.78) respectively.

In principle, one could calculate the number of accessible microstates and hence the entropy S in terms of U, V and N from the solutions of the Schrödinger time independent equation. In practice it is more usual to calculate thermodynamic quantities from atomic theory using the partition function. This approach will be developed in Chapter 5.

In practice it is easier to measure the intensive variables T and p rather than the extensive variables U and S. Experiments are often carried out at constant temperature and at either constant volume or constant pressure. If a thermodynamic system is kept at constant temperature during the approach to internal thermodynamic equilibrium, heat must either enter or leave the system from the surroundings to keep the temperature constant. In such conditions the entropy approach developed for a closed system is not always the most appropriate. It is possible to introduce new functions using Legendre transformations and still keep all the information about the system. (Reference: Callen [9].) The most important new functions are the Helmholtz free energy F, which is equal to $(U - TS)$, and can be expressed in terms of the natural variables T, V and N, and the Gibbs free energy G which is equal to $(U - TS + pV)$ and can be expressed in terms of the natural variables p, T and N. The Helmholtz free energy F is useful for changes taking place at constant T and V, whereas the Gibbs free energy is useful for changes taking place at constant p and T. The properties of the Helmholtz free energy will be developed from statistical mechanics in Chapter 6. Some of the properties of the Gibbs free energy G will also be discussed in Chapter 6.

REFERENCES

[1] Reif, F., *Fundamentals of Statistical and Thermal Physics*, page 624, McGraw-Hill, 1965.

[2] Kittel, C. and Kroemer, H., *Thermal Physics*, Second Edition, W. H. Freeman, 1980.

[3] Landau, L. D. and Lifshitz, E. M., *Statistical Physics*, pages 16–22, Pergamon Press, 1959.

[4] Simon, F. E., *Physics*, **4**, 1089 (1937).

[5] Griffiths, R. B., *Journ. of Math. Phys.*, **6**, 1447 (1965).

[6] ter Haar, D., *Elements of Thermostatics*, Chapter 9, Holt Rinehart and Winston, 1966.

[7] Wilks, J., *The Third Law of Thermodynamics*, Oxford University Press, Oxford 1961.

[8] Turner, R. E. and Betts, D. S., *Introductory Statistical Mechanics*, Chapter 3, Sussex University Press, 1974.

[9] Callen, H. B., *Thermodynamics*, Chapter 5, John Wiley, 1960.

PROBLEMS

Problem 3.1

Assume that, at very low temperatures, the molar heat capacity of copper is equal to $7 \times 10^{-4}T$ J K^{-1} mole^{-1} where T is the absolute temperature. Show that if one erg (10^{-7} J) of heat is added to a mole of copper, which is initially at the absolute zero, the temperature of the copper rises to 0.0169K and the increase in the entropy of the copper is 1.18×10^{-5} J K^{-1}. The volume of the copper is kept constant. What is the number of microstates accessible to the copper? (Boltzmann's constant $k = 1.3807 \times 10^{-23}$ J K^{-1}.)

Problem 3.2

The entropy S of an ideal monatomic gas is given by

$$S(U, V, N) = Nk[\log(gV/N) + (3/2)\log(mU/3N\pi\hbar^2) + (5/2)]$$

where V is the volume, N is the number of particles, g is the statistical weight (a constant), m is the mass of an atom and U is the total energy. (This is equation (5.79) of Chapter 5.) Use equations (3.23), (3.70) and (3.78) to determine the temperature, pressure and chemical potential of the ideal monatomic gas in terms of U, V and N.

Problem 3.3

The entropy S of a gas is given by $S = CV^{1/4}U^{3/4}$ where U is the energy, V is the volume and C is a constant. Determine the temperature of the gas. Show that $p = U/3V$. Show that the chemical potential of the gas is zero.

Chapter 4

The Boltzmann Distribution for a Small System in Thermal Equilibrium with a Heat Reservoir

---

4.1 INTRODUCTION

In Chapter 3, the general case of two subsystems in thermal equilibrium, making up the **closed composite system** shown in Figure 3.1(b), was considered. In practice, one often has the case of a thermodynamic system in thermal equilibrium with its surroundings, as shown previously in Figure 1.1(a). In this latter case, the thermal capacity of one of the subsystems is very much larger than the other, as shown in Figure 4.1(a). It will be assumed that any change in the energy of the smaller system in Figure 4.1(a) (which corresponds to subsystem 2 of Figure 3.1(b)) does not have any significant effect on the temperature of the larger system in Figure 4.1(a), (which corresponds to subsystem 1 of Figure 3.1(b)). The larger system in Figure 4.1(a) will act as a heat reservoir of constant absolute temperature T. The smaller system in Figure 4.1(a) can still be a macroscopic system, provided its thermal capacity is very much less than that of the heat reservoir. For example, one could have a lump of metal at the bottom of the Atlantic Ocean. The sea and the Earth would act as a heat reservoir for the lump of metal. A typical microscopic example would be a single atomic system in thermal equilibrium with the rest of the atoms in a crystal lattice.

4.2 THE BOLTZMANN DISTRIBUTION

Consider the small system in thermal equilibrium with a constant volume heat reservoir shown in Figure 4.1(a). (The adjective small will be used in this Chapter to remind the reader that the thermal capacity of the system is very much less than the thermal capacity of the heat reservoir. The 'small' system can still be a macroscopic system.) It will be assumed

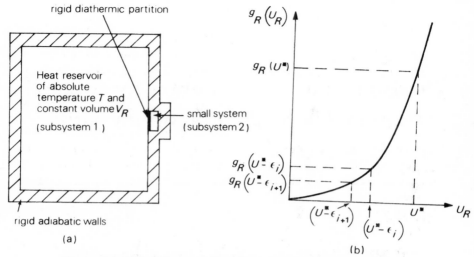

Figure 4.1—The Boltzmann distribution. (a) The small system is in thermal equilibrium with a heat reservoir of constant absolute temperature T. The small system and heat reservoir are inside rigid, outer, adiabatic walls and form a closed composite system. This is a special case of Figure 3.1(b). (b) The variation of $g_R(U_R)$, the number of microstates accessible to the heat reservoir, with U_R the energy of the heat reservoir. If the energy of the small system goes up from ε_i to ε_{i+1}, the energy of the heat reservoir goes down from $(U^* - \varepsilon_i)$ to $(U^* - \varepsilon_{i+1})$ and the number of microstates accessible to the heat reservoir goes down.

that the small system and heat reservoir in Figure 4.1(a) can exchange energy through the rigid diathermic partition, but they do not interchange particles, so that N, the number of particles in the small system, is constant. Both the heat reservoir and the small system are inside a rigid, adiabatic, outer enclosure. The heat reservoir plus small system can be treated as a *closed* isolated composite system of fixed total energy, fixed total volume and fixed total number of particles.

Since the *total* energy U^* of the heat reservoir plus small system is constant, the energy U_R of the heat reservoir, when the small system is in its ith microstate with energy eigenvalue ε_i, is

$$U_R = U^* - \varepsilon_i . \qquad (4.1)$$

The energy eigenvalue ε_i is the total energy of the small system and includes the kinetic energies and the potential energies of all the N particles making up the small system. The value of $\varepsilon_i(V, N)$ depends on the volume V of the small system through the boundary conditions used to solve Schrödinger's equation for the N particle system. The energy eigenvalue ε_i also depends on the number of particles in the small system.

According to the postulate of equal *a priori* probabilities, which is postulate 2 of Section 3.1, when the *closed* composite system made up of the small system in thermal contact with the heat reservoir in Figure 4.1(a) has had time to reach thermal equilibrium, one is equally likely to find the closed composite system in any one of the microstates accessible to the closed composite system, subject to the condition that the total energy is conserved. The volume V_R and the number of particles N_R in the heat reservoir are kept constant by the rigid diathermic partition and the rigid adiabatic outer walls. Hence the number of microstates accessible to the heat reservoir depends only on the energy of the heat reservoir. Let $g_R(U_R)$ be the number of microstates accessible to the heat reservoir, when it has energy U_R equal to $(U^* - \varepsilon_i)$. According to equation (2.19), since in this case $g_1 = g_R$ and $g_2 = 1$, the number of microstates accessible to the composite system, in which the small system is in its ith microstate, is equal to $g_R \times 1$. According to the principle of equal *a priori* probabilities, the probability P_i that at thermal equilibrium the small system is in its ith microstate is equal to the ratio of the number of microstates accessible to the composite system in which subsystem 2 is in its ith microstate to the total number g_T of microstates accessible to the composite system.

Hence
$$P_i = g_R(U_R)/g_T = Cg_R(U^* - \varepsilon_i) \tag{4.2}$$

where $C = 1/g_T$ is a constant. (Equation (4.2) is the same as equation (2.26) which was developed in Section 2.3.3 using a simple numerical example).

The function $\log[g_R(U^* - \varepsilon_i)]$ will be expanded in a Taylor series. According to equation (A1.28) of Appendix 1

$$f(a + h) = f(a) + hf'(a) + \ldots .$$

Let $a = U^*$ and let $h = -\varepsilon_i$. Let f stand for $\log g_R$. Neglecting terms of order ε_i^2 and higher order, we have

$$\log[g_R(U^* - \varepsilon_i)] = \log[g_R(U^*)] - \varepsilon_i(\partial \log g_R/\partial U_R) .$$

The differential coefficient in the above equation is evaluated at the energy $U_R = U^*$. From equations (3.14) and (3.19),

$$(\partial \log g_R/\partial U_R) = \beta = 1/kT \tag{4.3}$$

where β is the temperature parameter and T is the constant absolute temperature of the heat reservoir. The constant k is Boltzmann's constant. Hence

$$\log[g_R(U^* - \varepsilon_i)] = \log[g_R(U^*)] - \beta\varepsilon_i .$$

Rearranging and taking exponentials, we obtain

$$g_R(U^* - \varepsilon_i) = g_R(U^*)\exp(-\beta\varepsilon_i) . \tag{4.4}$$

In equation (4.4), $g_R(U^*)$ is equal to the number of microstates that would be accessible to the heat reservoir, if the heat reservoir had all the energy U^* of the composite system. Hence, $g_R(U^*)$ is a constant. Substituting in equation (4.2), we have

$$P_i = Cg_R(U^*)\exp(-\beta\varepsilon_i) = A\exp(-\beta\varepsilon_i) = A\exp(-\varepsilon_i/kT) \ . \quad (4.5)$$

The probability that the small system is in its ith microstate, which has energy eigenvalue ε_i, is proportional to $\exp(-\beta\varepsilon_i)$. The quantity $\exp(-\beta\varepsilon_i)$ is called the Boltzmann factor. Equation (4.5) is generally called the **Boltzmann distribution**, though it is sometimes referred to as the **canonical distribution**. [A slightly different way of developing the Boltzmann distribution was given in Section 3.9*, leading up to equation (3.100).]

It follows from equation (4.5) that the ratio of the probability P_{i+1} that the small system is in its $(i + 1)$th microstate, which has energy eigenvalue ε_{i+1}, to the probability P_i that it is in its ith microstate, which has energy eigenvalue ε_i, is

$$P_{i+1}/P_i = \exp(-\beta\varepsilon_{i+1})/\exp(-\beta\varepsilon_i) = \exp[-\beta(\varepsilon_{i+1} - \varepsilon_i)] \ . \quad (4.6)$$

If $\varepsilon_{i+1} > \varepsilon_i$, then, according to equation (4.6), P_{i+1} is less than P_i. If $\varepsilon_{i+1} > \varepsilon_i$, the energy of the heat reservoir is less, when the small system is in its $(i + 1)$th microstate, than when the small system is in its ith microstate. It was illustrated in Section 2.5.2 that for a normal macroscopic system, such as the heat reservoir, the number of microstates $g_R(U_R)$ accessible to the heat reservoir increases very rapidly as its energy U_R increases, so that if $\varepsilon_{i+1} > \varepsilon_i$, $g_R(U^* - \varepsilon_{i+1})$ is less than $g_R(U^* - \varepsilon_i)$, as shown in Figure 4.1(b). Hence, according to equation (4.2), P_{i+1} should be less than P_i. It is the decrease of $g_R(U^* - \varepsilon_i)$ with increasing ε_i, corresponding to a decrease in U_R, which leads to the decrease of P_i with increasing ε_i in equation (4.5).

Since the small system shown in Figure 4.1(a) must be in one of its possible microstates, ΣP_i must be equal to unity. Using equation (4.5) we have

$$\Sigma P_i = \Sigma A\exp(-\beta\varepsilon_i) = A\sum\exp(-\beta\varepsilon_i) = 1 \ .$$

Hence
$$A = \frac{1}{\underset{\text{(states)}}{\sum\exp(-\beta\varepsilon_i)}} \ . \quad (4.7)$$

In equation (4.7), the summation is over all the possible microstates of the small system. Equation (4.5) can be rewritten in the form

$$P_i = \frac{\exp(-\beta\varepsilon_i)}{Z} = \frac{\exp(-\varepsilon_i/kT)}{Z} \quad (4.8)$$

where
$$Z = \sum_{\text{(states)}} \exp(-\beta \varepsilon_i) = \sum_{\text{(states)}} \exp(-\varepsilon_i/kT) \qquad (4.9)$$

is the **partition function**. The summation in equation (4.9) is over all the microstates of the small system. In German, the partition function is called the Zustandssumme, which translated means 'sum over states'. If a particular energy level of the small system is degenerate, each of the associated microstates must be treated separately, when equation (4.9) is applied. In chemistry, the partition function is generally denoted by Q. The absolute temperature T in equation (4.9) is the absolute temperature of the heat reservoir.

Assume that, when the small system is in its ith microstate, the mean value of the variable α is α_i. Putting $x_i = \alpha_i$ in equation (A1.7) of Appendix 1 and substituting for P_i from equation (4.8), we find that the ensemble average of the variable α, when the small system is in thermal equilibrium with the heat reservoir, is

$$\bar{\alpha} = \sum P_i \alpha_i = \frac{\sum\limits_{\text{(states)}} \alpha_i \exp(-\beta \varepsilon_i)}{Z} . \qquad (4.10)$$

4.3 A TWO LEVEL SYSTEM

Consider a small system which can be in one of two non-degenerate microstates, which have energy eigenvalues ε_0 and ε_1 respectively as shown in Figure 4.2(a). The energy gap ε is given by

$$\varepsilon = \varepsilon_1 - \varepsilon_0 .$$

The two level system is in thermal equilibrium with a heat reservoir, which is at an absolute temperature T. This example will be applied to paramagnetism in Section 4.4. Since the small system has only two microstates, according to equation (4.9) the partition function (sum over states) is

$$Z = \exp(-\beta \varepsilon_0) + \exp(-\beta \varepsilon_1) .$$

Applying equation (4.8), we have

$$P_0 = \exp(-\beta \varepsilon_0)/Z = \exp(-\beta \varepsilon_0)/[\exp(-\beta \varepsilon_0) + \exp(-\beta \varepsilon_1)] \quad (4.11)$$

$$P_1 = \exp(-\beta \varepsilon_1)/Z = \exp(-\beta \varepsilon_1)/[\exp(-\beta \varepsilon_0) + \exp(-\beta \varepsilon_1)] . \quad (4.12)$$

It is useful to introduce the characteristic temperature θ, defined as

$$\theta = (\varepsilon_1 - \varepsilon_0)/k = \varepsilon/k . \qquad (4.13)$$

Multiplying both the numerators and the denominators in both equations

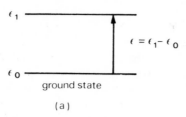

(a)

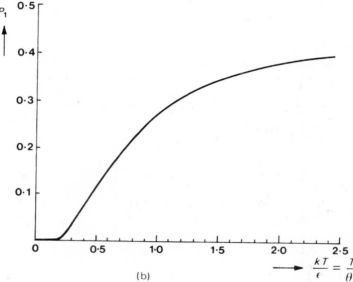

(b)

Figure 4.2—(a) A non-degenerate two level system, which is in thermal equilibrium with a heat reservoir of absolute temperature T. (b) The variation of the probability P_1, that the system is in the upper level with the ratio T/θ, where $\theta = \varepsilon/k$ is the characteristic temperature.

(4.11) and (4.12) by $\exp(\beta\varepsilon_0)$ and using equation (4.13), we obtain

$$P_0 = 1/[1 + \exp(-\beta\varepsilon)] = 1/[1 + \exp(-\theta/T)] \tag{4.14}$$

$$P_1 = \exp(-\beta\varepsilon)/[1 + \exp(-\beta\varepsilon)] = \exp(-\theta/T)/[1 + \exp(-\theta/T)] \tag{4.15}$$

It can be seen from equations (4.14) and (4.15) that both P_0 and P_1 depend on the absolute temperature T of the heat reservoir. The values of P_1 for various values of the ratio $T/\theta(= kT/\varepsilon)$ are plotted in Figure 4.2(b). If the absolute temperature T of the heat reservoir were zero, $P_1 = 0$ and $P_0 = 1$ and the system would be in its ground state. When $T \ll \theta$, that is when kT is much less than the energy gap ε, P_1 is negligible. For example, when $T = 0.01\theta$, 0.1θ and 0.2θ, P_1 is equal to 3.72×10^{-44}, 4.54×10^{-5} and 6.69×10^{-3} respectively. P_1 begins to increase significantly when

$T > 0.2\theta$. The maximum rate of increase of P_1 is in the temperature range $T = 0.3\theta$ to 0.8θ. As T increases further, P_1 tends towards a limiting value of 0.5. For example, if $T = 10\theta$, P_1 is equal to 0.475. [Reference: Problem 4.1].

According to equation (4.10), the ensemble average energy of the two level system is

$$\bar{\varepsilon} = \sum_{\text{(states)}} \varepsilon_i P_i = \frac{\varepsilon_0 \exp(-\beta\varepsilon_0)}{Z} + \frac{\varepsilon_1 \exp(-\beta\varepsilon_1)}{Z}$$

$$= \frac{[\varepsilon_0 \exp(-\beta\varepsilon_0) + \varepsilon_1 \exp(-\beta\varepsilon_1)]}{[\exp(-\beta\varepsilon_0) + \exp(-\beta\varepsilon_1)]} = \frac{[\varepsilon_0 + \varepsilon_1 \exp(-\theta/T)]}{[1 + \exp(-\theta/T)]} .$$

For example, if $\theta = T$, that is if kT is equal to the energy gap ε, we have

$$\bar{\varepsilon} = 0.731\varepsilon_0 + 0.269\varepsilon_1 .$$

It is sometimes convenient to calculate the partition function using a zero of energy equal to the energy of the ground state of the system. If Z' is the partition function measured on this energy scale

$$Z' = \exp(0) + \exp(-\beta\varepsilon) = 1 + \exp(-\beta\varepsilon)$$

Comparing with

$$Z = \exp(-\beta\varepsilon_0) + \exp(-\beta\varepsilon_1) = \exp(-\beta\varepsilon_0)[1 + \exp(-\beta\varepsilon)]$$

it can be seen that $Z = \exp(-\beta\varepsilon_0)Z'$ (4.16)

where Z' is the partition function calculated by taking the energy of the ground state as the zero of the energy scale, and Z is the partition function calculated using an energy scale on which the energy of the ground state is equal to ε_0. (It is left as an exercise for the reader to check that equation (4.16) is valid in the general case of a multilevel system.)

4.4 IDEAL PARAMAGNETISM

4.4.1 Introduction

Paramagnetism arises when some of the atoms in a crystal have a resultant magnetic dipole moment, associated with either electron orbital angular momentum or electron spin or both. When a magnetic field is applied, these magnetic moments tend to align themselves in the direction of the applied magnetic field. A typical example of a paramagnetic salt is chromium potassium alum which has the formula $Cr_2(SO_4)_3.K_2SO_4.24H_2O$. In this example the paramagnetism is due to the Cr^{+++} ions in the crystal. Each Cr^{+++} ion has three unpaired electrons, each of which has an atomic magnetic dipole moment of one Bohr magneton,

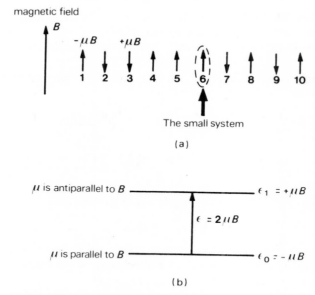

Figure 4.3—(a) N spatially separated independent atomic magnetic moments (an idealised spin system). Each magnetic dipole can point either parallel or anti-parallel to the magnetic field. One of the magnetic moments is chosen as the small system for the application of the Boltzmann distribution. (b) Energy level diagram for the chosen magnetic moment. When the magnetic moment is parallel to **B**, its potential energy is $(-\mu B)$ and when the magnetic moment is anti-parallel to **B** its potential energy is $(+\mu B)$. The energy gap ε is equal to $2\mu B$.

which is equal to 9.27×10^{-24} A m². (When the electron shells in atoms are built up, if two electrons have the same values of the quantum numbers n, l and m, but opposite spins, they are said to pair off, and in a magnetic field the effects of their magnetic moments compensate each other.) For each Cr^{+++} ion there are two sulphur atoms, one potassium atom, 20 oxygen atoms and 24 hydrogen atoms so that the Cr^{+++} ions are well separated from each other and, except at very low temperatures, it is reasonable, as a first approximation, to neglect the magnetic interactions between the Cr^{+++} ions.

To simplify the discussion, we shall consider a simplified model in which each paramagnetic atom has only one unpaired electron, which is in an s state of zero orbital angular momentum. It will be assumed that the magnetic interactions between these unpaired electrons can be neglected. When a magnetic field is applied to the specimen, the electron spin angular momentum of each unpaired electron can have a component of $\hbar/2$ either parallel to or anti-parallel to the magnetic field[†]and each unpaired electron has an atomic magnetic moment μ equal to one Bohr magneton pointing either anti-parallel to or parallel to the magnetic field, as shown in Figure 4.3(a). The potential energies of the atomic magnetic dipoles are different in these two states.

4.4.2 The potential energy of a magnetic dipole in a magnetic field

It is shown in text books on electromagnetism that the torque acting on a magnetic dipole of magnetic moment $\boldsymbol{\mu}$ in a magnetic field of induction **B** is given by the vector product $\boldsymbol{\mu} \times \mathbf{B}$. Consider the magnetic dipole shown in Figure 4.4. The magnetic dipole is pointing at an angle $(\pi/2 + \theta)$ to the

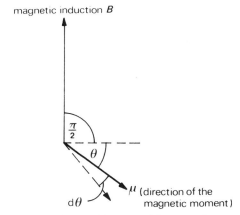

Figure 4.4—Calculation of the potential energy of a magnetic dipole in a magnetic field.

direction of the magnetic field. The numerical value of the torque T acting on the magnetic dipole is $\mu B \sin(\pi/2 + \theta)$, which is $\mu B \cos \theta$. The work dW done in rotating the magnetic dipole from an inclination of $(\pi/2 + \theta)$ to the direction of the magnetic field to an inclination of $(\pi/2 + \theta + d\theta)$ is

$$dW = T \, d\theta = \mu B \cos \theta \, d\theta \ .$$

The total work done in rotating the magnetic dipole from $\theta = 0$, when the dipole moment is pointing in a direction perpendicular to the magnetic field, to $\theta = \pi/2$, when the magnetic dipole is pointing in a direction oppo-

[†]Strictly this is the local magnetic field. If the magnetic interactions between the unpaired electrons can be ignored, the local macroscopic magnetic field is equal to the applied magnetic field. This approximation is reasonable for weak paramagnetic substances, whose relative permeability is only fractionally greater than unity.

site to the magnetic field, is

$$W = \int_0^{\pi/2} \mu B \cos \theta = \mu B \ .$$

If the potential energy of the magnetic dipole is assumed to be zero when $\theta = 0$, according to classical electromagnetism the potential energy of a magnetic dipole $\boldsymbol{\mu}$ pointing in a direction opposite to a magnetic field of magnetic induction $\mathbf{B}$ is $+\mu B$. Similarly, when the magnetic dipole is pointing in the direction of the magnetic field, according to classical electromagnetism, the potential energy is $-\mu B$.

4.4.3 Curie's law

In an applied magnetic field, the potential energy of each of the unpaired electrons in the idealised spin system shown in Figure 4.3(a) is either $-\mu B$ or $+\mu B$ depending on whether the magnetic moment points parallel to or anti-parallel to the magnetic field. This is illustrated in the energy level diagram for one of the unpaired electrons, shown in Figure 4.3(b). It is being assumed that the magnetic interactions between the unpaired electrons can be neglected. In the application of the Boltzmann distribution law in this Section, one of the unpaired electrons will be treated as the small system and the rest of the crystal will be treated as a heat reservoir at constant absolute temperature T. This problem is the same as the two level example treated in Section 4.3 and illustrated in Figure 4.2(a), provided $\varepsilon_0 = -\mu B$ and $\varepsilon_1 = +\mu B$, so that the energy gap ε is equal to $2\mu B$.

According to equation (4.9), if $\varepsilon_0 = -\mu B$ and $\varepsilon_1 = +\mu B$, the partition function is

$$Z = \exp(\mu B/kT) + \exp(-\mu B/kT) \ . \tag{4.17}$$

According to equation (4.11), the probability that, at thermal equilibrium, the atomic magnetic dipole moment is pointing parallel to the magnetic field is

$$P_0 = \exp(\mu B/kT)/[\exp(\mu B/kT) + \exp(-\mu B/kT)] \ . \tag{4.18}$$

Similarly, the probability that the atomic magnetic moment is pointing anti-parallel to the magnetic field is

$$P_1 = \exp(-\mu B/kT)/[\exp(\mu B/kT) + \exp(-\mu B/kT)] \ . \tag{4.19}$$

In this case the characteristic temperature, given by equation (4.13), is

$$\theta = \varepsilon/k = 2\mu B/k \ . \tag{4.20}$$

For a typical magnetic field of 1 tesla with $\mu = 9.27 \times 10^{-24}$ A m^2 (one Bohr magneton) and $k = 1.38 \times 10^{-23}$ J K^{-1}, we find that $\theta = 1.34$K.

For example, according to equations (4.18) and (4.19), when $T = 0.1\theta$, $T = \theta$ and $T = 10\theta$, P_0 is equal to 0.99995, 0.731 and 0.525 respectively, and P_1 is equal to 0.00005, 0.269 and 0.475 respectively. It can be seen that as the temperature decreases below the characteristic temperature θ, P_0 tends to unity as more and more of the magnetic moments are in the lower energy state of energy $-\mu B$, and point in the direction of **B**. As T increases above the characteristic temperature θ, the number of magnetic moments in the state of energy $+\mu B$ increases, and both P_0 and P_1 tend to limiting values of 0.5.

There are large number of paramagnetic atoms in a typical macroscopic sample. Each of these paramagnetic atoms can be treated as a small system in thermal equilibrium with the rest of the crystal. Equations (4.18) and (4.19) should be applicable to each of these, and it is reasonable to assume that the average number of unpaired electrons in states 0 and 1 with potential energies $\varepsilon_0 = -\mu B$ and $\varepsilon_1 = +\mu B$ are $N_0 = NP_0$ and $N_1 = NP_1$ respectively, where N is the total number of unpaired electrons. In such conditions, one often refers to NP_0 and NP_1 as the *populations* of the two states with energies $-\mu B$ and $+\mu B$. The term **relative populations of the states** is often used for the ratio of N_1 to N_0. Using equations (4.18) and (4.19) we have

$$N_1/N_0 = \exp(-\varepsilon/kT) = \exp(-2\mu B/kT) \ . \tag{4.21}$$

The magnetic dipole moments of atomic nuclei are generally of the order of a nuclear magneton, which is equal to 5.05×10^{-27} A m^2. For a nuclear magnetic moment of one nuclear magneton in a magnetic field of 1 tesla, the characteristic temperature is 7.3×10^{-4}K.

According to equation (4.10), for the case of a small system consisting of only one unpaired electron, the mean magnetic moment parallel to the magnetic field is

$$\bar{\mu} = \Sigma P_i\mu_i = P_0\mu - P_1\mu \ .$$

Using equations (4.18) and (4.19), we have

$$\bar{\mu} = \mu[\exp(\mu B/kT) - \exp(-\mu B/kT)]/[\exp(\mu B/kT) + \exp(-\mu B/kT)] \ .$$

Using the relation

$$\tanh y = [\exp(y) - \exp(-y)]/[\exp(y) + \exp(-y)]$$

with $y = \mu B/kT$, we obtain

$$\bar{\mu} = \mu \tanh(\mu B/kT) \ . \tag{4.22}$$

For N independent, spatially separated, non-interacting unpaired electrons the mean magnetic moment m of the sample is

$$m = N\bar{\mu} = N\mu \tanh(\mu B/kT) \tag{4.23}$$

Equation (4.23) gives the magnetic moment m of a specimen containing N independent, non-interacting unpaired electrons in a magnetic field of magnetic induction B at an absolute temperature T. The magnetisation M is defined as the magnetic moment per unit volume. If the magnetisation is uniform throughout the specimen, M is equal to (m/V), where V is the volume of the specimen.

If $y \gg 1$, $e^y \gg e^{-y}$ and $\tanh y$ is approximately equal to unity. Hence, when $(\mu B/kT) \gg 1$, corresponding to very low temperatures and/or very high magnetic fields, according to equation (4.23), $m \simeq N\mu$. This result was to be expected, since for large B or small T, nearly all the N atomic magnetic dipoles are in the lower energy state with their magnetic moments pointing in the direction of the magnetic field.

When $y \ll 1$, $e^y \simeq (1 + y)$ and $\tanh y \simeq y$. Hence, when $\mu B/kT \ll 1$, corresponding to high temperatures and/or weak magnetic fields, equation (4.23) reduces to

$$m = N\mu(\mu B/kT) = N\mu^2 B/kT \ . \qquad (4.24)$$

In this limit, the total magnetic moment is proportional to $1/T$. This is

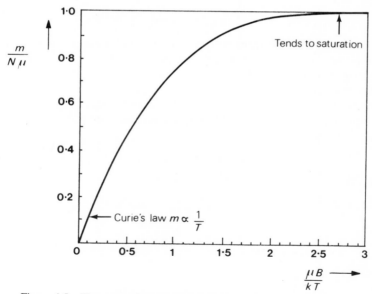

Figure 4.5—The magnetic moment m of a paramagnetic salt containing N unpaired electrons each of magnetic moment μ in a magnetic field of magnetic induction **B** is plotted against $(\mu B/kT)$. For large B and/or small T, $m/N\mu$ tends to unity, when all the magnetic moments tend to point in the direction of B. For high T and/or low B, m is proportional to $(1/T)$, if B is constant. This is Curie's law.

Curie's Law, which is generally valid experimentally for paramagnetic salts in weak magnetic fields at high temperatures.

The general form of the variation of m with $\mu B/kT$ is shown in Figure 4.5. A reader interested in the more general case when the total angular momentum of each paramagnetic atom is greater than $1/2$ is referred to Reif [1a]. The thermodynamic properties of paramagnetic salts will be developed in Section 5.8 of Chapter 5.

4.5 INTRODUCTION OF DEGENERACY

Generally, when calculating the partition function, we shall use equation (4.9) and sum over all the microstates of the system. If an energy level is degenerate, each of the degenerate microstates must be treated separately, when equations (4.8) and (4.9) are applied. Sometimes the reader may find these equations expressed as summations over the energy levels of the system, some of which may be degenerate.

Consider the example shown in Figure 4.6. The system has two energy levels having energies ε_0 and ε_1 respectively. The degeneracies (statistical weights) of energy levels 0 and 1 are $g_0 = 2$ and $g_1 = 3$ respectively. According to equation (4.9), when the system is in thermal equilibrium with a heat reservoir of absolute temperature T, the partition function, expressed as the sum over states, is

$$Z = \exp(-\beta\varepsilon_0) + \exp(-\beta\varepsilon_0) + \exp(-\beta\varepsilon_1) + \exp(-\beta\varepsilon_1) + \exp(-\beta\varepsilon_1) \ .$$

This can be rewritten in the form

$$Z = 2\exp(-\beta\varepsilon_0) + 3\exp(-\beta\varepsilon_1)$$

or $\qquad\qquad Z = g_0\exp(-\beta\varepsilon_0) + g_1\exp(-\beta\varepsilon_1) \ . \qquad\qquad$ (4.25)

The probability that at thermal equilibrium the small system is in any one of the 3 microstates with energy ε_1, which is the probability $P(U_1)$ that the small system has energy U_1 equal to ε_1 is

$$P(U_1) = 3\exp(-\beta\varepsilon_1)/Z = g_1\exp(-\beta\varepsilon_1)/Z \ . \qquad\qquad (4.26)$$

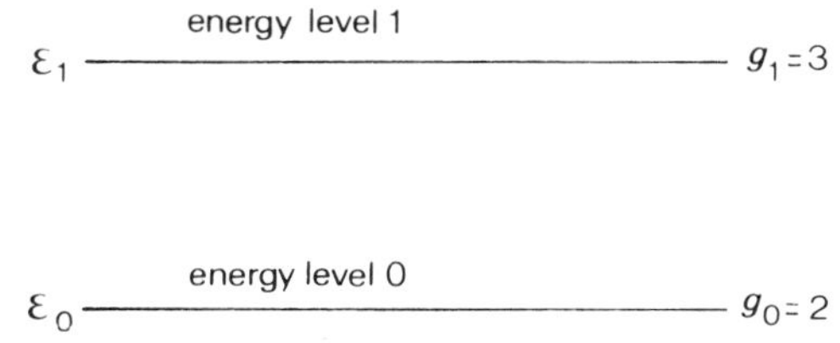

energy level 1

ε_1 ——————————————————— $g_1 = 3$

energy level 0

ε_0 ——————————————————— $g_0 = 2$

Figure 4.6—A two level system in which the lower energy level has a degeneracy of $g_0 = 2$ and the upper level has a degeneracy of $g_1 = 3$.

Equations (4.25) and (4.26) will now be generalized. Let g_j denote the degeneracy (statistical weight) of the jth energy level which has energy ε_j. By comparison with equation (4.25), we conclude that, when the small system is in thermal equilibrium with a heat reservoir of absolute temperature T, the partition function is

$$Z = \sum_{(\text{levels})} g_j \exp(-\beta \varepsilon_j) \ . \tag{4.27}$$

The summation in equation (4.27) is over all the energy levels of the small system. Generalising equation (4.26), we conclude that the probability $P(U_j)$ that at thermal equilibrium the small system is in any one of the g_j microstates which have energy $\varepsilon_j = U_j$ is

$$P(U_j) = g_j \exp(-\beta \varepsilon_j)/Z \ . \tag{4.28}$$

If the distribution of quantum eigenstates (microstates) is continuous, by comparison with equation (4.27) we conclude that if $D(\varepsilon)$ is the density of states

$$Z = \int_0^\infty D(\varepsilon) \exp(-\beta \varepsilon) \, d\varepsilon \ . \tag{4.29}$$

where $D(\varepsilon) \, d\varepsilon$ is the total number of microstates of the small system, which have energy eigenvalues in the range ε to $(\varepsilon + d\varepsilon)$. By comparison with equation (4.28), we conclude that the probability $P(\varepsilon) \, d\varepsilon$ that, at thermal equilibrium, the small system is in any one of its microstates, which have energy eigenvalues in the energy range ε to $(\varepsilon + d\varepsilon)$, is

$$P(\varepsilon) \, d\varepsilon = D(\varepsilon) \exp(-\beta \varepsilon) \, d\varepsilon/Z \ . \tag{4.30}$$

4.6 A MULTILEVEL SYSTEM

As an example of a multilevel system, consider a system which has six non-degenerate energy levels having energies 0, ε, 2ε, 3ε, 4ε and 5ε respectively, as shown in Figure 4.7(a). The characteristic temperature θ will be defined as $\theta = \varepsilon/k$ where ε is the energy gap between the first excited state and the ground state and k is Boltzmann's constant. According to equation (4.9), when the multilevel system is in thermal equilibrium with a heat reservoir of absolute temperature T, the partition function is

$$Z = \Sigma \exp(-\varepsilon_i/kT) = \exp(0) + \exp(-\theta/T) + \exp(-2\theta/T) + \exp(-3\theta/T)$$
$$+ \exp(-4\theta/T) + \exp(-5\theta/T) \ .$$
$$\tag{4.31}$$

The reader can check that, when $T = 0.2\theta$, $Z = 1.007$, when $T = \theta$, $Z = 1.578$ and when $T = 10\theta$, $Z = 4.741$. It can be seen that the value of

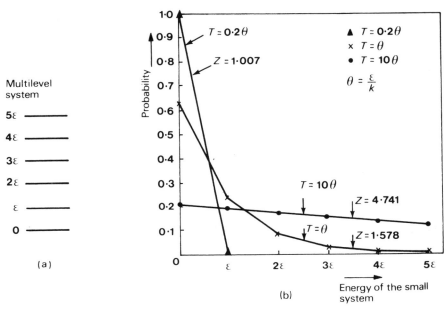

Figure 4.7—(a) The multilevel system has 6 non-degenerate equally spaced single particle energy levels. The energy of the ground state is zero and the separation of the energy levels is ε. The characteristic temperature θ is equal to ε/k. (b) The probability P_n that, when the multilevel system is in thermal equilibrium with a heat reservoir, it has energy $n\varepsilon$ is plotted against $n\varepsilon$. The results are presented for heat reservoir temperatures of 0.2θ, θ and 10θ.

the partition function Z increases from 1.007 to 4.741 as the temperature of the heat reservoir is increased from 0.2θ to 10θ. This illustrates that the partition function Z is a function of the absolute temperature of the heat reservoir. Since the energy eigenvalues of the small system, to be used in equation (4.9), depend on the volume V and the number of particles N in the small system, the partition function Z also depends on V and N. It will be shown in Chapter 5 how the other thermodynamic state variables of the small system can be derived from the equation relating the partition function Z to T, V and N.

According to equation (4.8) the probability that at thermal equilibrium the small system in Figure 4.7(a) is in the microstate, which has energy eigenvalue $n\varepsilon$, is

$$P_n = \exp(-n\varepsilon/kT)/Z = \exp(-n\theta/T)/Z \ . \tag{4.32}$$

The values of P_n are plotted against the energy of the small system in Figure 4.7(b) for values of T equal to 0.2θ, θ and 10θ. Notice the typical decrease of the probability with increasing energy above the ground state.

Notice also that, the higher the temperature of the heat reservoir, the larger is the probability that the small system is in one of its higher energy microstates.

It can be seen from Figure 4.7(b) that when $T = 0.2\theta$ the probability of finding the system in any of the states, other than the ground state and first excited state, is negligible. The low temperature behaviour of the system shown in Figure 4.7(a), when $T \ll \theta$ is dominated by the ground state and the first excited state.

4.7 ONE DIMENSIONAL HARMONIC OSCILLATOR IN THERMAL CONTACT WITH A HEAT RESERVOIR

Consider a one dimensional harmonic oscillator, which is in thermal equilibrium with a heat reservoir at constant absolute temperature T. According to quantum mechanics, the energy levels of the one dimensional harmonic oscillator are non-degenerate and have the energy eigenvalues

$$\varepsilon_n = (n + \tfrac{1}{2})h\nu = (n + \tfrac{1}{2})\hbar\omega \ , \tag{4.33}$$

where ν is the classical frequency of the harmonic oscillator and $\omega = 2\pi\nu$. The quantum number n can have the integer values $0, 1, 2, \ldots \infty$. [Reference: Section 2.1.3.]

According to equation (4.9) the partition function is

$$Z = \sum_{\text{(states)}} \exp[-\beta(n + 1/2)\hbar\omega] = \exp(-\beta\hbar\omega/2) \sum_{n=0}^{\infty} \exp(-n\beta\hbar\omega) \ .$$

Putting x equal to $\exp(-\beta\hbar\omega)$, we have

$$\sum_{n=0}^{\infty} \exp(-n\beta\hbar\omega) = 1 + x + x^2 + \ldots$$

This is a geometric series whose sum is $1/(1 - x)$. The reader can check this by expanding $(1 - x)^{-1}$ using the binomial theorem. Hence the partition function for a one dimensional harmonic oscillator is

$$Z = \exp(-\beta\hbar\omega/2)/[1 - \exp(-\beta\hbar\omega)] = \exp(-\theta/2T)/[1 - \exp(-\theta/T)] \ . \tag{4.34}$$

In this example the energy gap between the ground state and the first excited state is $\hbar\omega$. The characteristic temperature θ is defined by the equation

$$\theta = \hbar\omega/k = h\nu/k \ . \tag{4.35}$$

According to equation (4.8), the probability that, at thermal equilibrium, the one dimensional harmonic oscillator is in the microstate which has energy eigenvalue $(n + \tfrac{1}{2})\hbar\omega$ is $P_n = \exp[-\beta(n + \tfrac{1}{2})\hbar\omega]/Z$. According to equation (4.10) the mean energy of the harmonic oscillator is

$$\bar{U} = \Sigma P_n \varepsilon_n = \frac{\Sigma (n + \tfrac{1}{2})\hbar\omega \exp[-\beta(n + \tfrac{1}{2})\hbar\omega]}{\Sigma \exp[-\beta(n + \tfrac{1}{2})\hbar\omega]} \qquad (4.36)$$

$$\bar{U} = \hbar\omega\left[\tfrac{1}{2} + \frac{\Sigma nx^n}{\Sigma x^n}\right]$$

where x is equal to $\exp(-\beta\hbar\omega)$. We saw that Σx^n is equal to $(1 - x)^{-1}$. We also have

$$\Sigma nx^n = x \, \Sigma nx^{(n-1)} = x \frac{\mathrm{d}}{\mathrm{d}x}(\Sigma x^n) = x \frac{\mathrm{d}}{\mathrm{d}x}\left(\frac{1}{1 - x}\right) = \frac{x}{(1 - x)^2}$$

Hence, $\quad \bar{U} = \tfrac{1}{2}\hbar\omega + \dfrac{\hbar\omega}{[\exp(\beta\hbar\omega) - 1]} = \tfrac{1}{2}hv + \dfrac{hv}{[\exp(\beta hv) - 1]}$. $\qquad (4.37)$

This result will be derived from the partition function in Section 5.2 of Chapter 5. If $kT \gg \hbar\omega$, $[\exp(\hbar\omega/kT) - 1]$ is approximately equal to $\hbar\omega/kT$, and, in the high temperature limit, equation (4.37) becomes $\bar{U} \simeq \tfrac{1}{2}\hbar\omega + kT$. Thus in the classical limit, when $kT \gg \hbar\omega$, $\bar{U} \simeq kT$.

4.8 TYPICAL CHARACTERISTIC TEMPERATURES

According to equation (4.6), if a small system is in thermal equilibrium with a heat reservoir, the ratio of the probability that the small system is in its first excited state, having energy eigenvalue ε_1, to the probability that it is in its non-degenerate ground state, having energy eigenvalue ε_0, is

$$P_1/P_0 = \exp(-\beta\varepsilon_1)/\exp(-\beta\varepsilon_0) = \exp[-\beta(\varepsilon_1 - \varepsilon_0)] = \exp(-\beta\varepsilon) \ .$$

The characteristic temperature θ will be defined, in the general case, in terms of the energy gap ε between the ground state and the first excited state by the equation

$$\theta = \varepsilon/k = (\varepsilon_1 - \varepsilon_0)/k \ . \qquad (4.38)$$

We then have $\qquad\qquad P_1/P_0 = \exp(-\theta/T) \ . \qquad (4.39)$

When $T \ll \theta$, P_1/P_0 is very much less than unity, and the probability of finding the small system in one of its excited states is very small. When $T > \theta$ there is a significant probability of finding the system in one of its excited states. Some typical examples will now be considered.

IDEAL GASES

According to equation (2.5), for the case of a single atom of helium in a cubical box of volume 0.0224 m^3, $(\varepsilon_{211} - \varepsilon_{111})$ is 3.12×10^{-40} J. Since $k = 1.38 \times 10^{-23} \text{ J K}^{-1}$, the characteristic temperature in this case is equal to 2.26×10^{-17}K. According to equation (4.39) the *ratio* of the probability that the helium atom is in the 2,1,1 state to the probability that it is in the 1,1,1 state at a temperature of 2.26×10^{-17}K is 0.368. At 300K the ratio P_{211}/P_{111} is extremely close to unity.

SOLIDS

The energy gap in many processes taking place in solids is in the energy range 0.1 to 1 eV (corresponding to the energy range 1.6×10^{-20} to 1.6×10^{-19} J). For an energy gap in this range, the characteristic temperature θ is in the range 1.16×10^{3}K to 1.16×10^{4}K. At 300K, if θ is 5×10^{3}K corresponding to an energy gap of 0.43 eV, according to equation (4.39), P_{1}/P_{0} is equal to 5.8×10^{-8}. Such a probability might appear insignificant, but, if there is a total of 10^{20} such small (atomic) systems in the specimen, for a two level system approximately 5×10^{12} would be in the excited state at 300K, and could make a significant contribution to the properties of the specimen at 300K.

ATOMIC SYSTEMS

For many atomic processes, the excitation energy is of the order of 10 eV. For example, for atomic hydrogen the energy gap between one of the microstates with $n = 1$ and one of the microstates with $n = 2$ is 10.02 eV. For an energy gap ε equal to 10 eV, the characteristic temperature, given by equation (4.38), is 1.16×10^{5}K. The excitation of atomic electrons to excited states is generally negligible at room temperature. At temperatures of the order of 10^{4} to 10^{5}K, present in electrical discharges, some of the atomic electrons are in excited electronic states, and light is emitted during de-excitation. At a temperature of the order of 10^{6}K in the solar corona nearly all the atoms are ionised.

The excitation energies of the higher energy nuclear states of atomic nuclei are generally in the MeV range. The characteristic temperature for an energy gap of 1 MeV is 1.16×10^{10}K.

4.9 THE MAXWELL VELOCITY DISTRIBUTION FOR AN IDEAL GAS

Consider a *single* monatomic gas atom, such as a helium atom, inside a box of volume V, which is kept at a constant absolute temperature T. According to equation (A6.18) of Appendix 6, which is the same as equation (A3.18) of Appendix 3, the number of single particle quantum states (orbitals) which have energy eigenvalues in the energy range ε to $(\varepsilon + \mathrm{d}\varepsilon)$ is

$$D(\varepsilon) \, \mathrm{d}\varepsilon = (gV/4\pi^{2})(2m/\hbar^{2})^{3/2}\varepsilon^{1/2} \, \mathrm{d}\varepsilon. \qquad (4.40)$$

where $D(\varepsilon)$ is the density of states, and m is the mass of the gas atom.

According to the Boltzmann distribution, equation (4.8), the probability that the gas atom is in the ith single particle state (orbital) with energy eigenvalue ε_i is

$$P_i = \exp(-\varepsilon_i/kT)/Z \ . \tag{4.41}$$

The probability $P(\varepsilon)$ dε that the gas atom is in any one of the single particle quantum states (orbitals), which have energy eigenvalues in the range ε to $(\varepsilon + d\varepsilon)$, is equal to the product of the number of single particle quantum states (orbitals) in this energy range, which is given by equation (4.40), and the probability that the gas atom is in any particular quantum state in this energy range, which is given by equation (4.41). Hence

$$P(\varepsilon) \ d\varepsilon = D(\varepsilon)\exp(-\varepsilon/kT) \ d\varepsilon/Z \ . \tag{4.42}$$

Equation (4.42) is the same as equation (4.30). Substituting for $D(\varepsilon)$ from equation (4.40) we have

$$P(\varepsilon) \ d\varepsilon = (gV/4\pi^2Z)(2m/\hbar^2)^{3/2} \exp(-\varepsilon/kT)\varepsilon^{1/2} \ d\varepsilon \ , \tag{4.43}$$

If the potential energy of the gas atom inside the box is zero, all the energy of the gas atom is kinetic energy, so that

$$\varepsilon = \tfrac{1}{2}mv^2$$

where v is the velocity of the gas atom. Differentiating, we have $d\varepsilon = mv \ dv$. Hence

$$\varepsilon^{1/2} \ d\varepsilon = m^{3/2}v^2 \ dv/2^{1/2} \ .$$

Substituting in equation (4.43), we find that the probability that the velocity of the gas atom has a numerical value between v and $(v + dv)$ is

$$P(v) \ dv = Kv^2\exp(-mv^2/2kT) \ dv \ . \tag{4.44}$$

where K is a temperature dependent constant. (The temperature dependence of K comes via the partition function.) Since the gas atom must have some value of velocity

$$\int_0^\infty P(v) \ dv = 1 \ .$$

Hence
$$K\int_0^\infty v^2\exp(-mv^2/2kT) \ dv = 1 \ . \tag{4.45}$$

Strictly the upper limit to the velocity of the monatomic gas atom is the speed of light. Due to the exponential term, no significant error is introduced by taking the upper limit as infinity. According to equation (A1.39) of Appendix 1

$$I_2 = \int_0^\infty v^2\exp(-\alpha v^2)dv = \pi^{1/2}/4\alpha^{3/2} \ .$$

Putting $$\alpha = (m/2kT) \qquad (4.46)$$

and substituting in equation (4.45), we obtain

$$K = (4/\pi^{1/2})(m/2kT)^{3/2} \ . \qquad (4.47)$$

Substituting in equation (4.44), we obtain

$$P(v)\ dv = (4/\pi^{1/2})(m/2kT)^{3/2}v^2 \ \exp(-mv^2/2kT)\ dv \ . \qquad (4.48)$$

Equation (4.48) gives the probability that a single monatomic gas atom inside a box kept at an absolute temperature T, has a value of velocity between v and $(v + dv)$ in any direction of space. Equation (4.48) is generally called the Maxwell velocity distribution.

The nitrogen and oxygen molecules in air are diatomic. Though the vibrational and rotational internal energies of a diatomic molecule are temperature dependent, the precise value of the internal energy of a diatomic molcule and the precise value of its translational energy are, to a good approximation, independent of each other. Equation (4.48) is a reasonable approximation for the velocity distribution of a nitrogen or an oxygen molecule in a box. The calculated probability distribution for a nitrogen molecule at 273K is shown in Figure 4.8.

The most generally quoted speed is the root mean square velocity v_{rms}. Generalising equation (4.10), with $\alpha_i = v^2$, we find that the mean square

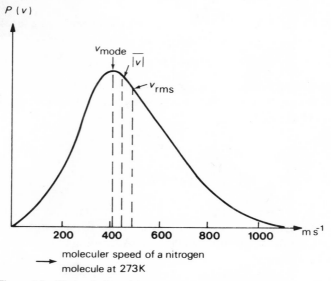

Figure 4.8—The variation with the speed v of the probability $P(v)$ that a nitrogen molecule has a speed v at 273K.

velocity is given by

$$\overline{v^2} = \int_0^\infty v^2 P(v)\, dv \ .$$

Using equation (4.44) we have

$$\overline{v^2} = \int_0^\infty v^2 K v^2 \exp(-mv^2/2kT)\, dv = K I_4 \ .$$

Using equation (A1.41) of Appendix 1 for I_4 with $\alpha = (m/2kT)$, and substituting for K from equation (4.47), we find that $\overline{v^2} = 3kT/m$. Hence

$$v_{\text{rms}} = (3kT/m)^{1/2} \ . \tag{4.49}$$

The mean kinetic energy of the monatomic gas atom is

$$\overline{\varepsilon} = \langle \tfrac{1}{2}mv^2 \rangle = \tfrac{1}{2}m\overline{v^2} = \tfrac{3}{2}kT \ . \tag{4.50}$$

The mass of a nitrogen molecule is 4.65×10^{-26} kg. Since $k = 1.38 \times 10^{-23}$ J K^{-1}, according to equation (4.49) the root mean square velocity of a nitrogen molecule at 273K is 493 m s^{-1}, as shown on Figure 4.8.

The most probable speed v_{mode} is the value of v at which the probability $P(v)$, given by equation (4.44), is a maximum. This is given by

$$dP(v)/dv = (d/dv)[K v^2 \exp(-\alpha v^2)] = 0$$

where α and K are given by equations (4.46) and (4.47) respectively. Carrying out the differentiation and solving the algebraic equation, we find that

$$v_{\text{mode}} = (2kT/m)^{1/2} = (\tfrac{2}{3})^{1/2} v_{\text{rms}} = 0.816 v_{\text{rms}} \ . \tag{4.51}$$

The most probable speed is less than v_{rms}, as shown in Figure 4.8.

The mean value of the speed of the gas atom, $\overline{|v|}$ is given by

$$\overline{|v|} = \int_0^\infty v P(v)\, dv = K \int_0^\infty v^3 \exp(-\alpha v^2)\, dv = K I_3 \ . \tag{4.52}$$

According to equation (A1.40) of Appendix 1,

$$I_3 = 1/2\alpha^2 = \tfrac{1}{2}(2kT/m)^2 \ .$$

Substituting for I_3 in equation (4.52) and for K from equation (4.47) we obtain

$$\overline{|v|} = (8kT/\pi m)^{1/2} = (8/3\pi)^{1/2} v_{\text{rms}} = 0.921 v_{\text{rms}} \ . \tag{4.53}$$

The mean speed is less than v_{rms}, as shown in Figure 4.8.

So far we have only considered one monatomic gas atom in a box. If there is more than one identical gas atom in the box, strictly, one should use either the Fermi-Dirac or the Bose–Einstein distribution. This approach will be discussed in Sections 12.1.7 and 12.5.1 respectively. In this Section we shall only consider the classical limit.

As an illustrative example of the classical limit, consider one helium gas atom in a cubical box of volume 0.0224 m³ kept at 273K. (Normally at STP, such a box would contain a mole of helium gas atoms.) According to equation (A3.17) of Appendix 3, the number of single particle quantum states (orbitals), which have energy eigenvalues less than or equal to ε is

$$\Phi(\leq \varepsilon) = (\pi/6)(2m/\hbar^2\pi^2)^{3/2}V\varepsilon^{3/2} \ . \tag{4.54}$$

where V is the volume of the box, and m is the mass of the gas atom. Equation (4.54) can also be obtained by integrating equation (4.40). According to equation (4.50) the mean energy of the helium gas atom is $3kT/2$, which at 273K is equal to 5.65×10^{-21} J. Using this value of ε in equation (4.54) to represent a typical helium gas atom at 273K with $m = 6.65 \times 10^{-27}$ kg and $V = 0.0224$ m³ we find that the number of single particle states having energy eigenvalues $\leq 3kT/2$ is

$$\Phi(\leq \tfrac{3}{2}kT) \simeq 2 \times 10^{29} \ .$$

This is very much greater than the number of helium atoms in a mole. The ratio $N_A/\Phi(\leq \tfrac{3}{2}kT)$ is equal to 3×10^{-6}. In the classical limit, a significant fraction of the helium atoms in a gas will be in single particle states having energy eigenvalues greater than $\tfrac{3}{2}kT$. For purposes of discussion, it will be assumed that, for a helium gas at STP, the mean occupancy of the single particle states is less than 3×10^{-6}. [This *very* rough estimate will be confirmed by taking the classical limit of the Bose–Einstein distribution function in Chapter 12. Reference: Problem 12.8.]

One of the monatomic gas atoms can be treated as the small system, and the rest of the gas atoms and the surroundings treated as the heat reservoir when the Boltzmann distribution is applied. If the gas atom is a fermion, it cannot occupy a single particle state that is already filled. However, if less than one in 3×10^5 of the single particle states is occupied at 273K, this does not limit the choice of states available to the fermion very much, since there are plenty of unoccupied single particle states in the near vicinity of every occupied state. The restriction to one fermion per single particle state is not a severe restriction and is not likely to affect the energy distribution of the fermion significantly in the classical limit. The probability $P(v) \, dv$ that the fermion has a velocity of magnitude between v and $(v + dv)$ should be given, in the classical limit, by equation (4.48). This argument applies to each of the gas atoms, so that $n(v) \, dv$, the number of gas atoms having velocities between v and $(v + dv)$, should be given, in the classical limit, by the product of N, the number of gas atoms, and $P(v)\,dv$. We then have

$$n(v)\,dv = (4N/\pi^{1/2})(m/2kT)^{3/2}v^2\,\exp(-mv^2/2kT)\,dv \qquad (4.55)$$

Equation (4.55) will be confirmed in Sections 12.1.7 and 12.5.1 of Chapter 12 by taking the classical limits of the Fermi–Dirac and Bose–Einstein distributions. Equation (4.55) has also been confirmed by experiments on molecular beams. (Reference Reif [1b].)

4.10 CLASSICAL STATISTICAL MECHANICS*

4.10.1* Introduction*

Before the advent of quantum mechanics, the approach to statistical mechanics was generally based on Liouville's theorem of classical mechanics. It was postulated that the number of accessible microstates was proportional to the accessible volume of phase space. To describe the motion of a particle in one dimension, we must use a phase space of two dimensions, using x and p_x as coordinates. For a particle moving in three dimensions, we must use a phase space of six dimensions, namely x, y, z, p_x, p_y, p_z. In the theory of classical statistical mechanics, it was assumed that, if the particle was restricted to spatial coordinates in the range x, y, z to $(x + dx)$, $(y + dy)$, $(z + dz)$ and to a momentum, which had components between p_x, p_y, p_z and $(p_x + dp_x)$, $(p_y + dp_y)$, $(p_z + dp_z)$, the number of accessible states $dN(\mathbf{r}, \mathbf{p})$ was proportional to the accessible volume $dx\,dy\,dz\,dp_x\,dp_y\,dp_z$ of phase space, that is

$$dN(\mathbf{r}, \mathbf{p}) \propto dx\,dy\,dz\,dp_x\,dp_y\,dp_z \ . \qquad (4.56)$$

The actual number of accessible microstates was not specified in classical statistical mechanics. [The vector $\mathbf{r}$ has components x, y, z and the momentum $\mathbf{p}$ has components p_x, p_y, p_z.]

The density of states given by equation (4.40) is the number of stationary wave solutions (normal modes) which have energies between ε and $(\varepsilon + d\varepsilon)$. In order to obtain progressive wave solutions of Schrödinger's equation, one can use periodic boundary conditions in the way described in Section A6.2* of Appendix 6. According to equation (A6.22), the number of progressive wave solutions of Schrödinger's equation for a single particle, of spin zero, satisfying periodic boundary conditions on the surface of a cubical volume V, is

$$N(\mathbf{p})\,d^3\mathbf{p} = V\,dp_x\,dp_y\,dp_z/h^3 \ . \qquad (4.57)$$

According to equation (4.57) there is one progressive wave solution per volume h^3 of phase space. This suggests that, in the classical limit, the constant of proportionality in equation (4.56) is $1/h^3$. The absence of the factor $1/h^3$ from equation (4.56) is, generally, not important in the classical limit. For example the constant h does not appear in equation (4.55). A reader interested in the full development of classical statistical mechanics is referred to Kittel [2].

4.10.2* An ideal gas in thermal contact with a heat reservoir*

Consider one of the gas atoms. The other gas atoms and the surroundings will be treated as a heat reservoir of absolute temperature T. According to the Boltzmann distribution, the probability that at thermal equilibrium the gas atom is in a single particle state with total energy (kinetic energy plus potential energy) equal to ε is proportional to $\exp(-\beta\varepsilon)$. The number of single particle states is given by equation (4.56). The probability that at thermal equilibrium the gas atom has total energy ε, is at a position having coordinates in the range x, y, z to $(x + dx)$, $(y + dy)$, $(z + dz)$ and has a momentum which has components between p_x, p_y, p_z and $(p_x + dp_x)$, $(p_y + dp_y)$, $(p_z + dp_z)$ is equal to the product of the number of single particle states in this range and the probability that the particle is in one of these states. Hence

$$P(\mathbf{r}, \mathbf{p})\, d^3\mathbf{r}\, d^3\mathbf{p} \propto \exp(-\beta\varepsilon)\, d^3\mathbf{r}\, d^3\mathbf{p} \qquad (4.58)$$

where
$$d^3\mathbf{r} = dx\, dy\, dz$$

and
$$d^3\mathbf{p} = dp_x\, dp_y\, dp_z \; .$$

The x component of momentum p_x is equal to mv_x, where m is the mass of the gas atom and v_x, which can be either positive or negative, is the x component of its velocity.

For the atoms of an ideal monatomic gas in a box, the variation of the gravitational potential energy inside the box can be ignored. The kinetic energy is equal to $\frac{1}{2}mv^2$. In this case equation (4.58) reduces to

$$P(\mathbf{v})\, d^3\mathbf{v} \propto \exp[-m(v_x^2 + v_y^2 + v_z^2)/2kT]\, dv_x\, dv_y\, dv_z \; .$$

Normalising for N particles in the box and using the value of $I_0 = \pi^{1/2}/2\alpha^{1/2}$, given by equation (A1.36) of Appendix 1, to evaluate the three integrals of the type

$$\int_{-\infty}^{\infty} \exp(-\alpha v_x^2)\, dv_x \; ,$$

in the classical limit we find that

$$n(\mathbf{v})\, d^3\mathbf{v} = n(\mathbf{v})\, dv_x\, dv_y\, dv_z = N\left(\frac{m}{2\pi kT}\right)^{3/2} \exp(-mv^2/2kT)\, dv_x\, dv_y\, dv_z \; .$$
$$(4.59)$$

where $n(\mathbf{v})$ $d^3\mathbf{v}$ is the number of atoms which have velocities, which have components between v_x, v_y, v_z and $(v_x + dv_x)$, $(v_y + dv_y)$, $(v_z + dv_z)$.

Writing equation (4.59) into components, we have

$$n(\mathbf{v})\, dv_x\, dv_y\, dv_z =$$

$$N\left(\frac{m}{2\pi kT}\right)^{3/2} \exp(-mv_x^2/2kT)\, \exp(-mv_y^2/2kT)\, \exp(-mv_z^2/2kT)\, dv_x\, dv_y\, dv_z \ .$$

Integrating over dv_y and dv_z and using $I_0 = \pi^{1/2}/2\alpha^{1/2}$ we find that

$$n(v_x)\, dv_x = N\left(\frac{m}{2\pi kT}\right)^{1/2} \exp(-mv_x^2/2kT)\, dv_x \qquad (4.60)$$

where $n(v_x)$ dv_x is the number of gas atoms which have velocities, which have a component in the x direction of magnitude between v_x and $(v_x + dv_x)$. If equation (4.59) is integrated over all possible directions of velocity, it gives equation (4.55). [Reference: Reif [1c].]

4.10.3* The law of isothermal atmospheres*

As an example, consider the Earth's atmosphere. One of the air molecules will be treated as the small system and the rest of the atmosphere as the heat reservoir. In this example, allowance must be made for the variation of the gravitational potential energy of the molecule with its altitude H above sea level. The sum of the potential and kinetic energy of the air molecule is $\varepsilon = mgH + \frac{1}{2}mv^2$. Substituting for ε in equation (4.58), for a volume element $dx\, dy\, dz$ at a position vector $\mathbf{r}$, using $m\, dv_x$ for dp_x etc., we have

$$P(\mathbf{r}, \mathbf{p})\, d^3\mathbf{r}\, d^3\mathbf{p} \propto \exp(-mgH/kT)\exp(-mv^2/2kT)\, dx\, dy\, dz\, dv_x\, dv_y\, dv_z \ .$$

At a fixed altitude H, $P(\mathbf{r}, \mathbf{p})$ is independent of the horizontal coordinates of the molecule. For an isothermal atmosphere T is constant, and the $\exp(-mv^2/2kT)$ term in the expression for $P(\mathbf{r}, \mathbf{p})$ is the same at all altitudes. Hence the probability that the air molecule is at an altitude between H and $(H + dH)$ is given by[†]

$$P(H)\, dH \propto \exp(-mgH/kT)\, dH \ . \qquad (4.61)$$

The above equation should be applicable to each of the air molecules.

[†]This result follows directly from the Boltzmann distribution. Consider one air molecule. The rest of the atmosphere acts as a heat reservoir of constant absolute temperature T. If the chosen air molecule is raised a height H from sea level, without a change of velocity, the extra energy mgH is taken from the heat reservoir. According to equation (4.4) this reduces the number of microstates accessible to the heat reservoir from g_R to $g_R \exp(-mgH/kT)$. Using equation (4.2) we obtain equation (4.61).

Hence the variation of the number of molecules per metre³ with altitude in an isothermal atmosphere is given by

$$n = n_0 \exp(-mgH/kT) \qquad (4.62)$$

where n_0 is the number of molecules per metre³ at sea level. According to the kinetic theory of gases, the pressure of an ideal gas is equal to $\frac{1}{3}\rho\overline{v^2}$ where ρ is the density of the gas. Since for an isothermal atmosphere, the velocity distribution is the same at all altitudes, $\overline{v^2}$ is independent of H and the pressure p is proportional to ρ at all altitudes. Since ρ is proportional to n, the variation of pressure with altitude is given by

$$p = p_0 \exp(-mgH/kT) = p_0 \exp(-H/H_0) \qquad (4.63)$$

where p_0 is the pressure at sea level. According to equation (4.63), the pressure of an isothermal atmosphere decreases exponentially with height. The scale height H_0 is given by

$$H_0 = kT/mg \ . \qquad (4.64)$$

Substituting the numerical values, $k = 1.38 \times 10^{-23}$ J K^{-1}, $T = 300$K, $g = 9.8$ m s^{-2} and $m = 4.9 \times 10^{-26}$ kg to represent an average air molecule (80% nitrogen, 20% oxygen) we find that H_0 is equal to 8.6 km. According to equation (4.63), the pressure of an isothermal atmosphere at 300K should be reduced to 0.368 of its sea level value at an altitude of 8.6 km. In practice H_0 is about 8 km near sea level. The difference arises from the significant decrease of temperature with altitude in the lower atmosphere. For example, at $H = 10$ km, the temperature is down to 223K. A lower average value of temperature should be used in equation (4.64) to estimate the scale height near sea level.

Equation (4.62) was used by Perrin to determine Avogadro's constant. (Reference: Problem 4.12.)

4.10.4* The equipartition theorem*
In many cases, the energy of a system can be expressed in a form which is quadratic in the position and momentum variables of the system. For example, the kinetic energy of a free particle of mass m, which has velocity v and momentum p equal to mv, can be expressed in the form $(p_x^2 + p_y^2 + p_z^2)/2m$. Another example is the total energy of a particle of mass m moving with simple harmonic motion along the x axis. In this example, the energy ε is

$$\varepsilon = (p_x^2/2m) + \tfrac{1}{2}Cx^2 \qquad (4.65)$$

where C is the restoring force, when the displacement from the mean position is 1 m. According to the theorem of equipartition of energy, when a system is in thermal equilibrium with a heat reservoir of absolute temperature T, the mean value of each contribution to the total energy, which is quadratic in either a position or a momentum coordinate, is $\frac{1}{2}kT$. The proof will be illustrated by finding the mean value of the term $\frac{1}{2}Cx^2$ in equation (4.65).

Using equation (4.58) it follows that the mean value of $\frac{1}{2}Cx^2$ at thermal equilibrium is

$$\langle \tfrac{1}{2}Cx^2 \rangle = \frac{\int \frac{1}{2}Cx^2 \exp[(-\frac{1}{2}Cx^2 - \varepsilon_0)/kT]\, dx\, dy\, dz\, dp_x\, dp_y\, dp_z}{\int \exp[(-\frac{1}{2}Cx^2 - \varepsilon_0)/kT]\, dx\, dy\, dz\, dp_x\, dp_y\, dp_z}.$$

The term ε_0 is the contribution to the total energy due to all the other coordinates and momenta except x. Separating the integral and cancelling the integration over the variables other than x, we have

$$\langle \tfrac{1}{2}Cx^2 \rangle = \frac{\int_{-\infty}^{+\infty} \frac{1}{2}Cx^2 \exp(-Cx^2/2kT)\, dx}{\int_{-\infty}^{+\infty} \exp(-Cx^2/2kT)\, dx}.$$

Putting $C/2kT = \alpha$, we obtain

$$\langle \tfrac{1}{2}Cx^2 \rangle = \frac{\frac{1}{2}C\int_{-\infty}^{+\infty} x^2\exp(-\alpha x^2)\, dx}{\int_{-\infty}^{+\infty} \exp(-\alpha x^2)\, dx} = \tfrac{1}{2}C\,\frac{2I_2}{2I_0}.$$

According to equations (A1.39) and (A1.36) of Appendix 1, $I_2 = \pi^{1/2}/4\alpha^{3/2}$ and $I_0 = \pi^{1/2}/2\alpha^{1/2}$ so that I_2/I_0 is equal to $1/2\alpha$, which with $\alpha = C/2kT$ is equal to kT/C. Hence $\langle \frac{1}{2}Cx^2 \rangle = \frac{1}{2}C(kT/C) = \frac{1}{2}kT$. A similar proof holds in the case of $p_x^2/2m$. This is left as an exercise for the reader.

In the case of the simple harmonic oscillator, according to the law of equipartition of energy, the mean value of both $\frac{1}{2}Cx^2$ and $p_x^2/2m$ in equation (4.65) is $\frac{1}{2}kT$, so that, in the classical limit, the mean energy of a harmonic oscillator is kT. This is in agreement with the high temperature limit of equation (4.37).

In the case of a free particle

$$\varepsilon = (p_x^2 + p_y^2 + p_z^2)/2m .$$

According to the equipartition theorem, the mean value of the contribution of each of the terms such as $(p_x^2/2m)$ to the total energy is $\frac{1}{2}kT$. Hence, the mean energy of the particle is $\frac{3}{2}kT$. This is in agreement with equation (4.50).

REFERENCES

[1] Reif, F., *Fundamentals of Statistical and Thermal Physics*, 1965, McGraw-Hill, (a) page 257 (b) page 274 (c) page 267.
[2] Kittel, C., *Elementary Statistical Physics*, 1958, John Wiley, New York.
[3] Gasser, R. P. H. and Richards, W. G., *Entropy and Energy Levels*, 1974, Clarendon Press, Oxford, page 38.

PROBLEMS

Boltzmann's constant $k = 1.3807 \times 10^{-23}$ J K^{-1}
Bohr magneton $= 9.273 \times 10^{-24}$ A m^2
Nuclear magneton $= 5.050 \times 10^{-27}$ A m^2
Acceleration due to gravity $= 9.80$ m s^{-2}
Electron volt $= 1.602 \times 10^{-19}$ J.

Problem 4.1

A system has two non-degenerate energy levels. The energy gap is 0.1 eV. Calculate the probability that the system is in the higher energy level, when it is in thermal equilibrium with a heat reservoir of absolute temperature (i) 300K, (ii) 600K, (iii) 1000K, (iv) 10 000K. At what temperature is the probability equal to (v) 0.25, (vi) 0.4, (vii) 0.49?

Problem 4.2

A solid is placed in an external magnetic induction field B of strength 3T. The solid contains non-interacting paramagnetic atoms of spin $\frac{1}{2}$, so that the potential energy of each atom in the magnetic field is $\pm\mu B$. The atoms are in thermal equilibrium with the lattice. If μ is equal to one Bohr magneton, below what temperature must the solid be cooled so that more than 60% of the atoms are polarised with their magnetic moments parallel to the magnetic field? What would the temperature need to be, if μ were equal to one nuclear magneton?

Problem 4.3

A paramagnetic salt contains 10^{24} magnetic ions per cubic metre, each with a magnetic moment of 1 Bohr magneton, which can be aligned either parallel or anti-parallel to a magnetic field. Calculate the difference between the number of ions in these two states, when a magnetic induction field of strength 1 T is applied, at a temperature of 3K. The volume of the sample is 1 m^3. Calculate the magnetic moment of the sample at this temperature.

Problem 4.4

The first excited energy level of a system lies at an energy of 20 eV above the ground state. If this excited level is three-fold degenerate while the ground state is non-degenerate, find the relative populations of the first excited level and the ground state, when the system is in thermal equilibrium with a heat reservoir at 10 000K.

Problem 4.5

The energy levels of a three-dimensional isotropic harmonic oscillator are given by

$$\varepsilon = (n_1 + n_2 + n_3 + \tfrac{3}{2})\hbar\omega \ .$$

Each of the quantum numbers n_1, n_2 and n_3 can have the integer values 0, 1, 2, ... ∞. Find the degeneracies of the levels with energies $\tfrac{5}{2}\hbar\omega$ and $\tfrac{7}{2}\hbar\omega$. Show that, for a system of such oscillators in thermal equilibrium with a heat reservoir of absolute temperature T, the latter level is more highly populated than the former if $T > (\hbar\omega/k\log 2)$.

Problem 4.6

The energy levels of atomic hydrogen are equal to $(-R_\alpha/n^2)$, where $R_\infty = 13.6$ eV is the Rydberg constant. The principal quantum number n can have the integer values $n = 1, 2, 3 \ldots \infty$. The degeneracy of the levels is $2n^2$. Use the Boltzmann distribution to calculate the relative populations of the energy levels, for which $n = 2$ and $n = 1$, at a temperature of 10^5K. Calculate the temperature at which the populations of the two energy levels are equal.

(*Comment*: Avoid calculating the partition function, which in this case is

$$\sum_{n=1}^{\infty} 2n^2 \exp[-R_\infty/n^2 kT] \ .$$

Due to the n^2 term in the summation, the partition function is infinite. In practice one cannot have a completely isolated hydrogen atom. Reference: Gasser and Richards [3].)

Problem 4.7

A single particle system has three non-degenerate energy levels of energy 0, ε and 2ε. The system is in thermal equilibrium with a heat reservoir. Calculate the partition function and the probability of finding the particle in each of the levels at absolute temperatures of (i) $0.1\varepsilon/k$, (ii) ε/k, (iii) $10\varepsilon/k$. Calculate the ensemble average energy at each temperature.

Problem 4.8

A single particle system has three non-degenerate energy levels 0, 0.5ε and 0.8ε. It is in thermal equilibrium with a heat reservoir of absolute temperature $T = \varepsilon/k$, where k is Boltzmann's constant. Calculate for each level the probability that the particle is in that level. Calculate also the partition function and the mean energy of the single particle system.

Problem 4.9

A system has energy levels at $\varepsilon = 0$, $\varepsilon = 300k$ and $\varepsilon = 600k$ where k is Boltzmann's constant. The degeneracies of the levels are 1, 3 and 5 respectively. Calculate the *partition function*, the *relative populations* of the energy levels and the *average energy*, all at a temperature of 300K. At what temperature is the population of the energy level at $600k$ equal to the population of the energy level at $300k$?

Problem 4.10

Calculate the rms, average and most probable speed of an argon atom of mass 6.64×10^{-26} kg at a temperature of 300K.

Problem 4.11

For purposes of discussion, assume that the mass of air molecules is 1 kg. Calculate the ratio of the probability of finding such a molecule 1 m above sea level to the probability of finding it at sea level, if the temperature is 300K. Calculate the scale height at 300K.

Problem 4.12*

In one of his experiments on gravitational sedimentation equilibrium, Perrin observed the number of gamboge particles in water at 293K. Perrin found that, when the microscope was raised by 100 μm, the mean number of particles in the field of view decreased from 203 to 91. If the molar gas constant is 8.32 J K^{-1} mole^{-1}, find a value for Avogadro's constant. Assume that the gamboge particles had a mean volume of 9.78×10^{-21} m^3 and a density of 1351 kg m^{-3}. The density of water is 1000 kg m^{-3}.

Problem 4.13

A strip of rubber, which is maintained at an absolute temperature T, is fastened at one end to a peg, and a mass m hangs from the other end. Assume the following simple model for the rubber band: It consists of a linked polymer chain of N segments joined end to end. Each segment has a length a and can be oriented either parallel or anti-parallel to the vertical direction. Show that the resultant mean length L of the rubber band is

$$Na \, \tanh(mga/kT)$$

(*Hint*: Apply the Boltzmann distribution to one segment to calculate the probabilities that it is in the up and the down positions. Then calculate the mean contribution of the segment to the length of the rubber. Neglect the weights of the segments themselves and any interactions between the segments.)

Problem 4.14*

The potential energy of a particle of mass m oscillating in one dimension in a potential well is equal to Cx^4, where x is the displacement of the particle from its equilibrium position and C is a constant. The oscillator is in thermal equilibrium with a heat reservoir, of absolute temperature T. Assume that equation (4.57) is applicable. Show that the mean energy (kinetic energy plus potential energy) is $3kT/4$.

Problem 4.15*

The total energy of a ballistic galvanometer suspension is equal to $(\frac{1}{2}C\theta^2 + \frac{1}{2}I\omega^2)$, where θ is the deflection from the mean position, C is the restoring couple when this deflection is 1 radian, I is the moment of inertia and ω is the angular velocity. The galvanometer suspension is in thermal equilibrium with a heat reservoir of absolute temperature T. Due to the interchange of energy between the suspension and the heat reservoir, there are fluctuations in the energy of the suspension and hence in θ. According to the theorem of the equipartition of energy, the mean value of $\frac{1}{2}C\theta^2$ is $kT/2$. The light from a light source, 1 m from a mirror on the galvanometer suspension is reflected by the mirror to give an image on a screen also 1 m from the mirror. Calculate the rms displacement of the image from its mean position, if $T = 300K$ and $C = 10^{-13}$ N m rad^{-1}.

Chapter 5

The Partition Function and Thermodynamics

5.1 INTRODUCTION

Consider the example of a small system, of volume V and consisting of N particles, which is in thermal equilibrium with a constant volume heat reservoir of absolute temperature T, as shown in Figure 4.1(a) of Chapter 4. There is no interchange of particles between the small system and the heat reservoir. According to the Boltzmann distribution, equation (4.8), the probability that, at thermal equilibrium, the small system is in its ith microstate, which has energy eigenvalue ε_i, is

$$P_i = \exp(-\beta\varepsilon_i)/Z = \exp(-\varepsilon_i/kT)/Z \tag{5.1}$$

According to equation (4.9) the partition function Z of the small N particle system is given by

$$Z = \sum_{\left(\substack{N \text{ particle} \\ \text{states}}\right)} \exp(-\beta\varepsilon_i) = \sum_{\left(\substack{N \text{ particle} \\ \text{states}}\right)} \exp(-\varepsilon_i/kT). \tag{5.2}$$

In equation (5.2) the summation is over all the N particle states of the small N particle system. The energy eigenvalues ε_i depend on both the volume V of the small system and N, the number of particles in the small system. The partition function also depends on the temperature of the heat reservoir. (An example of the dependence of Z on T is illustrated in Figure 4.7(b) of Section 4.6.) Summarising, the partition function defined by equation (5.2) is a function of V, the volume of the small system, N the number of particles in the small system and T the absolute temperature of the heat reservoir.

It is often convenient to use the temperature parameter β of the heat reservoir which is given by

$$\beta = 1/kT \tag{3.18}$$

and to express Z as a function of V, β and N. It follows from equation (3.18) that

$$\frac{\partial}{\partial \beta} = -kT^2 \frac{\partial}{\partial T} \tag{5.3}$$

$$\frac{\partial}{\partial T} = -k\beta^2 \frac{\partial}{\partial \beta} \quad . \tag{5.4}$$

In order to determine the mean values of the thermodynamic variables of the small system, consider an ensemble of identical small systems in thermal equilibrium with a heat reservoir of absolute temperature T. Following Gibbs, such an ensemble is generally called a *canonical ensemble*. The ensemble average is obtained by averaging over the members of the ensemble at a fixed time. According to equation (4.10) the ensemble average of any variable α is

$$\bar{\alpha} = \Sigma P_i \alpha_i = \Sigma \alpha_i \exp(-\beta \varepsilon_i)/Z \tag{5.5}$$

where α_i is the mean value of the variable α when the small system is in its ith microstate. It will now be shown how the ensemble averages of the other thermodynamic variables can be determined from the partition function Z, if Z is known as a function of V, T and N (or V, β and N). According to Postulate 4 of Section 3.2, these ensemble averages give the equilibrium values of the macroscopic variables of classical equilibrium thermodynamics.

5.2 MEAN ENERGY OF A SMALL SYSTEM IN THERMAL EQUILIBRIUM WITH A HEAT RESERVOIR

It follows from equation (5.5) that, if α_i is put equal to the energy eigenvalue ε_i, the ensemble average energy U of the small system, when it is in thermal equilibrium with the heat reservoir, is

$$U = \bar{\varepsilon}_i = \Sigma \varepsilon_i \exp(-\beta \varepsilon_i)/Z \quad .$$

Differentiating equation (5.2) partially with respect to β gives

$$\left(\frac{\partial Z}{\partial \beta}\right)_{V,N} = \frac{\partial}{\partial \beta} \Sigma \exp(-\beta \varepsilon_i) = \Sigma \frac{\partial}{\partial \beta} [\exp(-\beta \varepsilon_i)] = \Sigma(-\varepsilon_i)\exp(-\beta \varepsilon_i) \ . \tag{5.6}$$

Substituting into the expression for U, we have

$$U = \bar{\varepsilon}_i = -\frac{1}{Z}\left(\frac{\partial Z}{\partial \beta}\right)_{V,N} = -\left(\frac{\partial \log Z}{\partial \beta}\right)_{V,N} \ . \tag{5.7}$$

Using equation (5.3), we have

$$U = kT^2\left(\frac{\partial \log Z}{\partial T}\right)_{V,N} = -\left(\frac{\partial \log Z}{\partial \beta}\right)_{V,N} \ . \tag{5.8}$$

As an example of the use of equation (5.8), consider the one dimensional simple harmonic oscillator discussed in Section 4.7. According to equation (4.34), the partition function is

$$Z = \frac{\exp(-\beta\hbar\omega/2)}{[1 - \exp(-\beta\hbar\omega)]} \ . \tag{4.34}$$

Taking the natural logarithm of both sides, we have

$$\log Z = -\tfrac{1}{2}\beta\hbar\omega - \log[1 - \exp(-\beta\hbar\omega)] \ . \tag{5.9}$$

Using equation (5.8), we obtain

$$U = -\left(\frac{\partial \log Z}{\partial \beta}\right)_{V,N} = \tfrac{1}{2}\hbar\omega + \frac{\hbar\omega}{[\exp(\beta\hbar\omega) - 1]} \ . \tag{5.10}$$

According to equation (A1.8) of Appendix 1 the standard deviation σ in the energy of the small system is given by

$$\sigma^2 = \langle(\varepsilon_i - \bar{\varepsilon}_i)^2\rangle = \overline{\varepsilon_i^2} - 2\bar{\varepsilon}_i\bar{\varepsilon}_i + (\bar{\varepsilon}_i)^2$$

$$\sigma^2 = \overline{\varepsilon_i^2} - (\bar{\varepsilon}_i)^2 \ . \tag{5.11}$$

According to equation (5.5), with $\alpha_i = \varepsilon_i^2$ we have

$$\overline{\varepsilon_i^2} = \frac{\Sigma \varepsilon_i^2 \exp(-\beta\varepsilon_i)}{Z} \ . \tag{5.12}$$

Differentiating equation (5.6) partially with respect to β gives

$$\left(\frac{\partial^2 Z}{\partial \beta^2}\right)_{V,N} = \Sigma \varepsilon_i^2 \exp(-\beta\varepsilon_i) \ .$$

Substituting in equation (5.12), we have

$$\overline{\varepsilon_i^2} = \frac{1}{Z}\left(\frac{\partial^2 Z}{\partial \beta^2}\right)_{V,N} \ . \tag{5.13}$$

Squaring equation (5.7) gives

$$(\bar{\varepsilon}_i)^2 = \frac{1}{Z^2}\left(\frac{\partial Z}{\partial \beta}\right)_{V,N}^2 \ . \tag{5.14}$$

Substituting from equations (5.13) and (5.14) into equation (5.11), we obtain

$$\sigma^2 = \frac{1}{Z}\left(\frac{\partial^2 Z}{\partial \beta^2}\right)_{V,N} - \frac{1}{Z^2}\left(\frac{\partial Z}{\partial \beta}\right)_{V,N}^2 \ . \tag{5.15}$$

Differentiating equation (5.7) with respect to β gives

$$\left(\frac{\partial U}{\partial \beta}\right)_{V,N} = -\frac{\partial}{\partial \beta}\left[\frac{1}{Z}\left(\frac{\partial Z}{\partial \beta}\right)_{V,N}\right]$$

$$= -\left[\frac{1}{Z}\left(\frac{\partial^2 Z}{\partial \beta^2}\right)_{V,N} - \frac{1}{Z^2}\left(\frac{\partial Z}{\partial \beta}\right)^2_{V,N}\right] . \qquad (5.16)$$

Comparing equations (5.15) and (5.16), it can be seen that

$$\sigma^2 = -\left(\frac{\partial U}{\partial \beta}\right)_{V,N} = kT^2\left(\frac{\partial U}{\partial T}\right)_{V,N} = kT^2 C_V . \qquad (5.17)$$

where C_V is the heat capacity of the small system, defined by equation (1.28) of Chapter 1.

As an example consider a non-conducting solid. It will be shown in Chapter 10 that, in the classical limit, the heat capacity of an insulating solid, consisting of N atoms, is equal to $3Nk$, and that, in the classical limit, the internal energy of a non-conducting solid is $U = 3NkT$ where N is the number of atoms in the solid. Applying equation (5.17), we obtain $\sigma/U = (1/3N)^{1/2}$. This is in agreement with equation (A4.10) of Appendix 4*. If N were equal to 1, σ/U would be 0.58, showing that fluctuations are very important for microscopic systems. If N were equal to 10^{18}, corresponding to a small macroscopic system, σ/U would be equal to 5.8×10^{-10}. This shows that, when a small macroscopic system is in thermal equilibrium with a heat reservoir, the fluctuations in the energy of the small system from the ensemble average energy are small. In the context of classical equilibrium thermodynamics, it is reasonable to assume that the small macroscopic system has a precise value of energy, which, according to statistical mechanics, is equal to the ensemble average value of energy, given by equation (5.8).

According to equation (4.30), the probability that at thermal equilibrium the small system is in any one of its microstates which have energy eigenvalues between ε and $(\varepsilon + d\varepsilon)$ is

$$P(\varepsilon) \, d\varepsilon = \exp(-\beta\varepsilon)D(\varepsilon) \, d\varepsilon/Z . \qquad (4.30)$$

This probability distribution is proportional to the product of the Boltzmann factor $\exp(-\beta\varepsilon)$, which decreases exponentially as the energy of the small system increases, and the density of states D, which increases extremely rapidly as the energy of the small system increases. For macroscopic systems, this product gives an extremely sharp probability distribution. The effect on the ensemble average energy U of changing the temperature of the heat reservoir will be discussed in Section 7.1.3.

5.3 PRESSURE OF A SMALL SYSTEM IN THERMAL EQUILIBRIUM WITH A HEAT RESERVOIR

Using the principle of virtual work, it is shown in Appendix 5 that the pressure of the small system, when it is in its ith microstate, is

$$p_i = -(\partial \varepsilon_i / \partial V) \ . \tag{A5.2}$$

Using equation (5.5) with $\alpha_i = p_i$, we find that the mean (ensemble average) pressure p of the small system is given by

$$p = \bar{p}_i = \frac{1}{Z} \Sigma p_i \exp(-\beta \varepsilon_i) = \frac{1}{Z} \Sigma \left(-\frac{\partial \varepsilon_i}{\partial V} \right) \exp(-\beta \varepsilon_i) \tag{5.18}$$

Differentiating $\log Z$ with respect to V keeping β and N constant and using equation (5.2) for Z, we have

$$\left(\frac{\partial \log Z}{\partial V} \right)_{\beta, N} = \frac{1}{Z} \left(\frac{\partial Z}{\partial V} \right)_{\beta, N} = \frac{1}{Z} \left(\frac{\partial}{\partial V} [\Sigma \exp(-\beta \varepsilon_i)] \right)_{\beta, N} \ .$$

Since ε_i is a function of V and N only, carrying out the partial differentiation for constant β and N, and dividing by β, we have

$$\frac{1}{\beta} \left(\frac{\partial \log Z}{\partial V} \right)_{\beta, N} = \frac{1}{\beta Z} \Sigma (-\beta) \left(\frac{\partial \varepsilon_i}{\partial V} \right) \exp(-\beta \varepsilon_i) = \frac{1}{Z} \Sigma \left(-\frac{\partial \varepsilon_i}{\partial V} \right) \exp(-\beta \varepsilon_i) \ . \tag{5.19}$$

Comparing equations (5.18) and (5.19), we conclude that

$$p = \frac{1}{\beta} \left(\frac{\partial \log Z}{\partial V} \right)_{\beta, N} = kT \left(\frac{\partial \log Z}{\partial V} \right)_{T, N} \ . \tag{5.20}$$

5.4 THE ENTROPY OF A SMALL SYSTEM IN THERMAL EQUILIBRIUM WITH A HEAT RESERVOIR

Since $\log Z$ is a function of V, β and N, using equation (A1.31) of Appendix 1 and assuming that N is constant, we have

$$d(\log Z) = (\partial \log Z / \partial \beta)_{V, N} \, d\beta + (\partial \log Z / \partial V)_{\beta, N} \, dV \ .$$

Using equations (5.8) and (5.20), we have

$$d(\log Z) = -U \, d\beta + p\beta \, dV \ . \tag{5.21}$$

Differentiating the product βU gives

$$d(\beta U) = U \, d\beta + \beta \, dU \ . \tag{5.22}$$

Adding equations (5.21) and (5.22), we have

$$d(\log Z + \beta U) = \beta(dU + p \, dV) \ .$$

Using $\beta = 1/kT$, we obtain

$$dU + p \, dV = T \, d[k \log Z + (U/T)] \ . \tag{5.23}$$

According to the thermodynamic identity, equation (3.85), which was developed from statistical mechanics in Section 3.7.5, for constant N we have

$$dU + p \, dV = T \, dS \ . \tag{5.24}$$

Comparing equations (5.23) and (5.24), we conclude that

$$dS = d[k \log Z + (U/T)] \ .$$

Integrating, we obtain

$$S = k \log Z + (U/T) + S_0 \tag{5.25}$$

where S_0 is a constant of integration. Let the energy eigenvalues of the small system be equal to $\varepsilon_0, \varepsilon_1, \varepsilon_2 \ldots$ in order of increasing magnitude. The partition function is

$$Z = \exp(-\varepsilon_0/kT) + \exp(-\varepsilon_1/kT) + \ldots$$

$$= \exp(-\varepsilon_0/kT)\{1 + \exp[-(\varepsilon_1 - \varepsilon_0)/kT] + \ldots\} \ .$$

If the absolute temperature T of the heat reservoir tends to zero, $\exp[-(\varepsilon_1 - \varepsilon_0)/kT]$ is much less than unity, the partition function Z tends to $\exp(-\varepsilon_0/kT)$, and $\log Z$ tends to $(-\varepsilon_0/kT)$. Hence, in the limit when T tends to zero, since U tends to ε_0, equation (5.25) becomes

$$S = (-k\varepsilon_0/kT) + (\varepsilon_0/T) + S_0 = S_0 \ .$$

The constant of integration in equation (5.25) is equal to the entropy the small system would have at the absolute zero. If it is assumed that S is zero at $T = 0$, so as to be in accord with the third law of thermodynamics, equation (5.25) becomes

$$S = k \log Z + (U/T) = k[\log Z + \beta U] \tag{5.26}$$

where U is given by equation (5.8).

As an example of the use of equation (5.26), consider the case of a one dimensional linear harmonic oscillator in thermal equilibrium with a heat reservoir of absolute temperature T, which was discussed in Section 4.7. According to equation (5.9)

$$\log Z = -\tfrac{1}{2}\beta\hbar\omega - \log[1 - \exp(-\beta\hbar\omega)] \ .$$

According to equation (5.10)

$$U = \tfrac{1}{2}\hbar\omega + \hbar\omega/[\exp(\beta\hbar\omega) - 1] \ .$$

Substituting for $\log Z$ and for U into equation (5.26), we obtain

$$S = k\left\{\frac{\hbar\omega}{kT[\exp(\hbar\omega/kT) - 1]} - \log[1 - \exp(-\hbar\omega/kT)]\right\} .$$

Putting $\hbar\omega/k = \theta$, we have

$$S = k\left\{\frac{\theta}{T[\exp(\theta/T) - 1]} - \log[1 - \exp(-\theta/T)]\right\} . \qquad (5.27)$$

In the limit, when T tends to zero, $\exp(\theta/T) \gg 1$, $\exp(-\theta/T) \ll 1$ and equation (5.27) reduces to

$$S = (k\theta/T)\exp(-\theta/T) = (\hbar\omega/T)\exp(-\hbar\omega/kT)$$

confirming that the entropy given by equation (5.27) tends to zero near the absolute zero of temperature.

5.5 THE HELMHOLTZ FREE ENERGY

Equation (5.26) can be rewritten in the form

$$U - TS = -kT \log Z . \qquad (5.28)$$

In classical equilibrium thermodynamics the **Helmholtz free energy** F is defined by the equation

$$F = U - TS . \qquad (5.29)$$

Comparing equations (5.28) and (5.29), it can be seen that

$$F = -kT \log Z . \qquad (5.30)$$

Alternatively $\qquad\qquad Z = \exp(-F/kT) . \qquad (5\ 31)$

Since it is $\log Z$ which appears in the expressions for U, p and S, the algebraic forms of these equations can be simplified by using F instead of $\log Z$. This approach will be developed in Section 6.2.

5.6 THE CHEMICAL POTENTIAL

It follows from equation (5.28) that

$$-d(kT \log Z) = dU - T \, dS - S \, dT . \qquad (5.32)$$

According to the thermodynamic identity, equation (3.85), which was developed from statistical mechanics in Section 3.7.5, $dU - T \, dS = -p \, dV + \mu \, dN$. Substituting into equation (5.32) we have

$$-d(kT \log Z) = -p \, dV - S \, dT + \mu \, dN . \qquad (5.33)$$

If V and T are constant, it follows from equation (5.33) that the chemical potential μ is given by

$$\mu = -kT\left(\frac{\partial \log Z}{\partial N}\right)_{V,T} = -\frac{1}{\beta}\left(\frac{\partial \log Z}{\partial N}\right)_{V,\beta} . \tag{5.34}$$

It is straightforward to extend the result to a multi-component system consisting of $N_1, N_2 \ldots$ particles of types $1, 2 \ldots$ respectively. Using equation (3.86) instead of equation (3.85), it follows that, if T, V and all the N_i are constant except for N_j, the number of particles of the jth constituent present, then

$$\mu_j = -kT(\partial \log Z/\partial N_j)_{T,V,N_1,N_2\ldots \text{(except } N_j)} \tag{5.35}$$

where μ_j is the chemical potential of the jth constituent.

5.7 SUMMARY OF THE FORMULAE FOR ENSEMBLE AVERAGES

$$Z = \sum_{\substack{(N \text{ particle} \\ \text{states})}} \exp(-\beta\varepsilon_i) = \sum_{\substack{(N \text{ particle} \\ \text{states})}} \exp(-\varepsilon_i/kT) . \tag{5.36}$$

$$U = -\left(\frac{\partial \log Z}{\partial \beta}\right)_{V,N} = kT^2\left(\frac{\partial \log Z}{\partial T}\right)_{V,N} . \tag{5.37}$$

$$p = \frac{1}{\beta}\left(\frac{\partial \log Z}{\partial V}\right)_{\beta,N} = kT\left(\frac{\partial \log Z}{\partial V}\right)_{T,N} . \tag{5.38}$$

$$S = k[\log Z + \beta U] = k \log Z + \frac{U}{T} . \tag{5.39}$$

$$F = -\left(\frac{1}{\beta}\right)\log Z = -kT \log Z . \tag{5.40}$$

$$\mu = -\frac{1}{\beta}\left(\frac{\partial \log Z}{\partial N}\right)_{V,\beta} = -kT\left(\frac{\partial \log Z}{\partial N}\right)_{V,T} . \tag{5.41}$$

In the general case, the partition function given by equation (5.36) is obtained by summing over all the N particle states of the small system. In some idealised systems, the interactions between the N particles in the small systems can be neglected so that the small system can be treated as a system of N non-interacting particles. In Section 5.8 the case of N spatially separated particles will be considered, using the paramagnetic atoms of an ideal paramagnetic salt as an example. In Section 5.9, the partition function for N indistinguishable non-localised independent particles will be developed for the case of the ideal monatomic gas in the classical limit. In Section 5.10 the linear diatomic molecule will be discussed as an example of the factorisation of the partition function. For a fuller discussion of the calculation of the partition function, the interested reader is referred to *Introductory Statistical Mechanics* by Turner and Betts, (Sussex University Press, 1974).

5.8 THE PARTITION FUNCTION FOR A SYSTEM OF N SPATIALLY SEPARATED PARTICLES—A PARAMAGNETIC SALT

5.8.1 The partition function

As an example of a system of N distinguishable particles, consider the example of an idealised paramagnetic salt discussed previously in Section 4.4. In this example it is assumed that each of the spatially separated paramagnetic atoms has one unpaired electron. The wave functions of the unpaired electrons in the electron shells of different paramagnetic atoms do not overlap significantly, so that, even though the unpaired electrons are identical, the lattice sites occupied by the unpaired electrons can be distinguished. It makes no difference which unpaired electron is at which lattice site. What matters is whether, in the presence of a uniform magnetic field of magnetic induction **B**, the unpaired electron at a particular lattice site is in the state with potential energy $(-\mu B)$, when the magnetic moment μ of the unpaired electron is parallel to **B**, or in the state with potential energy $(+\mu B)$, when the magnetic moment of the unpaired electron is antiparallel to **B**, as illustrated in Figures 4.3(a) and 4.3(b). Such an idealised system is sometimes called an ideal spin system.

As an analogy, consider N coins in N spatially separated boxes. These coins can either have heads upwards or tails upwards. If the coins were all of different colours, in addition to whether the coin in a particular box had heads upwards or tails upwards, the microstate of the N coin system would also change if the coloured coins moved from box to box. To be analogous to the ideal spin system, the N coins would have to be indistinguishable, so that, if some of the identical coins moved from box to box, this in itself would not change the microstate of the N coin system. All that matters is whether the coin in a given box has heads upwards or tails upwards.

It will be assumed in this Section that each unpaired electron can be treated as an independent particle, which interacts only with the heat reservoir, which in this example is the rest of the crystal. If there is no interaction between the N unpaired electrons, the total energy of the N particle small system is

$$\varepsilon_i = E_1 + E_2 + \ldots + E_j + \ldots + E_N$$

where E_j is the energy of the unpaired electron at the jth lattice site. This can have the values $\pm \mu B$. The partition function of the N particle system is

$$Z = \sum_{\substack{(N \text{ particle} \\ \text{states})}} \exp(-\beta\varepsilon_i) = \Sigma \exp(-\beta E_1)\exp(-\beta E_2) \ldots \exp(-\beta E_N) \ .$$

In the absence of any interaction between the unpaired electrons, whether the unpaired electron at lattice site number 1 is in the state with potential energy $-\mu B$, or in the state with potential energy $+\mu B$, does not influence, and is not influenced by which single particle states the unpaired electrons at the other lattice sites are in. Both states of the unpaired electron at lattice site number 1 must be combined with every possible combination of the states of the unpaired electrons at the other lattice sites. Hence the partition function Z can be rewritten in the form

$$Z = \Sigma\exp(-\beta E_1)\Sigma\exp(-\beta E_2)\exp(-\beta E_3) \ldots \exp(-\beta E_N)$$

or $$Z = Z_1\Sigma\exp(-\beta E_2)\exp(-\beta E_3) \ldots \exp(-\beta E_N) \tag{5.42}$$

where $Z_1 = \Sigma\exp(-\beta E_1)$ is the single particle partition function, obtained by summing over all the single particle states of the unpaired electron at lattice site number 1. For this unpaired electron, there are only two single particle states. These have energies $\pm\mu B$. According to equation (4.17), the single particle partition function for one unpaired electron is

$$Z_1 = [\exp(\mu B/kT) + \exp(-\mu B/kT)] = 2\cosh(\mu B/kT) \ . \tag{5.43}$$

The summations in equation (5.42) can be carried out successively for each of the N lattice sites in turn, so that the partition function Z for the system of N spatially separated unpaired electrons is given in terms of the single particle partition function Z_1 by

$$Z = Z_1^N \ . \tag{5.44}$$

Taking the natural logarithm of both sides, we obtain

$$\log Z = N \log Z_1 \tag{5.45}$$

For the special case of the ideal paramagnetic salt illustrated in Figure 4.3(a), using equations (5.43) and (5.45), we have

$$\log Z = N \log[\exp(\mu B/kT) + \exp(-\mu B/kT)] = N \log[2\cosh(\mu B/kT)] \ . \tag{5.46}$$

Equations (5.44) and (5.45) also hold in the case of N distinguishable non-interacting particles.

5.8.2 Energy of the spin system

Substituting for $\log Z$ from equation (5.46) into equation (5.37) and carrying out the partial differentiation, we obtain

$$U = kT^2 \left(\frac{\partial \log Z}{\partial T}\right)_{V,N} = -N\mu B \left[\frac{e^{\mu B/kT} - e^{-\mu B/kT}}{e^{\mu B/kT} + e^{-\mu B/kT}}\right] \tag{5.47}$$

or
$$U = -N\mu B \tanh(\mu B/kT) \; . \tag{5.48}$$

The variation of $U/N\mu B$ with $kT/\mu B$ is shown in Figure 5.1(a). When the absolute temperature T tends to zero and/or the magnetic induction B tends to infinity $(\mu B/kT) \gg 1$, $\tanh(\mu B/kT)$ tends to unity and U tends to $(-N\mu B)$. This result was to be expected, since, when T tends to zero or B tends to infinity, all the magnetic moments tend to be in the state of lower energy $(-\mu B)$, and the energy U tends to $N(-\mu B)$. As T tends to infinity and/or B tends to zero, the populations of the two states with energies $\pm\mu B$ tend to equality and U tends to zero.

    According to equation (4.23) the magnetic moment m of the localised spin system is

$$m = N\mu \tanh(\mu B/kT) \; . \tag{4.23}$$

Comparing with equation (5.48), it can be seen that the energy of the

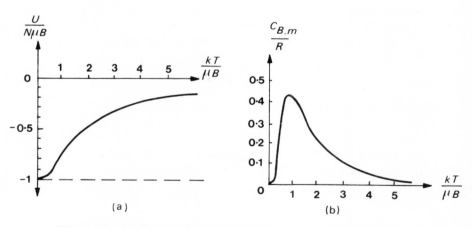

Figure 5.1—(a) The variation of the energy U of the N unpaired electrons in a paramagnetic salt with the ratio $(kT/\mu B)$, where B is the applied magnetic induction and T is the absolute temperature. The total energy U is negative since, at positive absolute temperatures, the majority of the magnetic moments point in the direction of the magnetic field, in which case they have potential energy $(-\mu B)$ each. (b) The variation of the molar heat capacity of the unpaired electrons in the paramagnetic salt with temperature. A maximum in the heat capacity of this type, associated with a two level system, is called a Schottky anomaly.

localised spin system can be rewritten in the form

$$U = -mB \; . \tag{5.49}$$

The *total* internal energy of a paramagnetic salt is the sum of the energy due to the lattice vibrations of the crystal and the 'magnetic energy' of the unpaired electrons given by equation (5.49).

5.8.3 The heat capacity of the spin system

Consider a mole of unpaired electrons, so that $N = N_A$, where N_A is Avogadro's constant. The molar heat capacity at constant B is

$$C_{B,m} = (\partial U / \partial T)_B \; .$$

Using equation (5.48), with $N_A k = R$, [reference: equation (5.77)], we have

$$C_{B,m} = N_A k (\mu B/kT)^2 \mathrm{sech}^2(\mu B/kT) = R(\mu B/kT)^2 \mathrm{sech}^2(\mu B/kT) \; . \tag{5.50}$$

The variation of $C_{B,m}/R$ with $kT/\mu B$ is shown in Figure 5.1(b). It can be seen that $C_{B,m}$ has a maximum value $\simeq 0.44R$ when $kT/\mu B \simeq 0.84$. It will be shown in Chapter 10 that, in the high temperature limit, the contribution of the lattice vibrations to the molar heat capacity of a solid tends to $3R$. It can be seen that, at the position of the maximum in Figure 5.1(b), the heat capacity of the localised spin system is about 15% of the value of $3R$. Both experiments and the theoretical predictions of the Debye T^3 law show that at $0.1\,\mathrm{K}$ the contribution of the lattice vibrations to the molar heat capacity of a solid is less than $10^{-5}R$. [Reference: Section 10.3 of Chapter 10.] Below $1\,\mathrm{K}$ the heat capacity of the spin system is very much bigger than the heat capacity associated with the lattice vibrations of the crystal. To quote Dugdale [1]: 'It has been pointed out that $1\,\mathrm{cm}^3$ of iron ammonium alum at $\sim 0.05\,\mathrm{K}$ (the temperature where its specific heat is a maximum) has a heat capacity equal to that of about 16 tons of lead at the same temperature'. This large difference in the two contributions to the total heat capacity is an important factor in the success of the method of cooling by adiabatic demagnetisation, which will be described in Section 5.8.5.

As the temperature T increases beyond the maximum of the curve in Figure 5.1(b), the molar heat capacity $C_{B,m}$ decreases. In the limit as T tends to infinity, $(\mu B/kT)$ tends to zero, $\mathrm{sech}(\mu B/kT)$ is very close to unity and $C_{B,m}$ is proportional to $1/T^2$. On the other hand, the contribution to the heat capacity of the lattice vibrations increases with increasing T and generally predominates at temperatures above $5\,\mathrm{K}$. The maximum in the heat capacity of a two level system at a particular value of T is known as a Schottky anomaly.

5.8.4* The entropy of the spin system*

Substituting the expression for U given by equation (5.47) and the expression of log Z given by equation (5.46) into equation (5.39), we find that the entropy S of the idealised spin system is

$$S/Nk = \log (e^x + e^{-x}) - x(e^x - e^{-x})/(e^x + e^{-x})$$

where
$$x = \mu B/kT \; . \tag{5.51}$$

Rearranging, we have

$$S/Nk = \log (1 + e^{-2x}) + 2x/(e^{2x} + 1) \; . \tag{5.52}$$

Alternatively, using hyperbolic functions we have

$$S/Nk = \log[2 \cosh(\mu B/kT)] - (\mu B/kT)\tanh(\mu B/kT) \; . \tag{5.53}$$

The variation of the entropy S with $(kT/\mu B)$ is shown in Figure 5.2(a).

If T tends to zero, so that $x \gg 1$, the term e^{-2x} in equation (5.52) can be neglected compared to unity and the term $(e^{2x} + 1)$ can be approximated by e^{2x}. Hence, when T tends to zero, the entropy is given approximately by

$$S = 2Nk(\mu B/kT)\exp[-2\mu B/kT] \; . \tag{5.54}$$

This shows that the entropy tends to zero exponentially as T tends to zero, and S is zero when T is zero, in agreement with the third law of thermodynamics.

If B tends to zero at a finite temperature, x tends to zero. In this limit both $\exp(-2x)$ and $\exp(2x)$ tend to unity, and equation (5.52) reduces to $S = Nk \log 2$. This is the same as equation (2.43).

5.8.5* Adiabatic demagnetisation*

The entropy S of the spin system, given by equation (5.53), is a function of (T/B). The abscissa in Figure 5.2(a) is $(kT/\mu B)$. Consider two values of magnetic induction B_1 and B_2, with $B_2 > B_1$. For a given value of T, the value of $(kT/\mu B)$ is less for B_2 than for B_1. Since S decreases with decreasing $(kT/\mu B)$ in Figure 5.2(a), the entropy S, for the same value of T, is less when the magnetic induction is equal to B_2 than when it is equal to the lower value of B_1. This is illustrated in Figure 5.2(b), where S is plotted against T not $(kT/\mu B)$, for both the magnetic induction values B_1 and B_2.

Let the magnetic induction be increased from B_1 to B_2 under isothermal conditions at a constant initial temperature T_i. This is illustrated in the change from state 1 to state 2 in Figure 5.2(b). Heat is evolved when some of the magnetic moments change their orientations from anti-parallel to parallel to the magnetic field, when the magnetic induction is increased from B_1 to B_2. The salt is then isolated thermally and the magnetic induc-

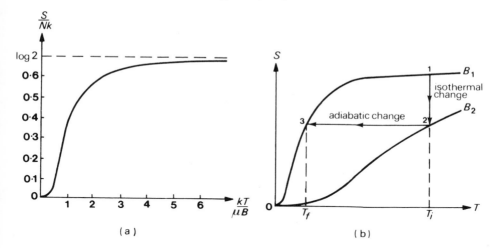

Figure 5.2—(a) The variation of the entropy S of the unpaired electrons in a paramagnetic salt with the ratio $kT/\mu B$, where B is the applied magnetic induction and T is the absolute temperature. (b) The entropy S is plotted against the absolute temperature T for two values of the magnetic induction B_1 and B_2, with $B_2 > B_1$. The principle of adiabatic demagnetisation is illustrated in the changes from 1 to 2 to 3.

tion is reduced very slowly, under adiabatic conditions, from B_2 to B_1. According to Ehrenfest's theorem (the adiabatic approximation of quantum mechanics), the probability that the spin system is in a particular microstate does not change in this reversible adiabatic process, and the entropy S of the spin system remains constant. (Reference: Section 7.1.5 of Chpater 7.) The spin system goes from state 2 to state 3 in Figure 5.2(b). Since S is a function of T/B, for the isentropic (adiabatic) change from state 2 to state 3 in Figure 5.2(b), we have

$$T_i/B_2 = T_f/B_1 \qquad (S \text{ is constant}) \qquad (5.55)$$

where T_f is the final temperature of the **spin system**. Since $B_1 < B_2$ it follows that $T_f < T_i$. Eventually, due to the interactions between the spins and the lattice, the whole paramagnetic crystal reaches a common temperature. It was shown in Section 5.8.3 that at temperatures < 0.1K the heat capacity of the spin system is very much bigger than the heat capacity of the rest of the crystal. Hence the final temperature of the paramagnetic salt will be only just above T_f.

A typical experimental arrangement is shown in Figure 5.3. In a typical experiment, one would use a value of B_2 equal to about 1T at an initial temperature T_i of about 1K. If B_1, the final value of the magnetic induction is approximately zero, it is possible to cool a typical paramagnetic salt to a

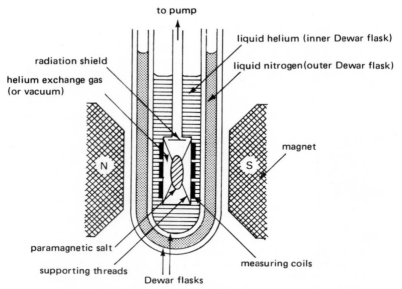

Figure 5.3—A typical experimental arrangement for cooling a paramagnetic salt by adiabatic demagnetisation. When the specimen is surrounded by helium exchange gas, the helium gas conducts heat to the liquid helium, which is boiling under reduced pressure in the inner Dewar flask. This cools the paramagnetic salt to an initial temperature T_i of about 1K. The outer Dewar flask contains liquid nitrogen to reduce the flow of heat from outside the apparatus to the liquid helium. The paramagnetic salt starts at position 1 in Figure 5.2(b), in a magnetic field B_1 at a temperature T_i. With the helium exchange gas still surrounding the paramagnetic salt, the magnetic field is increased from B_1 to B_2. The heat evolved, when the magnetisation of the paramagnetic salt increases, is conducted away to the liquid helium by the helium exchange gas. The paramagnetic salt is then in position 2 in Figure 5.2(b), in a magnetic field B_2 at a temperature T_i. The helium exchange gas is pumped away so that the paramagnetic salt is in a vacuum and hence thermally isolated. The magnetic field is reduced from B_2 to B_1. (In practice, the salt is demagnetised.) The paramagnetic salt undergoes an adiabatic reduction in its magnetisation to position 3 in Figure 5.2(b), and the temperature of the salt drops to a value of T_f of about 0.01K. (Copyright McGraw-Hill, 1971. Reproduced with permission from *Equilibrium Thermodynamics*, 2nd edition by C. J. Adkins.)

temperature of the order of 0.01K. For full details, the reader is referred to Zemansky [2a]. Using nuclear demagnetisation it is possible to reach temperatures of 10^{-5}K. (Reference: Zemansky [2b].)

5.9 THE PARTITION FUNCTION FOR *N* INDISTINGUISHABLE NON-LOCALISED INDEPENDENT PARTICLES—THE IDEAL MONATOMIC GAS IN THE CLASSICAL LIMIT

5.9.1 Introduction

The atoms of an ideal monatomic gas inside a box are not localised, so that in this case the identical particles cannot be distinguished. The ideal monatomic gas in the classical limit will be used as an example to develop the expression for the partition function of a system of *N* non-localised indistinguishable independent particles in terms of the single particle partition function for one particle. It will be assumed in this Section that the temperature of the heat reservoir is low enough for the possibility of exciting the higher energy electronic and the higher energy nuclear states of individual gas atoms to be negligible. Non-relativistic theory will be used throughout.

5.9.2 The partition function of a particle in a box

Consider one particle inside a cubical box of side L and volume V. The box is placed in thermal contact with a heat reservoir, which is at a constant absolute temperature T. According to equation (2.5) the energy eigenvalues of the possible *single* particle quantum states are

$$\varepsilon_i = (\pi^2 \hbar^2 / 2mV^{2/3})(n_x^2 + n_y^2 + n_z^2) \ . \tag{5.56}$$

Each of the quantum numbers n_x, n_y and n_z can have all the positive integer values from 1 to infinity. According to equation (5.56), the partition function of the single monatomic gas atom is

$$Z_1 = \sum_{(\text{states})} e^{-\beta\varepsilon_i} = \sum_{n_x} \sum_{n_y} \sum_{n_z} e^{-\alpha(n_x^2 + n_y^2 + n_z^2)} \ ,$$
$$= \sum_{n_x} \sum_{n_y} \sum_{n_z} e^{-\alpha n_x^2} e^{-\alpha n_y^2} e^{-\alpha n_z^2} \ , \tag{5.57}$$

where
$$\alpha = \beta\pi^2 \hbar^2 / 2mV^{2/3} \ . \tag{5.58}$$

To illustrate how we can sum over all the possible single particle states in equation (5.57), assume for example that $n_y = 5$ and $n_z = 7$. The quantum number n_x can have any positive integer value from 1 to ∞. This is true for every possible set of values for n_y and n_z, both of which can vary from 1 to ∞. Hence equation (5.57) can be rewritten in the form

$$Z_1 = \sum_{n_x=1}^{\infty} e^{-\alpha n_x^2} \sum_{n_y} \sum_{n_z} e^{-\alpha n_y^2} e^{-\alpha n_z^2} \ .$$

Generalising we have

$$Z_1 = \left(\sum_{n_x=1}^{\infty} e^{-\alpha n_x^2}\right)\left(\sum_{n_y=1}^{\infty} e^{-\alpha n_y^2}\right)\left(\sum_{n_z=1}^{\infty} e^{-\alpha n_z^2}\right) = Z_x Z_y Z_z \ , \qquad (5.59)$$

where, for example, $\qquad Z_x = \sum_{n_x=1}^{\infty} \exp(-\alpha n_x^2) \ . \qquad (5.60)$

It was shown in Section 2.1.4 that, in the special case of an ideal gas in the classical limit, we are dealing with quantum numbers n_x, n_y and n_z which are typically of the order of 10^{10}. In the classical limit, equation (5.60) can be approximated by

$$Z_x = \int_0^{\infty} \exp(-\alpha n_x^2) \, dn_x \ . \qquad (5.61)$$

This is the standard Gaussian integral I_0, which, according to equation (A1.36) of Appendix 1, has the numerical value of $\frac{1}{2}(\pi/\alpha)^{1/2}$. Hence $Z_x = \frac{1}{2}(\pi/\alpha)^{1/2}$. Substituting for α from equation (5.58), we obtain

$$Z_x = (m/2\beta\pi\hbar^2)^{1/2} V^{1/3} \ .$$

The expressions for Z_y and Z_z are the same. Equation (5.59) becomes

$$Z_1 = Z_x Z_y Z_z = (m/2\beta\pi\hbar^2)^{3/2} V \ . \qquad (5.62)$$

Equation (5.62) is the partition function associated with the translational degrees of freedom of a particle in a box. If the particle in the box is an electron or a proton of spin $s = \frac{1}{2}$, each set of quantum numbers n_x, n_y and n_z can be associated with either $s = +\frac{1}{2}$ or $s = -\frac{1}{2}$. This doubles the number of single particle states with a given value of n_x, n_y and n_z in equation (5.57) and doubles Z_1. If the particle in the box is an atom, the ground state of the orbital electrons may be degenerate. If the nucleus of the atom has a nuclear spin J, there is a nuclear spin degeneracy of $(2J + 1)$. To allow for the degeneracies associated with these internal degrees of freedom of the particle, equation (5.57) will be multiplied by g, the *total* degeneracy associated with the internal degrees of freedom. The single particle partition function then becomes

$$Z_1 = g[(m/2\beta\pi\hbar^2)^{3/2} V] \ . \qquad (5.63)$$

Taking the natural logarithm of both sides, we have

$$\log Z_1 = \log V + \log g - \tfrac{3}{2} \log \beta + \tfrac{3}{2} \log[m/2\pi\hbar^2] \ . \qquad (5.64)$$

Equation (5.64) was derived for the special case of a particle in a cubical box. It is valid whatever the shape of the box. [Reference: Problem 5.12*.] Substituting for $\log Z_1$ in equation (5.38), we find that the ensemble average pressure p is

$$p = (1/\beta)(\partial \log Z_1/\partial V)_{\beta,N} = 1/\beta V = kT/V .$$

Hence $$pV = kT .$$

Substituting for $\log Z_1$ in equation (5.37), we find that the ensemble average energy U_1 of the gas atom is

$$U_1 = -(\partial \log Z_1/\partial \beta)_{V,N} = 3/2\beta = \tfrac{3}{2}kT . \tag{5.65}$$

This is in agreement with equation (4.50), which was derived from the Maxwell velocity distribution in Section 4.9. It is also in agreement with the predictions of the equipartition theorem developed in Section 4.10*.

5.9.3 *N* atoms of an ideal monatomic gas in a box

Consider n moles of an ideal monatomic gas, consisting of $N = nN_A$ atoms, where N_A is Avogadro's constant. The gas is inside a cubical box of side L and volume V, which is placed in thermal contact with a heat reservoir of absolute temperature T. The partition function of the N-particle ideal monatomic gas system is

$$Z = \sum_{\left(\substack{N \text{ particle} \\ \text{states}}\right)} \exp(-\beta \varepsilon_i) . \tag{5.66}$$

In equation (5.66), $\varepsilon_i(V, N)$ is the energy eigenvalue of the ith quantum eigenstate (microstate) of the N particle ideal gas system. It will be assumed that there are no interactions between the atoms of the gas. It will also be assumed that the individual gas atoms are in single particle states, which can be determined by solving the Schrödinger equation for a single particle in a cubical box, in the way described in Appendix 3. The total energy of the N monatomic gas atoms is

$$\varepsilon_i = E_1 + E_2 + \ldots E_j + \ldots E_N$$

where E_j is the energy of the jth monatomic gas atom. Equation (5.66) can be rewritten in the form

$$Z = \sum \exp[-\beta(E_1 + E_2 + \ldots + E_N)]$$
$$= \sum \exp(-\beta E_1)\exp(-\beta E_2) \ldots \exp(-\beta E_N) .$$

If the N particles were *distinguishable*, there would be no restriction on the number of particles which could be in any single particle state (orbital) with a given set of quantum numbers. According to equation (5.44), the partition function for the system of N *distinguishable* particles would then be

$$Z = Z_1{}^N \tag{5.67}$$

where Z_1 is the single particle partition function given by equation (5.63).

The atoms of an ideal monatomic gas are *indistinguishable*. To illustrate the effect of the indistinguishability of the particles on the partition function, we shall consider an idealised system consisting of two particles inside a box, which is in thermal equilibrium with a heat reservoir. It will be assumed, initially, that the particles are distinguishable particles which are coloured red and white. In this case equations (5.66) and (5.67) give

$$Z = \Sigma \exp[-\beta(E_R + E_W)] = [\Sigma \exp(-\beta E_R)][\Sigma \exp(-\beta E_W)] \quad (5.68)$$

where E_R and E_W are the energy eigenvalues of the single particle states occupied by the red and the white particles respectively. The single particle states will be numbered. The ordering is not important. Each of the independent states of a degenerate energy level will be labelled with a different number. For fermions of spin $\frac{1}{2}$, the two states with the same values of n_x, n_y and n_z but having opposite spins are given different numbers. Consider a typical configuration such as 2;5 in which there is one particle in single particle state number 2 and one particle in single particle state number 5. In the case of two distinguishable particles there are two microstates, namely 2, 5 and 5, 2 where the first and second numbers are the numbers of the single particle quantum states occupied by the red and the white particles respectively. Both these microstates appear in equation (5.68), namely when $E_R = E_2$ with $E_W = E_5$ and when $E_R = E_5$ with $E_W = E_2$. If the two particles were indistinguishable, these two two-particle states would be the same state and the term $\exp[-\beta(E_2 + E_5)]$ should appear once not twice in the expression for the partition function, given by equation (5.68). This is always true, if the two indistinguishable particles are in different single particle states. Occasionally in Bose–Einstein statistics there is a possibility that the two indistinguishable particles are in the same single particle state. These cases only appear once in equation (5.68) when $E_R = E_W$, and we should not divide equation (5.68) by two in these cases. However, in the classical limit, the probability of double occupancy is less than 10^{-5} times the probability of single occupancy, and the rare cases of double occupancy can be ignored. Hence to obtain the partition function for two identical particles in the classical limit, the right hand side of equation (5.68) must be divided by 2.

For a system of three *distinguishable* particles, which are coloured red, white and blue (denoted R, W and B respectively), equations (5.66) and (5.67) give

$$Z = \Sigma \exp[-\beta(E_R + E_W + E_B)] = (\Sigma e^{-\beta E_R})(\Sigma e^{-\beta E_W})(\Sigma e^{-\beta E_B}) . \quad (5.69)$$

In this case, for any configuration in which the particles are in different single particle states there are 3! microstates. For example, in the 2;5;8 configuration the 3! microstates are 2,5,8; 2,8,5; 5,2,8; 5,8,2; 8,2,5 and 8,5,2 where the numbers are the numbers of the single particle states occupied by the red, white and blue particles respectively. If the three particles are *indistinguishable*, the six microstates are the same state. All the 3! microstates are included separately in equation (5.69). Hence to obtain the partition function for three indistinguishable particles, the right hand side of equation (5.69) must be divided by 3! The rare cases of multiple occupancy such as the 2;2;8 configuration can be ignored in the classical limit.

Generalising, for a system of *N* indistinguishable, non-localised, independent particles, *in the classical limit*, the partition function is[†]

$$Z = [\Sigma \exp(-\beta E_i)]^N/N! = Z_1{}^N/N! \quad \text{(classical limit)} . \quad (5.70)$$

Substituting for Z_1 from equation (5.63), for the special case of an ideal monatomic gas, we have

$$Z = g^N V^N (mkT/2\pi\hbar^2)^{3N/2}/N! \quad (5.71)$$

Taking the natural logarithm of both sides, we have

$$\log Z = N[\log(gV) + \tfrac{3}{2}\log(kT) + \tfrac{3}{2}\log(m/2\pi\hbar^2)] - \log N! .$$

According to Stirling's theorem, equation (A1.19) of Appendix 1, for large *N*

$$\log N! = N \log N - N .$$

Hence for a macroscopic system

$$\log Z = N[\log(gV/N) + \tfrac{3}{2}\log(kT) + \tfrac{3}{2}\log(m/2\pi\hbar^2) + 1] .$$
$$(5.72)$$

With $\beta = 1/kT$, we have

$$\log Z = N[\log(gV/N) - \tfrac{3}{2}\log\beta + \tfrac{3}{2}\log(m/2\pi\hbar^2) + 1] .$$
$$(5.73)$$

Equations (5.71), (5.72) and (5.73) were derived for the special case of an ideal monatomic gas inside a *cubical* box. It can be shown that these equations hold whatever the shape of the box. [Reference: Problem

---

[†]According to equation (A1.4) of Appendix 1, for a configuration of a system of *N* *distinguishable* particles in which there are n_j particles in the *j*th single particle state, the number of different microstates is $N!/(n_0! \, n_1! \, n_2! \ldots)$ not *N*!. If the probability of double occupancy of each single particle state were 10^{-6}, for a small macroscopic system consisting of 10^{18} particles, there would on average be 10^{12} cases of double occupancy, 10^6 of triple occupancy etc. The number of microstates in a typical configuration would be $N!/(10^{12} \times 2!)(10^6 \times 3!)(1 \times 4!) \ldots$ not *N*!. It can be seen from Section 5.7 that all the thermodynamic variables are related to log *Z*. Since $\log(10^{18}!)$ is about 4×10^{19}, whereas $\log(10^{12} \times 2!)$ is only 28.3 and, since we shall *always* work with log *Z*, we can always use *N*! in the denominator of equation (5.71) and neglect the effect of multiple occupancy in the classical limit.

5.12*.] In equations (5.72) and (5.73), g is the total degeneracy of one of the atoms due to its internal degrees of freedom, such as nuclear spin.

5.9.4 Internal energy of an ideal monatomic gas
Using equations (5.37) and (5.73), we have

$$U = -(\partial \log Z/\partial \beta)_{V,N} = 3N/2\beta = \tfrac{3}{2}NkT \ . \tag{5.74}$$

According to equation (5.74), the (internal) energy of an ideal monatomic gas is independent of the pressure and volume of the gas, and the (internal) energy is a function of only the absolute temperature of the gas. This statement is sometimes called Joule's law. The molar heat capacity of constant volume is

$$C_{V,m} = (\partial U/\partial T)_V = \tfrac{3}{2}N_A k \tag{5.75}$$

where N_A is Avogadro's constant.

5.9.5 Ideal monatomic gas equation
Using the value of $\log Z$ given by equation (5.73) in equation (5.38), we have

$$p = (1/\beta)(\partial \log Z/\partial V)_{\beta,N} = (1/\beta)(N/V) = NkT/V \ .$$

Rearranging and using equation (5.74), we have

$$pV = NkT = nN_A kT = \tfrac{2}{3}U \ . \tag{5.76}$$

where n is the number of moles in the ideal gas and N_A is Avogadro's constant. According to classical equilibrium thermodynamics, an ideal gas is defined as a gas which obeys the equation

$$pV = nRT \tag{1.2}$$

where R is the molar gas constant. Equations (5.76) and (1.2) are the same if

$$R = N_A k; \quad k = R/N_A \ . \tag{5.77}$$

The reader can check back that the absolute temperature T in equation (5.76) arises from the use of the abbreviation

$$\beta = 1/kT = (\partial \log g/\partial U)_{V,N}$$

in the deviation of the Boltzmann distribution in Section 4.2. Initially, Boltzmann's constant was introduced into our theory as the constant of proportionality in the equation

$$S = k \log g \ . \tag{3.22}$$

which relates the entropy S to the number of accessible microstates g.

Comparing equations (5.76) and (1.2) it can be seen that the definition of absolute temperature T by equation (3.19) in terms of the variation of the logarithm of the number of accessible microstates with energy gives precisely the same temperature scale as the ideal gas scale. The use of gas thermometers, corrected to the ideal gas scale, is a convenient way of determining absolute temperatures. (References: Section 1.4 of Chapter 1 and Zemanksy [2c].) An experimental value for Boltzmann's constant k can be determined from the experimental values of the molar gas constant R and Avogadro's constant N_A. According to experiments

$$k = 1.3807 \times 10^{-23} \text{ J K}^{-1} .$$

The reader should realise that equation (5.76) for an ideal gas was developed from Schrödinger's equation. The main steps were: (i) the solution of Schrödinger's equation in Appendix 3 leading up to equation (5.56) for the energy eigenvalues of the single particle states, (ii) the calculation of the single particle partition function Z_1 leading up to equation (5.63) in Section 5.9.2, (iii) the calculation of the N particle partition function Z in Section 5.9.3, leading up to equation (5.71), (iv) the application of equation (5.38) to log Z.

5.9.6 The entropy of an ideal monatomic gas
According to equation (5.39),

$$S = k \log Z + (U/T) . \qquad (5.39)$$

Substituting for $\log Z$ from equation (5.72) and for U from equation (5.74), we obtain

$$S = Nk[\log(gV/N) + \tfrac{3}{2}\log(mkT/2\pi\hbar^2) + \tfrac{5}{2}] . \qquad (5.78)$$

This is the *Sackur–Tetrode equation*. It gives the entropy of an ideal monatomic gas in terms of N, V and T. To express the entropy in terms of U, V and N, we can use the result, given by equation (5.76), that kT is equal to $2U/3N$. We have

$$S(U, V, N) = Nk[\log(gV/N) + \tfrac{3}{2}\log(mU/3N\pi\hbar^2) + \tfrac{5}{2}] . \qquad (5.79)$$

This is the **fundamental relation** for an ideal monatomic gas. The entropy S is expressed in terms of the extensive variables U, V and N. The temperature, pressure and chemical potential can be determined from equation (5.79) using equations (3.23), (3.70) and (3.78) respectively. [Reference: Problem 3.2.]

If T, V and N are given, the entropy of the ideal monatomic gas can be calculated using equation (5.78). For example, for neon at a temperature of 27.2K and a pressure of one atmosphere, the entropy calculated using equation (5.78) is 96.45 J mole^{-1} K^{-1}. The entropy of the gas can also be estimated by calculating the increase in the entropy of the neon when it is taken from the solid phase at $T = 0$ through the liquid phase to the gaseous phase at 27.2K and 1 atmosphere pressure, using the experimental values for the heat capacity and latent heats. The value obtained is 96.40 J mole^{-1} K^{-1}, which is in good agreement with the Sackur–Tetrode equation. (Reference: Kittel and Kroemer [3].)

5.9.7 The entropy of mixing

Consider two equal volumes V, both of which contain N atoms of the same ideal monatomic gas, and are separated, initially, by a solid partition, as shown in Figure 5.4(a). Both gases are in thermal equilibrium with a heat reservoir of absolute temperature T. According to equation (5.78) the entropy of the gas in each volume V is

$$S_1 = S_2 = Nk[\log(V/N) + C_1] \ .$$

where
$$C_1 = \tfrac{3}{2} \log(mkT/2\pi\hbar^2) + \log g + \tfrac{5}{2} \ .$$

If the absolute temperature T is constant, C_1 is a constant. The total initial entropy of the two *separate* volumes V of gas is

$$S_i = S_1 + S_2 = 2Nk[\log(V/N) + C_1] \ . \tag{5.80}$$

The partition is removed, so that there are $2N$ atoms in a volume $2V$, as shown in Figure 5.4(b). According to equation (5.78), the total entropy at thermal equilibrium is

$$S_f = (2N)k[\log(2V/2N) + C_1] = 2Nk[\log(V/N) + C_1] \ . \tag{5.81}$$

The same value of C_1 is used in equations (5.80) and (5.81), since the temperature is the same in both cases. Comparing equations (5.80) and (5.81) it can be seen that there is no change of entropy on mixing.

If the $N!$ term were absent from equation (5.71), equation (5.78) would have been

$$S = Nk[\log V + \log g + \tfrac{3}{2} \log(mkT/2\pi\hbar^2) + \tfrac{3}{2}] = Nk[\log V + C_1 - 1] \ .$$

In the absence of the $N!$ term, we would have had

$$S_i = 2Nk(\log V + C_1 - 1)$$

$$S_f = 2Nk(\log 2V + C_1 - 1)$$

so that
$$S_f - S_i = 2Nk(\log 2) \ .$$

But for the $N!$ term in equation (5.71), there would be an entropy of mixing, when the partition in Figure 5.4(a) was removed. This was the

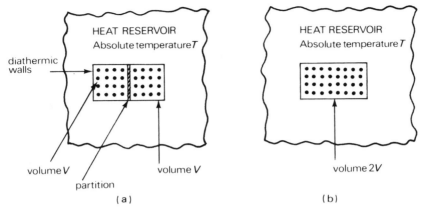

Figure 5.4—Example to illustrate the entropy of mixing. The partition in (a) is removed and the gases in the two separate volumes V are allowed to mix and fill the volume $2V$ in (b). It is shown in the text that, if the monatomic gas is the same in both the volumes V in (a), there is no change in the total entropy on mixing. If the atoms in the two volumes V are different, there is an increase in the total entropy, when the gases mix.

Gibbs paradox of the nineteenth century. It was only resolved by including the $N!$ term in equation (5.71) for the N particle partition function. The $N!$ term arose in our approach, when it was assumed that the atoms of a monatomic gas are indistinguishable.

If the volumes V in Figure 5.4(a) contained different monatomic gases, there would be an entropy of mixing. (Reference: Problem 5.10.)

If the concentration c is defined as the number of gas atoms per unit volume (N/V), equation (5.78) becomes

$$S = Nk\{-\log c + \log[g(mkT/2\pi\hbar^2)^{3/2}] + \tfrac{5}{2}\} . \qquad (5.82)$$

If the intensive variables c and T are kept constant in equation (5.82), the entropy S is proportional to the number of gas atoms N, showing that the entropy, given by equation (5.82), is an extensive variable. The calculated entropy of an ideal monatomic gas would not be an extensive variable, if the $N!$ term were omitted from equation (5.71).

When T tends to zero, the value of $\log T$ tends to $(-\infty)$, and the value for the entropy given by the Sackur–Tetrode equation tends to $(-\infty)$. The Sackur–Tetrode equation is not consistent with the third law of thermodynamics. It is only an approximation valid at high temperatures. At low temperatures either Fermi–Dirac or Bose–Einstein statistics must be used as appropriate.

5.9.8 The chemical potential of an ideal monatomic gas

Using equations (5.41) and (5.72) we find that, in the **classical limit**, the chemical potential of an ideal monatomic gas is given by

$$\mu = -kT\left(\frac{\partial \log Z}{\partial N}\right)_{V,T} = -kT \log\left[\frac{gV}{N}\left(\frac{mkT}{2\pi\hbar^2}\right)^{3/2}\right] . \qquad (5.83)$$

Taking exponentials, we have

$$\exp(-\mu/kT) = (gV/N)(mkT/2\pi\hbar^2)^{3/2} . \qquad (5.84)$$

As an example of a typical monatomic gas, consider a mole of helium gas at STP. Putting $V = 0.0224$ m^3, $N = 6.02 \times 10^{23}$, $m = 6.65 \times 10^{-27}$ kg, $T = 273$K, $k = 1.38 \times 10^{-23}$ J K^{-1}, $\hbar = 1.05 \times 10^{-34}$ J s and $g = 1$ in equation (5.84), we find that

$$\exp(-\mu/kT) = 2.56 \times 10^5 \qquad (5.85)$$
$$\mu/kT = -12.45$$

so that at $T = 273$K

$$\mu = -4.69 \times 10^{-20} \text{ J} .$$

This result shows that, in the classical limit, the chemical potential μ is negative. According to equation (5.85), $\exp(-\mu/kT)$ is very much greater than unity in the classical limit.

Equation (5.83) can be rewritten in the form

$$\mu = -kT \log(Z_1/N) \qquad (5.86)$$

where Z_1 is the single particle partition function, given by equation (5.63).

Equation (5.83) can be rewritten in the form

$$\mu = kT \log c - kT \log[g(mkT/2\pi\hbar^2)^{3/2}]$$

where $c = (N/V)$ is the particle concentration. This equation shows that μ increases when the particle concentration c is increased. (Actually, in the classical limit, μ gets less negative as c increases.) It was shown in Section 3.7.4 that particles tend to go from regions of high chemical potential to regions of low chemical potential, which, in the case of an ideal monatomic gas, is from regions of high concentration to regions of low concentration.

5.9.9 Discussion

The theory of the ideal monatomic gas was developed for a gas in a box in thermal equilibrium with a heat reservoir. It was shown in Section 5.2, that, for a small macroscopic system in thermal contact with a heat reservoir, the fluctuations in the energy of the small system are extremely small. In the context of classical equilibrium thermodynamics, it is reasonable to assume that the thermodynamic variables of a small macroscopic system in thermal equilibrium with a heat reservoir have precise values, which are equal to the ensemble averages of the variables. It is also reasonable to assume that the values of the thermodynamic variables of a small macroscopic system in thermal equilibrium with a heat reservoir are equal to the thermodynamic variables of an isolated system, consisting of the same number of atoms of the same type as the small system, but contained in an *adiabatic* enclosure of the same volume V, provided the absolute temperature of the isolated system is equal to the absolute temperature of the heat reservoir. For example, the entropy of an isolated monatomic gas system should be given in the classical limit by equation (5.78), the Sackur–Tetrode equation.

5.10 THE LINEAR DIATOMIC MOLECULE—AN EXAMPLE OF THE FACTORISATION OF THE PARTITION FUNCTION

5.10.1 Introduction

In some cases it is possible to separate the energy of a system into a number of independent contributions. As an example, consider one linear diatomic molecule inside a box kept at an absolute temperature T. As a first approximation, the diatomic molecule will be treated as two particles held together by an attractive force along the line joining the particles. To a good approximation the energy eigenvalue ε_i, of the ith microstate of the diatomic molecule in the box, can be rewritten in the form

$$\varepsilon_i = \varepsilon_t + \varepsilon_r + \varepsilon_v + \varepsilon_e + \varepsilon_n \qquad (5.87)$$

where ε_t is the contribution to the energy of the molecule associated with the transitional motion of the centre of mass of the molecule, ε_r is the contribution to the energy associated with the rotation of the two atoms in the molecule about the centre of mass of the molecule, ε_v is the contribution to the energy associated with the vibration of the two atoms in the molecule along the line joining them, ε_e is the energy of the atomic electrons and ε_n is the energy of the atomic nuclei. The partition function of the single diatomic molecule is

$$Z_1 = \sum_{\text{(states)}} \exp(-\beta\varepsilon_i)$$

$$= \sum_{\text{(states)}} \exp(-\beta\varepsilon_t)\exp(-\beta\varepsilon_r)\exp(-\beta\varepsilon_v)\exp(-\beta\varepsilon_e)\exp(-\beta\varepsilon_n) \ .$$

It will be assumed that the various contributions to the total energy, given by equation (5.87), can vary completely independently. Every translational state must be combined with every possible combination of the rotational, vibrational, electronic and nuclear states. Hence

$$Z_1 = [\sum \exp(-\beta\varepsilon_t)][\sum \exp(-\beta\varepsilon_r)\exp(-\beta\varepsilon_v)\exp(-\beta\varepsilon_e)\exp(-\beta\varepsilon_n)] \ .$$

Generalising, we have

$$Z_1 = [\Sigma \exp(-\beta\varepsilon_t)][\Sigma \exp(-\beta\varepsilon_r)][\Sigma \exp(-\beta\varepsilon_v)]$$
$$\times [\Sigma \exp(-\beta\varepsilon_e)][\Sigma \exp(-\beta\varepsilon_n)] \ .$$

This equation will be rewritten in the form

$$Z_1 = Z_t Z_r Z_v Z_e Z_n \tag{5.88}$$

where Z_t is equal to $\Sigma \exp(-\beta\varepsilon_t)$ etc. Taking the natural logarithm of equation (5.88), we have

$$\log Z_1 = \log Z_t + \log Z_r + \log Z_v + \log Z_e + \log Z_n \ . \tag{5.89}$$

Since it is $\log Z_1$ which appears in the formulae for the thermodynamic functions given in Section 5.7, the contributions of the various terms in equation (5.89) to the thermodynamic variables are additive. The calculations of each contribution to $\log Z_1$ will now be outlined.

5.10.2* The partition function for translational motion*

The partition function associated with the motion of the centre of mass of the molecule is given, *in the classical limit*, by equation (5.62) and is

$$Z_t = (m/2\beta\pi\hbar^2)^{3/2}V \tag{5.90}$$

where m is the total mass of the diatomic molecule. According to equation (5.65) the mean energy associated with the motion of the centre of mass of the molecule is

$$\bar{\varepsilon}_t = \tfrac{3}{2}kT \ . \tag{5.91}$$

5.10.3* The partition function for rotational motion*

According to classical mechanics, the kinetic energy of rotation of the two atoms in the molecule about their centre of mass is $\tfrac{1}{2}I\omega^2$, where ω is the angular velocity and I is the moment of inertia of the molecule about an axis through the centre of mass of the molecule perpendicular to the line joining the two atoms. According to quantum mechanics, the energy levels of a rigid rotator are

$$(\varepsilon_r)_j = j(j + 1)\hbar^2/2I \ . \tag{5.92}$$

The quantum number j can have the integer values 0, 1, 2 . . . . For a given value of j, the degeneracy is $(2j + 1)$ (Reference: French and Taylor [4]). Using equation (4.27), we find that the partition function for the rotational motion of heteronuclear molecule, such as HCl, which consists of two different types of atoms, is

$$Z_r = \sum_{j=0}^{\infty} (2j + 1)\exp\left[\frac{-j(j + 1)\hbar^2}{2IkT}\right] \ . \tag{5.93}$$

We have $Z_r = \left[1 + 3 \exp\left(-\dfrac{2\theta_r}{T} \right) + 5 \exp\left(-\dfrac{6\theta_r}{T} \right) + \ldots \right]$.   (5.94)

where θ_r is given by $\theta_r = \hbar^2 / 2Ik$.   (5.95)

When $T \ll \theta_r$ the overwhelming probability is that the heteronuclear diatomic molecule is in its ground state of rotational motion. When $T \simeq \theta_r$ the higher energy rotational states begin to be excited significantly. According to equation (5.95), θ_r is inversely proportional to the moment of inertia I, so that θ_r is higher for lighter molecules. For example, for the heteronuclear molecule HCl, θ_r is 15.2K, whereas for the lighter homonuclear molecule H_2, θ_r is 85.5K. At temperatures $T \gg \theta_r$, it is possible to approximate equation (5.93) by the integral

$$Z_r = \int_0^\infty (2j + 1)\exp\left[-\frac{j(j+1)\theta_r}{T} \right] dj .$$

Let $j(j+1) = z; \qquad dz = (2j+1)\, dj$.

Substituting and evaluating the definite integral, we have

$$Z_r = \int_0^\infty \exp\left[\frac{-z\theta_r}{T} \right] dz = T/\theta_r .$$

Using equation (5.95) we have

$$Z_r = 2IkT/\hbar^2 = 2I/\beta\hbar^2 \qquad (T \gg \theta_r) .$$ (5.96)

Hence $\log Z_r = -\log \beta + \log(2I/\hbar^2) \qquad (T \gg \theta_r)$.   (5.97)

Using equation (5.37), when $T \gg \theta_r$ we have

$$\bar{\varepsilon}_r = -(\partial \log Z/\partial\beta)_{V,N} = 1/\beta = kT .$$ (5.98)

Hence, when $T \gg \theta_r$, the mean energy of the rotational motion of the molecule is kT.

For a homonuclear diatomic molecule, such as H_2, which is composed of two identical atoms, the nuclei of the two atoms in the molecule are indistinguishable. We can allow for this, at a temperature $T \gg \theta_r$, by including the symmetry number σ in the denominator of equation (5.96) giving, to a very good approximation for temperatures $T \gg \theta_r$,

$$Z_r \simeq 2I/\sigma\beta\hbar^2 \qquad (\text{homonuclear molecule: } T \gg \theta_r) .$$ (5.99)

In the case of a linear homonuclear diatomic molecule, such as H_2, the symmetry number σ is equal to 2. [Equation (5.99) applies to a heteronuclear molecule, if we put $\sigma = 1$.] A reader interested in a full discussion of the symmetry number is referred to Gopal [5] and to Wilks [6].

5.10.4* The partition function for vibrational motion*

To a first approximation, the vibrational motion of the two atoms in a diatomic molecule along the line joining them can be treated as simple harmonic motion. According to equation (4.34), the partition function for vibrational simple harmonic motion is

$$Z_v = \exp(-\theta_v/2T)/[1 - \exp(-\theta_v/T)] \ . \tag{5.100}$$

The characteristic temperature for the vibrational motion is given by

$$\theta_v = hv/k = \hbar\omega/k \tag{5.101}$$

where v is the classical vibration frequency. Sometimes, particularly in Chemistry, the energies of the excited states are measured from the ground state. If the zero point energy $\frac{1}{2}\hbar\omega$ is omitted, the partition function is

$$Z_v' = [1 - \exp(-\theta_v/T)]^{-1} \ . \tag{5.102}$$

For an HCl molecule θ_v is 4130K, whereas for an H_2 molecule θ_v is 6140K. The higher energy vibrational states of diatomic molecules are not excited to any large extent at room temperatures, and do not contribute significantly to the heat capacity at room temperatures.

It was shown in Section (4.7) that the mean energy of a harmonic oscillator is

$$\bar{\varepsilon}_v = \frac{1}{2}k\theta_v + k\theta_v/[\exp(\theta_v/T) - 1] \ . \tag{4.37}$$

When $T \gg \theta_v$, the mean energy of the harmonic oscillator is

$$\bar{\varepsilon}_v \simeq kT \ . \tag{5.103}$$

5.10.5* The electronic partition function*

According to equation (4.27) the electronic partition function is

$$Z_e = \sum_{\text{(levels)}} g_e\exp(-\beta\varepsilon_e) \ . \tag{5.104}$$

For most diatomic molecules, the characteristic temperature θ_e for electronic excitation is very much higher than room temperature, so that the higher energy electronic states are not excited to any significant extent at room temperatures, and all the $\exp(-\beta\varepsilon_e)$ terms in equation (5.104) are negligible compared with $\exp(-\beta\varepsilon_0)$. Hence, at room temperatures when $T \ll \theta_e$, equation (5.104) reduces to

$$Z_e = g_0\exp(-\beta\varepsilon_0) \ . $$

where g_0 is the degeneracy (statistical weight) of the lowest energy level (ground state). If ε_0, the energy of the lowest energy level, is chosen as the zero of energy for measuring ε_e in equation (5.87), equation (5.104) reduces to

$$Z_e' = g_0 \; . \tag{5.105}$$

Since the higher electronic states are not excited significantly at room temperatures, they do not contribute significantly to the heat capacity of the diatomic molecule at room temperatures.

5.10.6* The nuclear partition function*

The characteristic temperature θ_n for nuclear excitation is very much higher than the characteristic temperature θ_e for electronic excitation. The higher energy nuclear states are not excited significantly at room temperatures.

A nucleus having a nuclear spin J has a nuclear spin degeneracy of $(2J + 1)$. If J_1 and J_2 are the nuclear spins of the ground states of the two nuclei in the molecule, since each of the $(2J_1 + 1)$ states of nucleus 1 can be associated with each of the $(2J_2 + 1)$ states of nucleus 2, the total degeneracy, associated with nuclear spin, is

$$g_n = (2J_1 + 1)(2J_2 + 1) \; . \tag{5.106}$$

If the ground state of the nucleus is chosen as the zero of energy for measuring ε_n in equation (5.87), at temperatures $T \ll \theta_n$, the nuclear partition function is

$$Z_n' = g_n = (2J_1 + 1)(2J_2 + 1) \; . \tag{5.107}$$

Since the higher energy nuclear states are not excited significantly at room temperatures, the atomic nuclei do not contribute, to any significant extent, to the heat capacity of a diatomic molecule at room temperatures.

The nuclear spin is closely associated with the symmetry number introduced into equation (5.99). The interested reader is referred to Gopal [5] and to Wilks [6].

5.10.7* The total partition function of a diatomic molecule*

According to equation (5.88)

$$Z_1 = Z_t Z_r Z_v Z_e Z_n \tag{5.88}$$

where Z_t, Z_r, Z_v, Z_e and Z_n are given by equations (5.90), (5.99), (5.100), (5.105) and (5.107) respectively. It is convenient to use the symbol Z_{int}, which is defined as

$$Z_{\text{int}} = Z_r Z_v Z_e Z_n \; , \tag{5.108}$$

to represent the contributions of all the internal degrees of freedom to the partition function. Since Z_r, Z_v, Z_e and Z_n are independent of the volume of the box, Z_{int} is a function of β only. Using equations (5.90) and (5.108), equation (5.88) can be rewritten in the form

$$Z_1 = (m/2\beta\pi\hbar^2)^{3/2}VZ_{int} . \tag{5.109}$$

5.10.8* The ideal diatomic gas in the classical limit*

Consider a system of N indistinguishable diatomic molecules in a box. If there is no interaction between the diatomic molecules, according to equation (5.70) the partition function of the N particle system, in the classical limit, is

$$Z = (Z_1)^N/N! = (1/N!)(m/2\beta\pi\hbar^2)^{3N/2}V^N Z_{int}^N . \tag{5.110}$$

Taking the natural logarithm of both sides and applying Stirling's theorem, equation (A1.19) of Appendix 1, we have

$$\log Z(\beta, V, N) = N[\log V - \tfrac{3}{2}\log \beta + \tfrac{3}{2}\log(m/2\pi\hbar^2)$$
$$+ \log Z_{int} - \log N + 1] \tag{5.111}$$

where Z_{int} is a function of β only. (For an ideal monatomic gas, Z_{int} reduces to $g_0 g_n$, which was represented by g in equation (5.63).) Applying equation (5.38) to equation (5.111), for an ideal diatomic gas we have

$$p = (1/\beta)(\partial \log Z/\partial V)_{\beta,N} = N/\beta V = NkT/V . \tag{5.112}$$

This is the equation of state of an ideal diatomic gas.

5.10.9* The heat capacity of a diatomic gas*

Consider a mole of gas, consisting of N_A diatomic molecules. At temperatures $T \ll \theta_r$, only the higher energy translational states are excited to any significant extent. In this limit, when according to equation (5.91) the mean translational energy of each molecule is approximately $3kT/2$, the total energy of the N_A diatomic molecules is approximately $3N_A kT/2$ and the molar heat capacity at constant volume

$$C_{V,m} = (\partial U/\partial T)_V \simeq \tfrac{3}{2}N_A k = \tfrac{3}{2}R \qquad (T \ll \theta_r)$$

as shown in Figure 5.5. When the temperature is of the order of θ_r, which for H_2 is 85.5K, the higher energy rotational states are beginning to be excited significantly. When $T \gg \theta_r$, according to equation (5.98), the mean rotational energy per molecule is kT, and the contribution to the total energy U due to the rotation of the molecules is $N_A kT$. In this temperature range, provided $T \ll \theta_v$, the molar heat capacity is

$$C_{V,m} \simeq N_A(\partial/\partial t)(\tfrac{3}{2}kT + kT) = \tfrac{5}{2}N_A k = \tfrac{5}{2}R$$

In the vicinity of $T = \theta_r$, $C_{V,m}$ increases from $3R/2$ to $5R/2$ as shown in Figure 5.5. Room temperature is generally between θ_r and θ_v, so that the

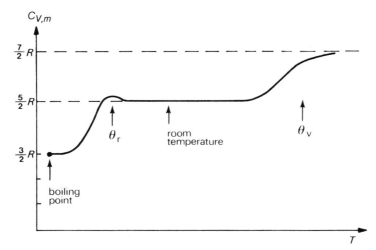

Figure 5.5—Sketch of the variation of the molar heat capacity of an ideal diatomic gas with temperature. The temperature scale is not linear. At room temperatures, the molar heat capacity is, generally, approximately equal to $\frac{5}{2}R$.

molar heat capacity of a diatomic gas, at constant volume, is generally approximately equal to $5R/2$ at room temperature.

As the temperature is increased further, the higher energy vibrational states are excited when $T \simeq \theta_v$. For hydrogen, $\theta_v = 6140K$. When $T \gg \theta_v$, according to equation (5.103) the mean vibrational energy per molecule is kT. When $T \gg \theta_v$, $C_{V,m}$ is given to a good approximation by

$$C_{V,m} = N_A(\partial/\partial T)(\tfrac{3}{2}kT + kT + kT) = \tfrac{7}{2}N_A k = \tfrac{7}{2}R \ .$$

In the vicinity of $T = \theta_v$, $C_{V,m}$ increases from $5R/2$ to $7R/2$, as shown in Figure 5.5.

Only a simplified theory of the variation of $C_{V,m}$ with T has been given. There are several corrections which should be applied. For example, the vibrations of the atoms along the line joining them is not exactly simple harmonic. A reader interested in a more comprehensive discussion is referred to Andrews [8] and to Hill [9].

5.10.10* The chemical potential of an ideal diatomic gas*

Using equations (5.41) and (5.111), we find that the chemical potential of a diatomic gas in the classical limit is

$$\mu = -kT\left(\frac{\partial \log Z}{\partial N}\right)_{V,T} = -kT \log\left[\frac{V}{N}\left(\frac{mkT}{2\pi\hbar^2}\right)^{3/2} Z_{\text{int}}\right] \ . \qquad (5.113)$$

Using equation (5.109), equation (5.113) can be rewritten in the form

$$\mu = -kT \log[Z_1/N] \qquad (5.114)$$

where Z_1 is the single particle partition function.

The ratio N/V is equal to the concentration c. Equation (5.113) can be rewritten in the form

$$\mu = kT \log c + \mu_1^{0}(T) \qquad (5.115)$$

where $$\mu_1^{0}(T) = -kT \log[(mkT/2\pi\hbar^2)^{3/2}Z_{int}] \qquad (5.116)$$

is the value of the chemical potential, when the concentration is equal to unity. The higher the concentration the larger the chemical potential. It was shown in Section 3.7.4 that particles tend to go from regions of high chemical potential to regions of lower chemical potential, which, for an ideal diatomic gas, is from regions of high concentration to regions of lower concentration.

Substituting for (V/N) from equation (5.112) into equation (5.113) we have

$$\mu(p, T) = kT \log p + \mu_2^{0}(T) \ . \qquad (5.117)$$

In this case, $\mu_2^{0}(T)$ is the chemical potential at unit pressure and is given by

$$\mu_2^{0}(T) = -kT \log[kT(mkT/2\pi\hbar^2)^{3/2}Z_{int}] \ . \qquad (5.118)$$

REFERENCES

[1] Dugdale, J. S., *Entropy and Low Temperature Physics*, Hutchinson, 1966, page 191.
[2] Zemanksy, M. W., *Heat and Thermodynamics*, 5th Edition, McGraw-Hill, New York 1957, (a) page 461, (b) page 484, (c) page 14.
[3] Kittel, C. and Kroemer, H., *Thermal Physics*, 2nd Edition, W. H. Freeman, 1980, page 167.
[4] French, A. P. and Taylor, E. F., *An Introduction to Quantum Physics*, Thomas Nelson, 1979, page 487.
[5] Gopal, E. S. R., *Statistical Mechanics and Properties of Matter*, Ellis Horwood, 1974, page 235.
[6] Wilks, J., *The Third Law of Thermodynamics*, Clarendon Press, Oxford 1961, page 93.
[7] Gasser, R. P. H. and Richards, W. G., *Entropy and Energy Levels*, Clarendon Press, Oxford 1974, page 74.
[8] Andrews, F. C., *Equilibrium Statistical Mechanics*, 2nd edition, Wiley, 1975, Chapter 11.
[9] Hill, T. L., *An Introduction to Statistical Thermodynamics*, Addison-Wesley, 1960, Chapter 8.

PROBLEMS

Planck's constant $h = 6.626 \times 10^{-34}$ J s
Boltzmann's constant $k = 1.3807 \times 10^{-23}$ J K^{-1}
Avogadro's constant $N_A = 6.022 \times 10^{23}$ mole^{-1}

Problem 5.1
A thermodynamic system consists of N spatially separated subsystems. Each subsystem has non-degenerate energy levels of energy 0, ε, 2ε and 3ε. The system is in thermal equilibrium with a heat reservoir of absolute temperature T equal to ε/k. Calculate the partition function, the mean energy and the entropy of the thermodynamic system.

Problem 5.2
Repeat problem 5.1 for the case when the degeneracies of the energy levels of energy 0, ε, 2ε and 3ε are 1, 2, 4 and 4 respectively.

Problem 5.3
The partition function of a system is given by

$$\log Z = aT^4V$$

where a is a constant, T is the absolute temperature and V is the volume. Calculate the internal energy, the pressure and the entropy.

Problem 5.4
A paramagnetic salt contains a total of 10^{24} magnetic ions, each with a magnetic moment of 1 Bohr magneton (9.273×10^{-24} A m^2) which can be aligned either parallel or anti-parallel to a magnetic field. A magnetic field of magnetic induction 1T is applied at a temperature of 3K. Calculate the partition function, the magnetic energy and the heat capacity.

Problem 5.5
Calculate the contribution to the total heat capacity of a diatomic gas of the vibration of the molecules, at an absolute temperature T equal to θ_v, where θ_v is given by equation (5.101). The gas consists of N molecules.
[*Hint*: Compare with equation (10.5) of Chapter 10.]

Problem 5.6
Calculate the contribution of the vibrations of the molecules to the total molar heat capacity of a diatomic gas at 300K, if the vibration frequency is 2×10^{14} Hz. What is the contribution at 1000K and 10^4K?

Problem 5.7
Calculate the entropy of a mole of helium gas at STP ($T = 273$K, $p = 101\ 325$ N m^{-2}). The mass of a helium atom is 6.65×10^{-27} kg and $g = 1$.

Problem 5.8
Calculate the chemical potential of a mole of argon gas at STP ($T = 273$K, $p = 101\ 325$ N m^{-2}). The degeneracy g is equal to 1 in equation (5.83). The mass of an argon atom is 6.64×10^{-26} kg.

Problem 5.9
An ideal monatomic gas consists of a mole of atoms in a volume V. The ideal gas is allowed to expand to fill a volume $2V$. Show that the increase in entropy is equal to $R \log 2$. (Assume that the temperature of the gas does not change.)

Problem 5.10
Assume that the ideal monatomic gas in the left hand side volume V in Figure 5.4(a) consists of N atoms of type 1 and the ideal monatomic gas in the other volume V consists of N atoms of type 2. Show that the increase in entropy, after the partition is removed and the gases allowed to come to thermodynamic equilibrium, is $2Nk \log 2$. The temperature remains constant.

Problem 5.11
Consider a mixture of ideal gases consisting of N_1 diatomic molecules of type 1 and N_2 diatomic molecules of type 2. Show that the total partition function is

$$Z = (Z_1^{N_1}/N_1!)(Z_2^{N_2}/N_2!) \ .$$

where Z_1 and Z_2 are the single particle partition functions of molecules of types 1 and 2 respectively. Show that

$$pV = NkT$$

where $p = (p_1 + p_2)$ is the total pressure and N is equal to $(N_1 + N_2)$.

Problem 5.12*
Assume that equation (A6.17) of Appendix 6 is valid whatever the shape of the box. Show that the partition function for a single monatomic gas atom in the box is

$$Z_1 = \frac{gV}{2\pi^2\hbar^3} \int_0^\infty p^2 \exp[-p^2/2mkT]\, dp \ .$$

Use equation (A1.39) of Appendix 1 to evaluate the integral, and show that Z_1 is given by equation (5.63). Check that for an ideal monatomic gas, this leads to equations (5.71) and (5.72) whatever the shape of the box. [Alternatively, the reader can start with equation (A6.22) of Appendix 6, remembering that in this case dp_x, dp_y and dp_z can be negative as well as positive.]

Chapter 6

Thermodynamic Functions and Equilibrium

6.1 INTRODUCTION

In the approach to thermodynamics based on entropy, which is developed from statistical mechanics in Chapter 3, the natural variables for expressing the entropy S of a thermodynamic system are the extensive variables: the energy U, the volume V and the number of particles N in the system. The intensive variables: the pressure p, the absolute temperature T and the chemical potential μ can then be determined from $S(U, V, N)$ using equations (3.70), (3.23) and (3.78) respectively. In practice, experiments are often carried out under conditions of constant temperature and/or constant pressure. In these cases, it is more convenient to use the enthalpy H for changes at constant pressure, the Helmholtz free energy F when V and T are kept constant and the Gibbs free energy G when p and T are kept constant. The Helmholtz free energy F is related to the partition function Z by equation (5.30). The properties of these thermodynamic functions will be outlined in this Chapter. A reader interested in an account of the development of the properties of H, F and G from the viewpoint of classical equilibrium thermodynamics is referred to Callen [1].

6.2 THE HELMHOLTZ FREE ENERGY

6.2.1 Introduction

In classical equilibrium thermodynamics, the Helmholtz free energy F of a thermodynamic system is defined by the equation

$$F = U - TS \tag{6.1}$$

where U is the internal energy, T the absolute temperature and S the entropy of the system. (Books on chemistry generally use the symbol A instead of F for the Helmholtz free energy.) Since U, T and S are functions of the thermodynamic state of the system, F is also a function of the state of the system. Since U and S are extensive variables and T is an intensive variable, F is doubled if both U and S are doubled at a fixed temperature T, showing that F is an extensive variable.

According to equation (5.30), in statistical mechanics we have

$$F(V, T, N) = -kT \log Z(V, T, N) \qquad (5.30)$$

where $Z(V, T, N)$ is the partition function of an N particle system, which has a volume V and is in thermal equilibrium with a heat reservoir of absolute temperature T. Since it is log Z which appears in the expressions for U, p and μ given by equations (5.37), (5.38) and (5.41) respectively, these equations can be simplified by using $(-F/kT)$ for log Z. We then have

$$U = -T^2 \left(\frac{\partial(F/T)}{\partial T} \right)_{V,N} \qquad (6.2)$$

$$p = -(\partial F/\partial V)_{T,N} \qquad (6.3)$$

$$\mu = (\partial F/\partial N)_{V,T} . \qquad (6.4)$$

Differentiating equation (5.30) with respect to T, at constant V and N, we have

$$(\partial F/\partial T)_{V,N} = -(\partial/\partial T)(kT \log Z) = -k \log Z - kT(\partial \log Z/\partial T)_{V,N} .$$

Using equation (5.37) and then comparing with equation (5.39), we have

$$S = -(\partial F/\partial T)_{V,N} . \qquad (6.5)$$

If we can determine Z in terms of V, T and N, the Helmholtz free energy F can be expressed in terms of V, T and N using equation (5.30). The other thermodynamic variables can then be calculated using equations (6.2), (6.3), (6.4) and (6.5). Some readers may prefer to use the Helmholtz free energy F rather than the partition function in all the discussions given in Chapter 5. It is straightforward to develop equations (6.3), (6.4) and (6.5) from the thermodynamic identity, equation (3.85), whenever the reader needs these equations. It follows from equation (6.1) that

$$dF = dU - T \, dS - S \, dT . \qquad (6.6)$$

According to equation (3.85), for a one component system

$$dU - T \, dS = -p \, dV + \mu \, dN .$$

Substituting in equation (6.6) we obtain

$$dF = -S \, dT - p \, dV + \mu \, dN . \qquad (6.7)$$

The natural variables for F are T, V and N. Using equation (A1.31) of

Appendix 1, we have

$$dF = (\partial F/\partial T)_{V,N} \, dT + (\partial F/\partial V)_{T,N} \, dV + (\partial F/\partial N)_{V,T} \, dN \; . \quad (6.8)$$

Since equations (6.7) and (6.8) must hold for independent variations of T, V and N, it follows that

$$S = -(\partial F/\partial T)_{V,N}; \qquad p = -(\partial F/\partial V)_{T,N}; \qquad \mu = (\partial F/\partial N)_{V,T} \; .$$

These equations are in agreement with equations (6.5), (6.3) and (6.4) respectively. Substituting for S in equation (6.1), we have

$$U = F - T(\partial F/\partial T)_{V,N} \; . \quad (6.9)$$

This is one of the Gibbs–Helmholtz equations. Equation (6.9) relates U to F and T. [Alternatively, equation (6.9) can be derived by carrying out the differentiation in equation (6.2).]

For an n component system, using equation (3.86) instead of equation (3.85), we have

$$dF = -S \, dT - p \, dV + \sum_{i=1}^{n} \mu_i \, dN_i \; . \quad (6.10)$$

If T, V and all the N_i except N_j are kept constant, it follows from equation (6.10) that the chemical potential of the jth component is

$$\mu_j = (\partial F/\partial N_j)_{T,V,N_1 \ldots N_n(\text{except } N_j)} \; . \quad (6.11)$$

6.2.2 Conditions for the equilibrium of a system kept at constant T and V (thermodynamic approach)

Consider a small macroscopic system of constant total volume V, in thermal contact with a heat reservoir of constant absolute temperature T_R, as shown in Figure 6.1(a). (The adjective small is used to emphasise the fact that the thermal capacity of the small system is very much less than the termal capacity of the heat reservoir.) The small system may consist, for example, of two subsystems separated by a partition, as shown in Figure 6.1(b). Alternatively, the small system may be a multicomponent system consisting of N_1 particles of type 1, N_2 particles of type 2 etc. The small system is surrounded by rigid, diathermic walls, which can maintain a pressure difference between the small system and the heat reservoir, and prevent the exchange of particles between the small system and the heat reservoir.

It will be assumed that an internal constraint of the small system is removed. For example, the partition between the subsystems in Figure 6.1(b) could be changed from a fixed, rigid, adiabatic partition, to a movable, diathermic partition with holes in it. According to equation (1.25) of

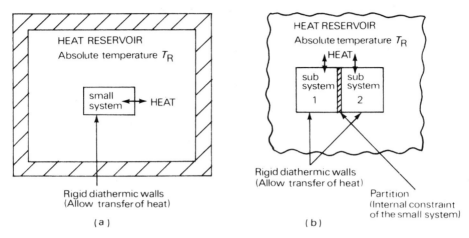

Figure 6.1—(a) The small system is in thermal equilibrium with a heat reservoir of absolute temperature T_R, which keeps the temperature of the small system constant. An example of a small system used in the text is a system composed of n different components filling the whole volume of the small system. (b) Enlargement of another typical small system. In this case, the small system consists of two subsystems separated by a partition. Heat can flow across the diathermic walls from the heat reservoir to both subsystems to keep their temperatures constant and equal to T_R.

Chapter 1, during the approach to the new state of equilibrium following the removal of the internal constraint of the small system

$$dS + dS_R \geqslant 0 \qquad (6.12)$$

where dS_R is the increase in the entropy of the heat reservoir and dS is the increase in the total entropy of the small system. The equality sign in equation (6.12) holds for a reversible change. Let heat $đQ$ flow *from* the heat reservoir *to* the small system during the approach to equilibrium, so as to keep the temperature of the small system constant. According to equation (1.13) of Chapter 1, the increase in the entropy of the heat reservoir is

$$dS_R = -(đQ/T_R)$$

where T_R is the absolute temperature of the heat reservoir. Substituting in equation (6.12), we have

$$-(đQ/T_R) + dS \geqslant 0 \ . \qquad (6.13)$$

According to the first law of thermodynamics, equation (1.7),

$$dU = đQ + đW \qquad (6.14)$$

where dU is the increase in the internal energy of the small system, $đQ$ is the heat added to the small system and $đW$ is the work done *on* the small system. Since the volume V of the small system is kept constant by the rigid diathermic walls surrounding the small system, $đW$ is zero in equation (6.14). Substituting dU for $đQ$ in equation (6.13), and multiplying by T_R we have

$$T_R \, dS - dU \geqslant 0 \ . \tag{6.15}$$

If the temperature of the small system is kept constant at the temperature of the heat reservoir, dT is zero in equation (6.6), which with $T = T_R$ becomes

$$dF = dU - T_R \, dS \ .$$

Substituting in equation (6.15), we have

$$-dF \geqslant 0 \ .$$

Hence, according to classical equilibrium thermodynamics, when a constant volume small system in thermal contact with a heat reservoir, of the type shown in Figures 6.1(a) or 6.1(b), approaches equilibrium after an internal constraint of the small system is removed

$$dF \leqslant 0 \qquad (T = T_R; \quad V \text{ is constant}) \ . \tag{6.16}$$

This shows that the total Helmholtz free energy F of a system of constant volume, kept at a constant temperature by a heat reservoir, decreases in the approach to internal thermodynamic equilibrium, reaching a *minimum* value at equilibrium.

6.2.3* Conditions for the equilibrium of a system of constant volume in thermal contact with a heat reservoir (statistical mechanical approach)*

In this subsection, the conclusion that the Helmholtz free energy F of a small system, kept at constant V and T, is a minimum at thermodynamic equilibrium will be interpreted in terms of statistical mechanics. The heat reservoirs plus the small systems in Figures 6.1(a) and 6.1(b) form *closed* composite systems. According to equation (3.43), which was developed from statistical mechanics in Section 3.7.2, the mathematical recipe, which enables one to calculate the equilibrium values of the thermodynamic variables for the cases shown in Figures 6.1(a) and 6.1(b), is

$$S_R(U_R, V_R, N_R) + S(U, V, N) \text{ is a maximum} \tag{6.17}$$

subject to the restriction $U_R + U = U^*$ and the restriction that V_R and V are constant. It is being assumed that there is no interchange of particles between the heat reservoir and the small system. In equation (6.17), S_R and S are the entropies the heat reservoir and small system would have, if they were isolated systems having energies U_R and U, volumes V_R and V and consisted of N_R and N particles respectively. Equation (6.17) corresponds to the most probable division of energy between the heat reservoir and the small system, which according to equation (3.59), is when the product

$$g_R(U_R, V_R, N_R)g(U, V, N) \text{ is a maximum .} \qquad (6.18)$$

In equation (6.18), g_R and g are the number of microstates accessible to the heat reservoir and the small system respectively. For macroscopic systems, the probability distribution given by equation (3.59) is extremely sharp. It is being assumed that the heat reservoir is large enough to have the same absolute temperature T_R, whatever the energy U of the small system. In this special case, according to equation (4.4), or equation (3.100),

$$g_R = g_R^* \exp(-U/kT_R) \qquad (6.19)$$

where g_R^* is the number of microstates accessible to the heat reservoir when it has all the energy U^*, which is equal to $(U_R + U)$. Substituting for g_R in equation (6.18), the condition for equilibrium becomes

$$g_R g = g_R^*[\exp(-U/kT_R)]g(U, V, N) \text{ is a maximum .} \qquad (6.20)$$

Since g_R^* is a constant, taking the natural logarithm, the condition of equilibrium of the small system becomes

$$[-(U/kT_R) + \log g] \text{ is a maximum .}$$

Using equation (3.22), the condition of equilibrium of the small system becomes

$$[S - (U/T_R)] \text{ is a maximum} \qquad (6.21)$$

or $\qquad\qquad (U - T_R S) \text{ is a minimum .} \qquad (6.22)$

Consider a small system consisting of one component only. Differentiating partially with respect to U, since T_R is constant, for constant V and N we have

$$1 - T_R(\partial S/\partial U)_{V,N} = 0 .$$

Using equation (3.23) this reduces to $T = T_R$. Hence at equilibrium the absolute temperature of the small macroscopic system must be equal to the absolute temperature of the heat reservoir.

Using the definition of the Helmholtz free energy F given by equation (6.1), the condition for the equilibrium of the small system in Figure 6.1(a), given by equation (6.22), becomes

$$F = (U - TS) \text{ is a minimum} \qquad (6.23)$$

(subject to the conditions: $T = T_R$; V and N are constant).

In equation (6.23), F, U, T and S are the Helmholtz free energy, the (internal) energy, the absolute temperature and the entropy of the small system respectively. We were able to allow for the effect of the possible exchange of heat between the small system and the heat reservoir in Figure 6.1(a) by using equation (6.19) in equation (6.18).

Since, according to equation (5.30),

$$F = -kT \log Z \qquad (5.30)$$

for constant T, $\log Z$ is a maximum when F is a minimum. The partition function Z of the small N particle system is a maximum when $\log Z$ is a maximum. The condition for the equilibrium of the small system in Figure 6.1(a) can be expressed in the following alternative forms:

$$\left.\begin{array}{c} Z \text{ is a maximum} \\[6pt] \log Z \text{ is a maximum} \end{array}\right\} \qquad (6.24)$$

(subject to the conditions: $T = T_R$; V and N are constant).

The application of equation (6.23) to the case shown in Figure 6.1(b) is developed in Problem 6.6. As a further illustrative example of the application of equation (6.23), assume that the small system shown in Figure 6.1(a) is a multicomponent system. It will be assumed that an exothermic chemical reaction takes place between the components. Since energy is released in such a reaction, if the temperature of the small system is to remain equal to the temperature of the heat reservoir, energy must flow from the small system to the heat reservoir, reducing U, the energy of the small system, and increasing the energy U_R of the heat reservoir. The increase in the energy of the heat reservoir increases the value of g_R in equation (6.18), tending to increase the product $(g_R g)$ and to make it a more probable situation. The other factor in equation (6.18) is g, the number of microstates accessible to the small system. If the transfer of energy from the small system to the heat reservoir, and the changes in the amounts of the various chemical constituents in the chemical reaction reduce g, this tends to reduce the product $(g_R g)$ in equation (6.18). An exothermic chemical reaction proceeds if the increase in g_R more than compensates any decrease in g in the product $(g_R g)$.

In thermodynamic terms, the decrease in the internal energy U of the small system in Figure 6.1(a) in an exothermic chemical reaction reduces the Helmholtz free energy F of the small system, and according to equation (6.23), this favours the chemical reaction. Due to the negative sign in the $(-TS)$ term in equation (6.23), a decrease in the entropy S of the small system, associated with a decrease in g, increases F and does not favour the chemical reaction. The chemical reaction proceeds if the decrease in U predominates and leads to an overall decrease in F, the Helmholtz free energy of the small system.

Some endothermic reactions proceed even though in these cases, if the temperature of the small system is to remain constant, energy must be taken from the heat reservoir leading to a decrease in g_R in equation (6.18). In these cases there is an increase in g, the number of microstates accessible to the small system, associated with the energy taken from the heat reservoir and the changes in the amounts of the various chemical constituents following the chemical reaction. For the endothermic reaction to proceed, this increase in g must be large enough to compensate for the decrease in g_R so that the product $(g_R g)$ is increased following the chemical reaction. In thermodynamic terms, U is increased in equation (6.23) increasing F, but the increase in S, the entropy of the small system, makes the $(-TS)$ term in equation (6.23) more negative and is sufficient to lead to an overall decrease in F the Helmholtz free energy of the small system.

The detailed discussion of the application of statistical mechanics to chemical reactions is deferred until Section 6.6. See also Problems 6.8 and 6.9.

6.3 ENTHALPY

Chemists carry out many of their experiments at constant pressure, which is generally atmospheric pressure. When the pressure is constant, it is more convenient to use the change in enthalpy in a chemical reaction than the change in internal energy. According to the first law of thermodynamics

$$\Delta U = \Delta Q + \Delta W \qquad (6.25)$$

where ΔW is the work done on the hydrostatic system. For a reversible expansion at constant pressure p, if the increase in volume is ΔV, the work done on the system is

$$\Delta W = -p\Delta V \ . \qquad (1.8)$$

For chemical reactions in solids, the $p\Delta V$ term is small, but it can be important in gaseous reactions, when there are large changes in volume. Substituting for ΔW in equation (6.25) for constant pressure p, we have

$$\Delta Q = \Delta U + p \Delta V = \Delta(U + pV) = \Delta H \ , \qquad (6.26)$$

where the *enthalpy H* is defined by the equation

$$H = U + pV \ . \qquad (6.27)$$

Since U, p and V are functions of the thermodynamic state of the system, H is also a function of state. Since U and V are extensive variables and p is an intensive variable, H is doubled if U and V are doubled at constant p, showing that H is an extensive variable.

According to equation (6.26), the increase ΔH in the enthalpy of a system is equal to ΔQ, the heat absorbed by the system **when the pressure of the system remains constant**. For a full account of the role of enthalpy in chemistry, the reader is referred to Smith [2a].

A typical chemical reaction is written in the form

$$CS_2 + 3O_2 \rightarrow CO_2 + 2SO_2 \qquad \Delta H = -1108 \text{ kJ} \ .$$

According to this equation, when one mole of carbon disulphide reacts with three moles of oxygen to give one mole of carbon dioxide and two moles of sulphur dioxide at a temperature of 298K and a constant pressure of one atmosphere, the total enthalpy of the system is decreased by 1108 kJ. This is equal to the heat liberated in the exothermic reaction. In an endothermic reaction, ΔH is positive and heat is absorbed from the surroundings, if the temperature is kept constant.

According to equation (6.27),

$$dH = dU + p \, dV + V \, dp \ . \tag{6.28}$$

According to the thermodynamic identity, equation (3.85), $dU + p \, dV = T \, dS + \mu \, dN$. Substituting in equation (6.28) we obtain

$$dH = T \, dS + V \, dp + \mu \, dN \tag{6.29}$$

It can be seen that the natural variables for H are S, p and N. Since equation (6.29) holds for independent variations in S, p and N, it follows that

(a) $T = (\partial H/\partial S)_{p,N}$;   (b) $V = (\partial H/\partial p)_{S,N}$;   (c) $\mu = (\partial H/\partial N)_{S,p}$.
$$\tag{6.30}$$

6.4 THE GIBBS FREE ENERGY

6.4.1 Introduction

The Gibbs free energy G is defined by the equation

$$G = U - TS + pV \ . \tag{6.31}$$

Using equation (6.27), equation (6.31) can be rewritten in the form

$$G = H - TS \tag{6.32}$$

where H is the enthalpy. Since U, T, S, p and V are functions of the thermodynamic state of the system, G is also a function of state. Since U, S and V are extensive variables and T and p are intensive variables, G is doubled if U, S and V are doubled at constant p and T, showing that G is an extensive variable.

It follows from equation (6.31) that

$$dG = dU - T \, dS - S \, dT + p \, dV + V \, dp \ . \tag{6.33}$$

According to the thermodynamic identity, equation (3.85), for a one component system $dU - T\,dS + p\,dV = \mu\,dN$. Substituting in equation (6.33), we have

$$dG = -S\,dT + V\,dp + \mu\,dN \ . \tag{6.34}$$

It can be seen that the natural variables for G are T, p and N. Using equation (A1.31) of Appendix 1

$$dG = (\partial G/\partial T)_{p,N}\,dT + (\partial G/\partial p)_{T,N}\,dp + (\partial G/\partial N)_{T,p}\,dN \ . \tag{6.35}$$

Since equations (6.34) and (6.35) must hold for independent variations of T, p and N, it follows that

$$S = -(\partial G/\partial T)_{p,N} \tag{6.36}$$

$$V = (\partial G/\partial p)_{T,N} \tag{6.37}$$

$$\mu = (\partial G/\partial N)_{T,p} \ . \tag{6.38}$$

Substituting for S in equation (6.32) we have

$$H = G - T(\partial G/\partial T)_{p,N} \ . \tag{6.39}$$

This the second of the Gibbs–Helmholtz equations.

Since both $G(T, p, N)$ and N are extensive variables, if N is doubled, keeping the intensive variables T and p constant, G is doubled, showing that G is proportional to N if p and T are constant. For a one component system, we have

$$G = Nf(T, p) \tag{6.40}$$

where f is a function of T and p. Differentiating equation (6.40) with respect to N keeping T and p constant, we have

$$f = (\partial G/\partial N)_{T,p} \ .$$

Comparing with equation (6.38), it can be seen that for a one component system, f is equal to the chemical potential μ. For a one component system, equation (6.40) can be rewritten as

$$G = N\mu(p, T). \tag{6.41}$$

The chemical potential μ of a one component system can be expressed as a function of the pressure p and the temperature T. According to equations (6.38) and (6.41), the chemical potential of a one component system is equal to the increase in the Gibbs free energy G when one extra particle is added to the system, provided p and T are kept constant. For an n component system, using equation (3.86) instead of equation (3.85), we obtain

$$dG = -S\,dT + V\,dp + \sum_{i=1}^{n} \mu_i\,dN_i \tag{6.42}$$

so that $\qquad\qquad \mu_j = (\partial G/\partial N_j)_{T,p,N_1\ldots N_n \text{ (except } N_j)}$. $\qquad\qquad$ (6.43)

In the general case of a multicomponent system, the chemical potential μ_j of the jth component depends on T, p and the concentrations of all the other components. [It is only in the special case of a mixture of ideal gases, when each component can be treated as a completely independent gas filling the whole volume V, that μ_j is independent of the concentrations of the other components and given by equation (5.114).]

If T and p are constant, dT and dp are zero in equation (6.42), so that, for an n component system, we have

$$dG = \sum_{i=1}^{n} \mu_i \, dN_i \qquad (T, p \text{ constant}) .$$

Let N_1, $N_2 \ldots N_i$ etc. all be increased by a factor of $(1 + \alpha)$, so that $dN_1 = \alpha N_1$, $dN_2 = \alpha N_2$, $\ldots dN_i = \alpha N_i$ etc. Substituting for dN_i, we have

$$dG = \alpha \sum_{i=1}^{n} \mu_i N_i \qquad (T, p \text{ are constant}) .$$

Since the Gibbs free energy is an extensive variable, if N_1, $N_2 \ldots$ are all increased by a factor of $(1 + \alpha)$, keeping the intensive variables p and T constant, G is also increased by a factor of $(1 + \alpha)$, so that dG is equal to αG. Hence for an n component system at constant T and p

$$dG = \alpha \sum_{i=1}^{n} \mu_i N_i = \alpha G .$$

Cancelling α, we have $\qquad\qquad G = \sum_{i=1}^{n} \mu_i N_i$ $\qquad\qquad$ (6.44)

where N_i is the number of molecules of the ith constituent present and μ_i is the chemical potential of the ith constituent.

6.4.2 Conditions for the equilibrium of a system kept at constant T and p (thermodynamic approach).

Consider a small system of variable total volume, which is in contact with a heat and pressure reservoir, as shown in Figure 6.2. The small system may, for example, be a multicomponent system or consist of two or more subsystems. Proceeding as in Section 6.2.2, and substituting for $đQ$ from equation (6.14) into equation (6.13), during the approach to thermodynamic equilibrium, we have

$$T_R \, dS - dU + đW \geqslant 0$$

According to equation (1.8), $đW$ the work done on the system in a reversible process is equal to $(-p \, dV)$. According to equation (6.33),

$$dU + p \, dV - T \, dS = dG + S \, dT - V \, dp$$

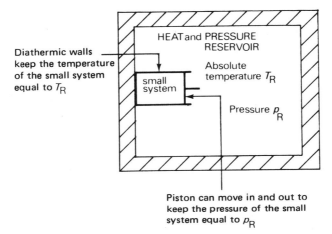

Figure 6.2—The small system is in thermodynamic equilibrium with a heat and pressure reservoir, which keeps the temperature of the small system equal to T_R and the pressure of the small system equal to p_R.

Substituting in the inequality, with $T_R = T$ we have

$$dG + S \, dT - V \, dp \leqslant 0.$$

If the changes take place at the temperature and pressure of the reservoir, dT and dp are zero. Hence during the approach of the small system in Figure 6.2 to thermodynamic equilibrium at constant T and p,

$$dG \leqslant 0 \qquad (T = T_R; \ p = p_R) \ . \qquad (6.45)$$

This shows that in the approach to equilibrium, the Gibbs free energy of a system kept at constant temperature and constant pressure decreases, reaching a *minimum* value at thermodynamic equilibrium.

6.4.3* Conditions for the equilibrium of a system in contact with a heat and pressure reservoir (statistical mechanical approach)*

The heat reservoir and the small system in Figure 6.2 form a *closed* system. According to equation (3.64), which was derived from statistical mechanics in Section 3.7.2, the condition of equilibrium, in the case shown in Figure 6.2, is

$$S(U_R, V_R, N_R) + S(U, V, N) \text{ is a maximum} \qquad (3.64)$$

subject to

$$U_R + U = U^* \quad \text{and} \quad V_R + V = V^* \ .$$

It is being assumed that there is no interchange of particles between the reservoir and the small system. Using equation (3.22), equation (3.64) can be rewritten in the form

$$g_R(U_R, V_R, N_R)g(U, V, N) \text{ is a maximum} . \qquad (6.46)$$

Compared with the case considered in Section 6.2.3, both the energy and the volume of the small system can vary in the present case, so that the energy and the volume of the reservoir can also vary, but it is being assumed that the changes in the reservoir are small enough for the pressure and temperature of the reservoir to remain constant. If U and V are the energy and volume of the small system, expanding $\log[g_R(U^* - U, V^* - V)]$ in a Taylor expansion, we have

$$\log[g_R(U^* - U, V^* - V)] \simeq \log[g_R(U^*, V^*)]$$

$$- U\left(\frac{\partial \log g_R}{\partial U_R}\right) - V\left(\frac{\partial \log g_R}{\partial V_R}\right) . \qquad (6.47)$$

Since S_R, the entropy of the heat reservoir, is equal to $k(\log g_R)$ it follows from equations (3.14) and (3.70) that

$$\left(\frac{\partial \log g_R}{\partial U_R}\right) = \beta_R = \frac{1}{kT_R} ; \qquad \left(\frac{\partial \log g_R}{\partial V_R}\right) = \frac{p_R}{kT_R} .$$

Substituting in equation (6.47), and taking exponentials, we obtain

$$g_R = g_R^*(U^*, V^*)\exp[-(U + p_RV)/kT_R] . \qquad (6.48)$$

In equation (6.48), g_R^* is the number of microstates accessible to the reservoir, when the reservoir has all the energy U^* and all the volume V^*. If, when the volume of the small system in Figure 6.2 is kept constant, its energy U is increased by a flow of heat from the reservoir, the energy of the reservoir U_R decreases and g_R decreases. The variation of g_R with increasing U (decreasing U_R) at constant volume V is proportional to $\exp(-U/kT_R)$. If the volume of the small system in Figure 6.2 is increased at a fixed value of U, the volume V_R of the reservoir is decreased and the energy eigenvalues of the microstates of the reservoir are all increased in the way illustrated in Figure 2.1. When V_R decreases, the constant energy U_R of the heat reservoir is only sufficient to make accessible microstates of the reservoir which previously had energy eigenvalues less than U_R and were in an energy range where the degeneracy g_R was less. The variation of g_R with increasing V (decreasing V_R) for fixed U_R is proportional to $\exp(-p_RV/kT_R)$.

Substituting for g_R from equation (6.48) into equation (6.46), since g_R^*

is a constant, the condition for equilibrium becomes

$$g \exp[-(U + p_R V)/kT_R] \text{ is a maximum} . \qquad (6.49)$$

Taking the natural logarithm and replacing $\log g$ by S/k, the condition for equilibrium becomes

$$S - (U/T_R) - (p_R V/T_R) \text{ is a maximum} . \qquad (6.50)$$

It is left as an exercise for the reader to show, by differentiating equation (6.50) partially with respect to U keeping V and N constant (remembering that S is a function of U, V and N) and using equation (3.23), that at thermodynamic equilibrium $T = T_R$. Differentiating equation (6.50) partially with respect to V, keeping U and N constant and using equation (3.70), it can also be shown that at equilibrium $p = p_R$. Hence equation (6.50) can be rewritten as $(TS - pV - U)$, is a maximum (subject to the conditions: $T = T_R$; $p = p_R$), where T, S, p, V and U are the thermodynamic variables of the small system. Using equation (6.31), the condition for the small system in Figure 6.2 to be in thermodynamic equilibrium is

$$G \text{ is a minimum} \qquad (6.51)$$

(subject to the conditions: $T = T_R$; $p = p_R$).

In equation (6.51), G is the Gibbs free energy of the small system in Figure 6.2. This result is in agreement with the discussion based on classical equilibrium thermodynamics given in Section 6.4.2.

6.5 CHANGE OF PHASE

The case of a liquid in equilibrium with its vapour will be used to illustrate the conditions of equilibrium under various external conditions. In Figure 6.3(a) the liquid and its vapour are inside a rigid adiabatic enclosure. The liquid and vapour form a closed system of fixed total energy, fixed total volume and fixed total number of particles. According to equation (3.64), at equilibrium

$$S_1(U_1, V_1, N_1) + S_2(U_2, V_2, N_2) \text{ is a maximum}$$

Throughout this Section, the suffixes 1, 2 and 3 refer to the vapour, liquid and solid phases respectively. It was shown in Sections 3.7.3 and 3.7.4 that at equilibrium, according to equations (3.68), (3.72) and (3.80), $T_1 = T_2$; $p_1 = p_2$; $\mu_1 = \mu_2$. Enough of the liquid phase vapourises, such that at equilibrium the intensive variables, T, p and μ are equal in both phases.

In practice, if one wants to study the effects of changing some of the

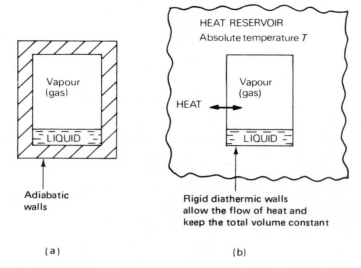

Figure 6.3—A liquid is in thermodynamic equilibrium with its vapour.
(a) The liquid and vapour are inside rigid adiabatic walls, and make up a
closed system of constant total energy, constant total volume and constant total number of particles. Sufficient vapour will liquefy, or sufficient liquid will vapourise until the total entropy $(S_1 + S_2)$ of vapour plus
liquid is a maximum at equilibrium, when the temperatures, pressures
and chemical potentials of the two phases are equal. (b) The liquid and
its vapour are separated from a heat reservoir by rigid diathermic walls,
which keep the total volume of liquid plus vapour constant. The diathermic walls allow the exchange of heat with the heat reservoir. This
keeps the temperature of the liquid and its vapour at the temperature of
the heat reservoir. In this case, the Helmholtz Free Energy $(F_1 + F_2)$ of
the vapour plus liquid is a minimum, when the liquid is in thermodynamic equilibrium with its vapour.

parameters, it is more convenient to use the arrangement shown in Figure
6.3(b). In this case the liquid and vapour are in a constant volume diathermic enclosure in thermal equilibrium with a heat reservoir of absolute
temperature T. It was shown in Section 6.2.3 that in these conditions, at
equilibrium according to equation (6.23)

$$F_1(T_1, V_1, N_1) + F_2(T_2, V_2, N_2) \text{ is a minimum} \qquad (6.23)$$

where F_1 and F_2 are the Helmholtz free energies of the vapour and liquid
phases respectively. It is shown in Problem 6.6, that this leads to the
conditions

$$T_1 = T_2 = T \qquad (6.52)$$

$$p_1 = p_2 \qquad (6.53)$$

$$\mu_1 = \mu_2 \ . \qquad (6.54)$$

For a fixed temperature T, provided there is enough of the substance present, the amounts of the substance in the liquid and vapour phases adjust themselves until at equilibrium p_1 is equal to p_2 and μ_1 is equal to μ_2. (Since the diathermic enclosure in Figure 6.3(b) is rigid, p_1 and p_2 are not in general equal to the pressure of the heat reservoir.) It was shown in Section 6.4.1 that for a one component system the chemical potential is a function of pressure and temperature, so that, according to equation (6.54), at equilibrium

$$\mu_1(T,p) = \mu_2(T,p) \tag{6.55}$$

Since T is the fixed temperature of the heat reservoir, for a fixed total amount of substance, the pressure $p\,(= p_1 = p_2)$ of the liquid and vapour phases at equilibrium is the only variable in equation (6.55). If we knew the expressions for μ_1 and μ_2 in terms of p and T, which in general we do not, equation (6.55) could be used to determine p for the fixed value of T. The equilibrium pressures and temperatures are plotted in Figure 6.4(a). If the temperature of the heat reservoir is increased from T to $(T + dT)$, the equilibrium values of pressure and chemical potentials are changed such that

$$(p_1 + dp_1) = (p_2 + dp_2) \tag{6.56}$$

$$(\mu_1 + d\mu_1) = (\mu_2 + d\mu_2) \ . \tag{6.57}$$

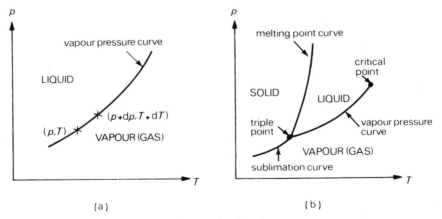

(a)   (b)

Figure 6.4—(a) A phase diagram showing the vapour pressure co-existence curve relating the pressure p and the absolute temperature T of a liquid in thermodynamic equilibrium with its own vapour. (b) A phase diagram showing the three co-existence curves, namely the vapour pressure, melting point and sublimation curves, for the three phases of a pure substance. The three curves intersect at the triple point. At temperatures above the temperature of the critical point, the vapour (gas) will not liquefy whatever the pressure.

The new values of pressure and temperature are shown in Figure 6.4(a). The line joining such points is called a coexistence curve. In the special case of equilibrium between the liquid and vapour phases, it is called a vapour–pressure curve.

Subtracting equation (6.53) from equation (6.56) gives $dp_1 = dp_2$. Subtracting equation (6.54) from equation (6.57) gives

$$d\mu_1 = d\mu_2 \ . \tag{6.58}$$

It follows from equations (6.41) and (6.34) that, for a one component system,

$$dG = N \, d\mu + \mu \, dN = -S \, dT + V \, dp + \mu \, dN \ .$$

Rearranging, $$d\mu = -(S/N) \, dT + (V/N) \, dp \ . \tag{6.59}$$

Using equation (6.59) in equation (6.58) with $dT_1 = dT_2 = dT$ and with $dp_1 = dp_2 = dp$, after rearranging we have

$$\frac{dp}{dT} = \frac{(S_1/N_1) - (S_2/N_2)}{(V_1/N_1) - (V_2/N_2)} = \frac{s_1 - s_2}{v_1 - v_2} \ , \tag{6.60}$$

where s_1 and s_2 are the entropies per particle and v_1 and v_2 are the volumes per particle at the pressure p and the absolute temperature T.

Let one mole (N_A molecules) change from the liquid to the vapour phase at a constant temperature T and at the saturation vapour pressure of the liquid. The volume of vapour is increased in this process. Since the intensive variables p and T are kept constant in this process, and the entropy and volume are both extensive variables the entropy per particle (s_1 and s_2) and the volume per particle (v_1 and v_2) in both the vapour and the liquid phases are unchanged. After multiplying both the numerator and denominator of the extreme right hand side by N_A, equation (6.60) can be rewritten in the form

$$\frac{dp}{dT} = \frac{N_A(s_1 - s_2)}{N_A(v_1 - v_2)} = \frac{\Delta S}{\Delta V} \tag{6.61}$$

where ΔS is the increase in entropy and ΔV is the increase in volume, when one mole of substance changes from the liquid to the vapour phase at the temperature T and the pressure p. The latent heat L is defined as the quantity of heat ΔQ that must be supplied, when a mole of liquid is changed to the vapour phase under its own saturation vapour pressure at the temperature T. (According to equation (6.26) since p is constant, ΔQ is equal to the change in enthalpy ΔH.) Since T is constant,

$$\Delta S = \Delta Q/T = \Delta H/T = L/T \ .$$

Substituting for ΔS in equation (6.61) we have

$$\frac{dp}{dT} = \frac{L}{T\,\Delta V} \qquad (6.62)$$

This is the Clausius–Clapeyron equation. It gives the slope of the co-existence curve in Figure 6.4(a) in terms of the measurable quantities L, T and ΔV. Equation (6.62) also holds for the phase changes from the solid to liquid and from the solid to vapour phases.

In the case of the change of phase from the liquid to vapour phase $\Delta V_1 \gg \Delta V_2$, so that $\Delta V \simeq \Delta V_1$. If it is assumed that the vapour obeys the ideal gas equation, for one mole we have $p\,\Delta V_1 = RT \simeq p\,\Delta V$. Substituting (RT/p) for ΔV in equation (6.62), and rearranging we have,

$$dp/p = (L/RT^2)\,dT \ .$$

Integrating, assuming that L is constant, we have

$$\log p = -(L/RT) + \text{constant} \ . \qquad (6.63)$$

Taking exponentials we obtain

$$p = p_0 \exp(-L/RT) \qquad (6.64)$$

where p_0 is a constant. Equation (6.63) is a useful interpolation formula over a limited region of the vapour pressure curve.

If the experimental arrangement, shown previously in Figure 6.2 is used, with the liquid and its vapour making up the small system, according to equation (6.51) the Gibbs free energy $(G_1 + G_2)$ of the small system is a minimum at equilibrium subject to the conditions, $T_1 = T_2 = T$, and $p_1 = p_2 = p$, where T and p are the absolute temperature and pressure of the reservoir respectively. The condition that $(G_1 + G_2)$ is a minimum at equilibrium leads to the condition that $\mu_1 = \mu_2$. (Reference: Problem 6.7.) In general, the applied values of p and T do not lie on the coexistence curve in Figure 6.4(a), and one cannot satisfy the condition that the chemical potentials of the two phases must be equal at equilibrium. For values of p and T above the coexistence curve in Figure 6.4(a), the piston in Figure 6.2 will move inwards until all the vapour liquefies. For values of p and T giving a point below the coexistence curve in Figure 6.4(a), only the vapour phase is present at equilibrium. [At temperatures above the temperature of the critical point, only the vapour phase is present, whatever the pressure.]

At a temperature below the melting point, the solid phase is in equilibrium with the vapour phase. The coexistence curve in this case is labelled the sublimation curve in Figure 6.4(b). At a sufficiently high pressure, the liquid and solid phases can be in equilibrium without the vapour phase present. This coexistence curve is labelled the melting point curve in Figure

6.4(b). The three coexistence curves intersect at the triple point, when

$$\mu_1(p, T) = \mu_2(p, T) = \mu_3(p, T) \ . \tag{6.65}$$

The triple point of water is defined by international agreement to have the precise temperature of 273.16K. The pressure of the triple point of water is 4.58 mm of mercury (The special case of helium will be discussed in Section 12.5.5 of Chapter 12).

6.6* CHEMICAL EQUILIBRIUM*

6.6.1* General condition for equilibrium*

Consider an n component system, consisting of N_1 molecules of type 1, N_2 molecules of type 2 etc. The system is in thermal equilibrium with a heat reservoir of absolute temperature T and a pressure reservoir of pressure p. The Gibbs free energy G is a function of p, T, N_1, ... N_n. Using equation (A1.31) of Appendix 1, if p and T are constant, we have

$$dG = \sum_{i=1}^{n} (\partial G/\partial N_i) \, dN_i \ .$$

Using equation (6.43), we have

$$dG = \sum \mu_i \, dN_i \qquad (T, p \text{ constant})$$

where μ_i is the chemical potential of the ith chemical component. [Alternatively, we could have put dT and dp equal to zero in equation (6.42).] According to equation (6.51), the Gibbs free energy G of a system kept at constant p and T is a minimum at equilibrium, so that for small variations away from equilibrium dG is zero. Hence at equilibrium

$$\sum \mu_i \, dN_i = 0 \ . \tag{6.66}$$

Consider the chemical reaction:

$$a A_1 + b A_2 + \ldots \rightleftharpoons l A_l + m A_m + \ldots \tag{6.67}$$

This can be rewritten in the general form

$$v_1 A_1 + v_2 A_2 + \ldots + v_n A_n \rightleftharpoons 0 \tag{6.68}$$

where $v_1, v_2 \ldots v_n$ are the stoichiometric coefficients, which are taken to be positive for the products of the reaction and negative for the reactants. The changes dN_1, dN_2 etc. in the numbers of particles of type 1, 2 ... must satisfy equation (6.68). If dN_1 is equal to αv_1, where α is an integer, dN_2 must be equal to αv_2 etc. Putting dN_i equal to αv_i in equation (6.66) and cancelling α, the condition for equilibrium becomes

$$\sum \mu_i v_i = 0 \ . \tag{6.69}$$

Since many chemical experiments are performed at constant p and T, we chose to develop equation (6.69) from the Gibbs free energy. (Equation (6.69) can also be derived using the property that the Helmholtz free energy F is a minimum at equilibrium (or that $\log Z$ is a maximum), if V and T are constant. Reference: Problem 6.11.)

6.6.2* Reactions between ideal gases*

In the case of an ideal gas mixture, each chemical component can be treated as an *independent* ideal gas filling the whole volume V. If the components are all completely independent, the chemical potential of the ith component of the ideal gas mixture is independent of the concentrations of the other components. According to equation (5.86), which is valid for atoms, and equation (5.114), which is valid for diatomic molecules, the chemical potential of the ith component of a mixture of *ideal* gases is

$$\mu_i = -kT \log(Z_i/N_i) \qquad (5.114)$$

where Z_i is the single particle partition function of one of the atoms (or molecules) of the ith chemical component and N_i is the number of atoms (or molecules) of the ith chemical component present in the system. Substituting for μ_i in equation (6.69), the condition of equilibrium of a mixture of *ideal* gases is

$$-kT \sum_{i=1}^{n} v_i \log(Z_i/N_i) = 0 \ .$$

Cancelling kT and rearranging, for equilibrium at an absolute temperature T, we have

$$\log[(Z_1/N_1)^{v_1} \ldots (Z_n/N_n)^{v_n}] = 0 \ . \qquad (6.70)$$

Since zero is equal to $\log 1$, equation (6.70) can be rewritten in the form

$$N_1^{v_1} N_2^{v_2} \ldots N_n^{v_n} = Z_1^{v_1} Z_2^{v_2} \ldots Z_n^{v_n} = K_n \ . \qquad (6.71)$$

where K_n is an equilibrium constant. As an example, consider an isomeric transition of the type

$$A \rightleftharpoons B \ . \qquad (6.72)$$

For example, A and B in equation (6.72) could stand for the *cis* and *trans* forms of dichloro-ethylene. (Reference: Guggenheim [3].) According to equation (6.71), at equilibrium, since $v_B = 1$ and $v_A = -1$, we have

$$N_B/N_A = Z_B/Z_A \ . \qquad (6.73)$$

where N_A and N_B are the mean numbers of particles of types A and B respectively at equilibrium and Z_A and Z_B are the single particle partition functions of the particles of types A and B. (An alternative derivation of equation (6.73) is developed in Problem 6.8.) It can be shown that the ratio of the standard deviation in the number of particles of type A at equilibrium to N_A, the ensemble average number of particles of type A at equilibrium, is of the order of $1/(N_A + N_B)^{1/2}$. This shows that fluctuations from the ensemble average values of N_A and N_B are small for macroscopic systems. (Reference: Hill [4].)

In the case of the reaction

$$a\text{A} + b\text{B} \rightleftharpoons l\text{L} + m\text{M} \qquad (6.74)$$

equation (6.71) gives

$$\frac{N_L^l N_M^m}{N_A^a N_B^b} = \frac{Z_L^l Z_M^m}{Z_A^a Z_B^b} = K_n \qquad (6.75)$$

where N_L is the ensemble average number of atoms (molecules) of type L at equilibrium and Z_L is the *single* particle partition function of the atoms (molecules) of type L etc. According to equations (5.63) and (5.109), for ideal gases Z_L, Z_M, Z_A and Z_B depend on both V and T so that K_n is a function of V and T.

It is more usual to use an equilibrium constant expressed in terms of the concentration c_i of the ith chemical component, where c_i is defined by the equation

$$c_i = N_i/V \; . \qquad (6.76)$$

Substituting $c_i V$ for N_i in equation (6.71), at equilibrium we have

$$c_1^{v_1} c_2^{v_2} \ldots c_n^{v_n} V^{\Delta v} = K_n \qquad (6.77)$$

where

$$\Delta v = \Sigma v_i \; .$$

Let

$$K_c = c_1^{v_1} c_2^{v_2} \ldots c_n^{v_n} \; . \qquad (6.78)$$

Equation (6.77) becomes

$$K_c = K_n V^{-\Delta v} \; . \qquad (6.79)$$

For the special case of the reaction given by equation (6.74), we have

$$K_c = \frac{c_L^l c_M^m}{c_A^a c_B^b} = \left(\frac{Z_L^l Z_M^m}{Z_A^a Z_B^b}\right) V^{-(l+m-a-b)} \; . \qquad (6.80)$$

The equilibrium constant K_c can be calculated from the single particle partition functions of the reactants and the products. According to equations (5.63) and (5.109), the single particle partition functions for an ideal monatomic gas and an ideal diatomic gas are proportional to V. Hence in the case of chemical reactions between *ideal* gases, K_c depends only on the absolute temperature T.

Equation (6.80) is the *law of mass action*. If the equilibrium constant K_c is calculated from statistical mechanics using equation (6.79), or if K_c is measured experimentally at a given value of temperature, these values of K_c can be used to calculate the final molecular concentrations of all the reacting gases, if they are mixed in any known proportion and allowed to come to equilibrium at that fixed temperature. As an example consider the following reaction between ideal gases:

$$A_2 + B_2 \rightleftharpoons 2AB \ .$$

Assume that we start with $N_A{}^0$ molecules of type A and $N_B{}^0$ molecules of type B but no molecules of type AB. Let x molecules of types A and B react to give $2x$ molecules of type AB at equilibrium. If V is the volume of the system at equilibrium, the concentrations at equilibrium are $(N_A{}^0 - x)/V$, $(N_B{}^0 - x)/V$ and $2x/V$. The equilibrium constant K_c is given by

$$K_c = c_{AB}{}^2/c_A c_B = 4x^2/(N_A{}^0 - x)(N_B{}^0 - x) \ .$$

This equation allows us to calculate x for any given initial values of $N_A{}^0$ and $N_B{}^0$ at the fixed temperature at which K_c is known.

6.6.3* Example of a chemical reaction*

Consider the dissociation of a homonuclear diatomic molecule A_2 into two atoms of type A according to the equation

$$A_2 \rightleftharpoons A + A \ . \tag{6.81}$$

Let the mean number of A_2 molecules at equilibrium be denoted by N_2, and let the mean number of atoms of type A at equilibrium be denoted by N_1. According to equation (6.71), at equilibrium we have

$$N_1{}^2/N_2 = Z_1{}^2/Z_2 \tag{6.82}$$

where Z_1 and Z_2 are the single particle partition functions of atoms of type A and diatomic molecules of type A_2 respectively. Equation (6.82) can also be developed using the condition that the partition function Z of the complete system, is a maximum at equilibrium if V and T are constant. (Reference: Problem 6.9.)

If the concentrations c_1 and c_2 are used, according to equation (6.80), since, for the chemical reactions given by equation (6.81), $\Delta v = 1$

$$K_c = c_1{}^2/c_2 = (Z_1{}^2/Z_2)V^{-1} \ . \tag{6.83}$$

When the expression for the single particle partition function of a diatomic molecule was developed in Section 5.10 of Chapter 5, the zero of the energy scale for ε_e, the energy of the electrons in equation (5.87), was the ground state of the molecule. The binding energy of the molecule was ignored. To illustrate the importance of the binding energy in chemical reactions, we shall return to the derivation of the Boltzmann distribution in

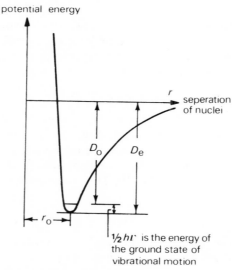

Figure 6.5—Variation of the potential energy of the two atoms in a homonuclear diatomic molecule with r, the separation of the two atomic nuclei in the molecule. When the potential energy is a minimum at $r = r_0$, the potential energy is negative due to the attractive forces holding the molecule together. At shorter separations the potential energy gets less negative, and eventually becomes positive corresponding to a repulsive force. Also shown is the ground state of the vibrational motion of the two atoms, measured from the minimum of the potential energy curve.

Section 4.2. [The Boltzmann distribution led to the expression for the partition function, which was then used in Chapter 5 to derive the expressions for the chemical potentials to be used in equation (6.69).] According to equation (4.2), the probability that, at thermal equilibrium, the small system in Figure 4.1(a) is in a particular microstate is proportional to the number of microstates accessible to the heat reservoir, when the small system is in that microstate. Consider a small system made up of two atoms in thermal equilibrium with a heat reservoir. If the atoms combine to form a diatomic molecule in its ground state, energy is released and at thermal equilibrium, this energy is given to the heat reservoir. This increases the energy of the heat reservoir and the number of microstates accessible to the heat reservoir is increased, so that a microstate in which the two atoms form a molecule in its ground state is more probable than a microstate in which the small system consists of two spatially separated atoms in their ground states. This illustrates how the binding energy of the molecule can have a profound effect on the equilibrium state of the system.

The variation of the potential energy of a typical homonuclear diatomic molecule with the separation r of the atomic nuclei is sketched in Figure

6.5. The equilibrium separation of the atomic nuclei is when the potential energy is a minimum at $r = r_0$. According to the theory of the homopolar chemical bond, the negative potential energy at $r = r_0$, which is equal to D_e, arises from the electrostatic forces between the electrons themselves and the atomic nuclei in the molecule, after allowing for the indistinguishability of the electrons, when Schrödinger's equation is used. To allow for the dissociation (binding) energy of the molecule, it will be assumed that the potential energy of the 'molecule' is zero when the atoms in the 'molecule' are at rest in their ground states at an infinite separation. When the atoms are at a separation r_0 in a molecule inside a box of volume V, one must add the potential energy $(-D_e)$ to the right hand side of equation (5.87) to give

$$\varepsilon_i = \varepsilon_t + \varepsilon_r + \varepsilon_v + \varepsilon_e + \varepsilon_n - D_e \ .$$

The energy ε_e of the electrons in the molecule is still measured from the ground state of the molecule. The energies ε_r, ε_v and ε_n are unchanged. If the volume V is the same, the energy of translational motion ε_t is also unchanged. The extra term $(-D_e)$ on the right hand side leads to an extra term $\exp(\beta D_e)$ in equation (5.88), and the expression for the partition function Z_2 of the diatomic molecule becomes

$$Z_2 = Z_t Z_r Z_v Z_e Z_n \exp(\beta D_e) \tag{6.84}$$

where
$$Z_t = (m_2/2\beta\pi\hbar^2)^{3/2} V \ .$$

The mass m_2 of the molecule is double the mass m_1 of an atom. According to equation (5.99), if $T \gg \theta_r$, putting $\sigma = 2$ for a homonuclear molecule, we have $Z_r = I/\beta\hbar^2$. The approximation $(m_1 r_0^2/2)$ is sometimes used for the moment of inertia I of the molecule (This expression holds for two point masses m_1, a distance r_0 apart, rotating about their centre of mass). According to equation (5.100), the vibrational partition function Z_v is

$$Z_v = \exp(-\theta_v/2T)/[1 - \exp(-\theta_v/T)]$$

where θ_v is equal to $(h\nu/k)$. According to equation (5.105), if the absolute temperature T is very much less than the characteristic electron temperature θ_e, the electron partition function Z_e of the molecule is equal to g_{02}, which is degeneracy of the electronic ground state of the molecule. Most diatomic molecules have a $^1\Sigma$ electronic ground state, which has a degeneracy g_{02} equal to unity. An exception is the O_2 molecule which, at room temperatures, has an electronic ground state degeneracy g_{02} equal to 3.

   According to equation (5.106), if the absolute temperature T is much less than the nuclear characteristic temperature θ_n, the nuclear partition function Z_n is equal to $(2J + 1)^2$, where J is the nuclear spin of one of the atomic nuclei. Substituting for Z_r, Z_v, Z_e and Z_n in equation (6.84), we obtain

$$Z_2 = \left(\frac{m_1 kT}{\pi \hbar^2}\right)^{3/2} \frac{VkTI \exp(-\theta_v/2T)g_{02}(2J + 1)^2 \exp(D_e/kT)}{\hbar^2[1 - \exp(-\theta_v/T)]} . \quad (6.85)$$

The product $\exp(-\theta_v/2T) \exp(D_e/kT)$ in equation (6.85) can be rewritten as $\exp[(D_e - hv/2)/kT]$, which can be rewritten as $\exp(D_0/kT)$, where $D_0 = D_e - \frac{1}{2}hv$. The dissociation energy of a molecule is generally defined as the work that must be done to separate the two atoms in a molecule, which is initially at rest in its ground state, until the atoms are at rest in their ground states at an infinite separation. According to equation (5.92), the energy of the ground state of rotational motion is zero. According to equation (2.5) of Chapter 2 the energy of the ground state of translational motion of a molecule in a box is very small compared with D_e. However, in its ground state, the molecule must have the zero point energy $\frac{1}{2}hv$ of the ground state of vibrational motion, and this can be significant. [For example the characteristic temperature $\theta_v(=hv/k)$ of vibrational motion is typically 5000K. The corresponding value of $\frac{1}{2}hv$ is 3.45×10^{-20} J or 0.22 eV. For one mole this amounts to 20.8 kJ.] The zero point energy of vibrational motion is zero, when the atoms are an infinite distance apart, since there is then no attractive force between the atoms. Since it starts with the zero point energy $\frac{1}{2}hv$ of vibrational motion, the work that must be done in separating the atoms in the molecule is not D_e but $(D_e - \frac{1}{2}hv)$, which is equal to D_0. Most chemistry and physics text books define the dissociation energy to be equal to D_0. It is D_0 which is determined experimentally. However, some physics text books call D_e the binding energy, or the strength of the chemical bond, since it is D_e which represents the contribution of the homopolar chemical bond to the total energy of the molecule. The $\frac{1}{2}hv$ term arises in a different way, namely from the vibration of the molecule. The reader should always check carefully which definition of binding or dissociation energy is used. We shall continue to use both D_e and D_0.

According to equation (5.63), if $T \ll \theta_e$ and $T \ll \theta_n$, the partition function of an atom is

$$Z_1 = (m_1 kT/2\pi \hbar^2)^{3/2} V g_{01}(2J + 1) . \quad (6.86)$$

In equation (6.86), g_{01} is the degeneracy of the electronic ground state of the atom. The nuclear spin degeneracy of an atom is equal to $(2J + 1)$, where J is the nuclear spin quantum number.

Substituting for Z_1 and Z_2 from equations (6.85) and (6.86) into equation (6.83), with $h = 2\pi \hbar$, we obtain

$$K_c = \frac{c_1^2}{c_2} = \frac{m_1^{3/2}(kT)^{1/2}g_{01}^2[1 - \exp(-\theta_v/T)]\exp(-D_e/kT)}{4h\pi^{1/2}Ig_{02}\exp(-\theta_v/2T)} \quad (6.87)$$

$$K_c = \frac{m_1^{3/2}(kT)^{1/2}g_{01}^2[1 - \exp(-\theta_v/T)]\exp(-D_0/kT)}{4h\pi^{1/2}Ig_{02}} \quad . \qquad (6.88)$$

The nuclear spin terms cancel out in equation (6.87). The quantities I, θ_v and D_0 can be determined by spectroscopic methods, and then used to estimate the value of the equilibrium constant K_c at the absolute temperature T. This calculated value of K_c is generally in reasonable agreement with the experimental results.

6.6.4* An equilibrium constant expressed in terms of the partial pressures of the ideal gases*

Another equilibrium constant used frequently in chemistry is expressed in terms of the partial pressures of the ideal gases. According to the equation of state of an ideal gas, $N_i = p_i V/kT$. Substituting in equation (6.71), we obtain

$$K_n = p_1^{v_1}p_2^{v_2}\ldots p_n^{v_n}(V/kT)^{\Delta v} \quad , \qquad (6.89)$$

where p_i is the partial pressure of the ith chemical component. Let

$$K_p = p_1^{v_1}p_2^{v_2}\ldots p_n^{v_n} \qquad (6.90)$$

where K_p is an equilibrium constant expressed in terms of the partial pressures of the gases at equilibrium. Substituting for K_p in equation (6.89), we obtain

$$K_p = K_n(kT/V)^{\Delta v} \quad . \qquad (6.91)$$

Since K_n can be calculated from the single particle partition functions using equation (6.71), K_p can also be estimated from the single particle partition functions using statistical mechanics.

Substituting for K_n from equation (6.79) into equation (6.91), we obtain

$$K_p = K_c(kT)^{\Delta v} \quad . \qquad (6.92)$$

For ideal gases K_c is a function of T only. Hence, according to equation (6.92), for ideal gases, K_p is also a function of T only. With real gases both K_c and K_p vary slightly with pressure at a fixed temperature T.

Another equilibrium constant used extensively in chemistry is K_x, which is expressed in terms of the mole fractions of the constituents. The properties of K_x are developed in Problem 6.10.

6.6.5* Discussion*

The theory of chemical equilibrium was only worked out for ideal gases. The theory works reasonably well for real gases and for reactions between condensed components, because changes in energy associated with the formation and breaking of chemical bonds are much larger than the changes in energy associated with changes in the small attractive forces between molecules, such as the van der Waals' forces. (For example, the latent heat of fusion of ice is ~6 kJ mole^{-1}, whereas the change in enthalpy when one mole of water is formed from hydrogen and oxygen at 298K is 285 kJ.) Generally, the error in ignoring the attractive forces between molecules is small. Only a brief insight has been given into the interpretation of chemical equilibrium based on statistical mechanics. For further details the reader is referred to text books on Physical Chemistry.

Our discussion of chemical equilibrium has illustrated the important role of the chemical potential in chemical equilibrium. The chemical potential is also used widely in the theory of solutions and electrolytes. (References: Smith [2b], ter Haar and Wergeland [5].) The reader should appreciate the important role the chemical potential plays in Physical Chemistry, so that the chemical potential does not appear to be quite such a strange new quantity, when we come to discuss the grand canonical distribution in Chapter 11 and the Fermi–Dirac and Bose–Einstein distributions in Chapter 12.

References

[1] Callen, H. B., *Thermodynamics*, 1960, Wiley, New York, Chapter 6.
[2] Smith, E. B., *Basic Chemical Thermodynamics*, Clarendon Press, Oxford, 1973, (a) page 18, (b) page 74.
[3] Guggenheim, E. A., *Boltzmann's Distribution Law*, North-Holland, Amsterdam, 1963, page 32.
[4] Hill, T. L., *An Introduction to Statistical Thermodynamics*, Addison–Wesley, 1960, page 181.
[5] ter Haar, D. and Wergeland, H., *Elements of Thermodynamics*, Addison–Wesley, 1966, Chapters 6 and 7.

PROBLEMS

Problem 6.1

A thermodynamic system consists of N spatially separated subsystems. Each subsystem has energy levels of energy 0, ε and 2ε. The degeneracies of these energy levels are 1, 2 and 4 respectively. The system is in thermal equilibrium with a heat reservoir of absolute temperature T equal to ε/k. Calculate the Helmholtz free energy of the system.

Problem 6.2

Outline the importance of the Helmholtz free energy F in classical equilibrium thermodynamics. How is the free energy F related to the partition function? Indicate how other quantities such as internal energy, entropy and pressure can be calculated from the free energy.

The energy levels of a one dimensional harmonic oscillator are given by

$(n + \frac{1}{2})h\nu$, where n is a positive integer. Show that at high temperatures, when $kT \gg h\nu$, the Helmholtz free energy of the oscillator approximates to $kT \log(h\nu/kT)$. Use equation (6.2) to show that the mean energy of the oscillator is kT.

Problem 6.3
The partition function of an ideal monatomic gas containing N atoms each of mass m is

$$Z = (V^N/N!)(2\pi mkT/h^2)^{3N/2}$$

where V is the volume of the gas and T is its temperature. Comment on the origin of the $N!$ term. Calculate the Helmholtz free energy F of the gas for large values of N. Use this expression for F to determine expressions for the pressure, chemical potential and the entropy of the gas. Calculate the heat capacity at constant volume from the expression for the entropy. Check your results with the formulae derived in Section 5.9. [In this example, $g = 1$.]

Problem 6.4
According to equation (A1.1) of Appendix 1,

$$\frac{\partial^2 f}{\partial T \, \partial V} = \frac{\partial^2 f}{\partial V \, \partial T} .$$

Apply this result to equations (6.3) and (6.5) to show that

$$(\partial S/\partial V)_{T,N} = (\partial p/\partial T)_{V,N} .$$

Apply equation (A1.1) to equations (3.23) and (A5.7), to equations (6.36) and (6.37), and to equations (6.30a) and (6.30b) to show that

$$(\partial V/\partial T)_{S,N} = -(\partial S/\partial p)_{V,N}; \qquad (\partial V/\partial T)_{p,N} = -(\partial S/\partial p)_{T,N}$$

and

$$(\partial p/\partial T)_{S,N} = (\partial S/\partial V)_{p,N} .$$

These are the Maxwell thermodynamic relations.

Problem 6.5
At the absolute zero, all the N atoms in a crystal occupy a lattice site of a simple cubic lattice with no vacancies. At higher temperatures, it is possible for an atom to move from a lattice site to an interstitial site in the centre of the cube. (This is a Frenkel defect.) An atom needs an energy ε to make this transition. If at an absolute temperature T there are n atoms in interstitial sites leaving n holes (vacancies) at empty lattice sites, show that there are $[N!/n!(N - n)!]^2$ ways of doing this. (Hint: use equation (A1.3) of Appendix 1 and apply it to both the arrangement of the n atoms in N interstitial sites and to n holes in N lattice sites.) Show that the entropy S is

$$S = 2k \log[N!/n! \, (N - n)!] \; .$$

Use equation (6.1) to show that the Helmholtz free energy F is

$$F = n\varepsilon - 2kT \log[N!/n! \, (N - n)!] \; .$$

Assume that both n and N are much greater than unity. By minimising F with respect to variations in n show that, provided $N \gg n$, which implies $\varepsilon \gg 2kT$, the equilibrium value of n is $N \exp(-\varepsilon/2kT)$.

Problem 6.6*

The small system in Figure 6.1(b) is made up of two subsystems 1 and 2. The total volume $(V_1 + V_2)$ and the total number of particles $(N_1 + N_2)$ are constant. Both subsystems can interchange heat with the heat reservoir, so that it is $(U_1 + U_2 + U_R)$ which is constant. Corresponding to equation (6.18), the condition of equilibrium is that $g_R g_1 g_2$ is a maximum. Using equation (6.19), the condition of equilibrium becomes

$$g_R^* \{\exp[-\beta(U_1 + U_2)]\} g_1(U_1) g(U_2) \text{ is a maximum} \; .$$

Take the natural logarithm and use the equation $S = k \log g$ to show that at equilibrium

$$(U_1 - T_R S_1) + (U_2 - T_R S_2) \quad \text{is a minimum} \; .$$

Differentiate partially with respect to U_1, and use the condition that, since it is $(U_1 + U_2 + U_R)$ which is a constant, U_1 and U_2 can vary independently, to show that $T_1 = T_R$ at equilibrium. By differentiating with respect to U_2, show that at equilibrium $T_2 = T_R$. The condition of equilibrium becomes

$$(F_1 + F_2) \quad \text{is a minimum}$$

where F_1 and F_2 are the Helmholtz free energies of subsystems 1 and 2 respectively. By differentiating partially with respect to V_1 at constant T_1, T_2, N_1 and N_2 and using the condition that $(V_1 + V_2)$ is a constant, show that, if the partition between subsystems 1 and 2 is a movable partition, $p_1 = p_2$ at equilibrium. By differentiating with respect to N_1 at constant T_1, T_2, V_1 and V_2 and using the condition that $(N_1 + N_2)$ is a constant, show that, if the two subsystems are in diffusive equilibrium, $\mu_1 = \mu_2$.

Problem 6.7*

Assume that the small system in Figure 6.2 is made up of two subsystems, both of which are in thermal and mechanical equilibrium with the heat and pressure reservoir. In this case $(U_1 + U_2 + U_R)$, $(V_1 + V_2 + V_R)$ and $(N_1 + N_2)$ are constant. Using equation (6.48), the condition of equilibrium becomes

$$g_R g_1 g_2 = g_R^* \exp[-(U_1 + U_2 + p_R V_1 + p_R V_2)/kT_R] g_1(U_1) g_2(U_2)$$
$$\text{is a maximum}$$

Take the natural logarithm and use the equation $S = k \log g$ to show that at equilibrium

$$(U_1 - T_R S_1 + p_R V_1) + (U_2 - T_R S_2 + p_R V_2) \text{ is a minimum}.$$

By differentiating with respect to U_1, keeping U_2, V_1, V_2, S_2 constant show that at equilibrium $T_1 = T_R$. By differentiating partially with respect to U_2 show that $T_2 = T_R$. By differentiating partially with respect to V_1 and with respect to V_2 show that at equilibrium $p_1 = p_2 = p_R$. The condition of equilibrium becomes

$$(G_1 + G_2) \text{ is a minimum}.$$

By differentiating with respect to N_1 at constant p and T, show that if the two subsystems are in diffusive equilibrium, $\mu_1 = \mu_2$.

Problem 6.8*

Consider the simple isomeric change

$$A \rightleftharpoons B$$

between ideal gas molecules A and B. Show that, if there are N_A molecules of type A and N_B molecules of type B in a container of fixed volume, which is in thermal equilibrium with a heat reservoir of absolute temperature T, the partition function is

$$Z = Z_A{}^{N_A} Z_B{}^{N_B}/N_A! N_B!$$

where Z_A and Z_B are the single particle partition functions of molecules of type A and type B respectively. By maximising $\log Z$ or minimising F with respect to changes in N_A, subject to the condition that $(N_A + N_B)$ is constant, show that at equilibrium

$$N_B/N_A = Z_B/Z_A .$$

(*Hint:* Use Stirling's theorem to show that $d \log N!/dN \simeq \log N$, for large N.)

Problem 6.9*

Consider the dissociation of a homonuclear diatomic molecule $A_2 \rightleftharpoons A + A$ discussed in Section 6.6.3. Show that, if there are N_1 atoms of type A and N_2 molecules of type A_2, the partition function is

$$Z = Z_1{}^{N_1} Z_2{}^{N_2}/N_1! N_2!$$

where Z_1 and Z_2 are the single particle partition functions of atoms of type A and molecules of type A_2 respectively. By maximising $\log Z$ with respect to variations in N_1, subject to the condition that $(N_1 + 2N_2)$, the total number of atoms, is a constant show that at equilibrium

$$N_1{}^2/N_2 = Z_1{}^2/Z_2 .$$

Problem 6.10*
The mole fraction x_i of the ith component of a multicomponent ideal gas system is given by

$$x_i = n_i/(n_1 + n_2 + \dots) = N_i/N$$

where n_i is the number of moles of the ith component, N_i is the number of molecules of the ith component and N is the total number of molecules in the system. Substitute $(x_i N)$ for N_i in equation (6.71), and derive the following relations

$$K_x = x_1^{\nu_1} x_2^{\nu_2} \dots = K_n N^{-\Delta\nu} = K_c (kT/p)^{\Delta\nu} = K_p p^{-\Delta\nu} \ .$$

Problem 6.11*
Consider an n component system, composed of N_1 molecules of type 1, N_2 molecules of type 2 etc., making up the small system shown in Figure 6.1(a). Use equations (A1.31) and (6.11) to show that for chemical equilibrium at constant V and T,

$$\mathrm{d}F = \sum_{i=1}^{n} (\partial F/\partial N_i)\, \mathrm{d}N_i = \sum_{i=1}^{n} \mu_i\, \mathrm{d}N_i = 0.$$

Then follow the method of Section 6.6.1 to derive equation (6.69).

Chapter 7*

Heat Work and Cycles*

7.1* HEAT AND WORK*

7.1.1* Introduction

Consider the hydrostatic system shown in Figure 7.1. At one end, the system makes thermal contact with a heat reservoir through a fixed, rigid, diathermic partition. At the other end of the system there is an adiabatic piston, which can be moved in and out any desired distance without friction and then fixed rigidly in the position required.

Let the system have a volume V when it is in thermal equilibrium with a heat reservoir, which is at an absolute temperature T. According to equation (2.28) the ensemble average energy of the system is

$$U = \sum_{\text{(states)}} P_i \varepsilon_i \qquad (7.1)$$

where P_i is the probability that the system is in its ith microstate having energy eigenvalue ε_i, when it is in thermal equilibrium with the heat reservoir. According to equation (4.2), which is the same as equation (2.26),

$$P_i = g_R(U^* - \varepsilon_i)/g_T \qquad (7.2)$$

where $g_R(U^* - \varepsilon_i)$ is the number of microstates accessible to the heat reservoir, when the heat reservoir has energy $(U^* - \varepsilon_i)$, U^* is the total energy of the system plus heat reservoir and g_T is the total number of microstates accessible to the closed composite system made up of the system in thermal equilibrium with the heat reservoir. It was shown in Section 4.2 of Chapter 4 that, if the heat reservoir in Figure 7.1 has a large enough thermal capacity for its temperature to remain constant whatever the value of ε_i, equation (7.2) reduces to the Boltzmann distribution, namely equation (4.5).

Let the fixed, rigid, diathermic partition between the system and the heat reservoir in Figure 7.1 be changed to a fixed, rigid, *adiabatic* partition. Let the adiabatic piston on the right hand side of the system in Figure 7.1

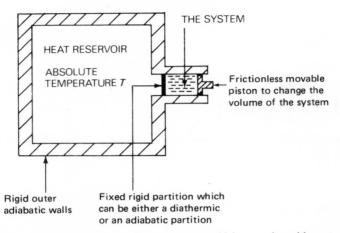

THE SYSTEM

HEAT RESERVOIR

ABSOLUTE
TEMPERATURE T

Frictionless movable
piston to change the
volume of the system

Rigid outer
adiabatic walls

Fixed rigid partition which
can be either a diathermic
or an adiabatic partition

Figure 7.1—An experimental arrangement which can give either a reversible change of volume or add heat to a system kept at constant volume.

be moved inwards very slowly, until the volume of the system changes to $(V + dV)$, where dV is negative. Let dU_1 be the change in the ensemble average energy of the system. Let the piston be fixed rigidly in this new position. According to the adiabatic approximation of quantum mechanics (Ehrenfest's principle), when the hydrostatic system is compressed *quasistatically* under adiabatic conditions, the probability that the system makes a transition to a different quantum state is negligible. [This result follows directly from the time dependent perturbation theory of quantum mechanics. Reference: Landau and Lifshitz [1]. A reader interested in a full discussion is referred to Wannier [2].] According to equation (7.1.), for constant P_i, which implies that all the dP_i are zero, the change in the ensemble average energy of the system in the reversible adiabatic change is

$$dU_1 = \sum_{(states)} P_i \, d\varepsilon_i . \qquad (7.3)$$

Using equation (A5.5) of Appendix 5, equation (7.3) becomes

$$dU_1 = \Sigma P_i \, d\varepsilon_i = -p \, dV = dW . \qquad (7.4)$$

Hence the term $\Sigma P_i \, d\varepsilon_i$ is equal to the mechanical work dW done *on* the system in a reversible adiabatic change. Summarising, the values of the energy eigenvalues ε_i change in a reversible *adiabatic* change of volume, but the probability P_i that the system is in a particular microstate does not change.

Let the temperature of the heat reservoir in Figure 7.1 be raised to

$(T + dT)$, and let the adiabatic partition separating the system from the heat reservoir be changed back to a fixed, rigid, *diathermic* partition. The volume of the system is kept constant at $(V + dV)$. Heat flows from the heat reservoir to the system until, at thermal equilibrium, the temperature of the system is increased to $(T + dT)$. Let dU_2 be the change in the ensemble average energy of the system due to the heat which flows to the system from the heat reservoir. If dU is the total change in the ensemble average energy of the system due to both the reversible adiabatic change of volume and the addition of heat to the system at constant volume, according to equation (7.1)

$$dU = dU_1 + dU_2 = \Sigma\, P_i\, d\varepsilon_i + \Sigma\, \varepsilon_i\, dP_i \; . \qquad (7.5)$$

It is being assumed that the number of particles in the system is constant. Since, according to equation (7.4), the $\Sigma\, P_i\, d\varepsilon_i$ term is equal to dU_1, equation (7.5) gives

$$\Sigma\, \varepsilon_i\, dP_i = dU_2 \; . \qquad (7.6)$$

Hence the $\Sigma\, \varepsilon_i\, dP_i$ term in equation (7.5) is equal to the heat added to the system in Figure 7.1, when its temperature is raised from T to $(T + \Delta T)$ keeping the volume of the system constant. (A simple numerical application of equation (7.6) is developed in Problem 7.1.)

Summarising, equations (7.6) and (7.4) illustrate the essential difference between the effect of heat and work on a system. When heat is added to a system, kept at constant volume, the values of the energy eigenvalues of the system do not change, but the probability that the system is in its ith microstate does change. On the other hand, in a reversible adiabatic change of volume, the values of the energy eigenvalues ε_i change, but the probability P_i that the system is in its ith microstate does not change. It will be shown in Section 7.1.7 that this difference between the effect of heat and work on a system leads to the Kelvin–Planck statement of the second law of thermodynamics, according to which: 'No process is possible whose sole result is the absorption of heat from a reservoir and the conversion of all of this heat into work'.

7.1.2* The Boltzmann definition of entropy*

This subsection can be omitted without loss of continuity. According to the thermodynamic identity, equation (3.85), which was developed from statistical mechanics in Section 3.7.5 of Chapter 3, if N and V are constant, the increase in the (internal) energy of a system is equal to $T\, dS$. Hence equation (7.6) can be rewritten in the form

$$dU_2 = T\, dS = \Sigma\, \varepsilon_i\, dP_i \; . \qquad (7.7)$$

This relation shows that there is a close connection between the entropy S

of the system and the probabilities P_i. To illustrate this, assume that the heat reservoir in Figure 7.1 has a large enough heat capacity for the Boltzmann distribution to be applicable to the system. According to equation (4.8),

$$P_i = \exp(-\varepsilon_i/kT)/Z$$

where Z is the partition function, which is a constant for given values of V, T and N. Taking the natural logarithm of P_i, and rearranging we have

$$\varepsilon_i = -kT[\log P_i + \log Z] \ .$$

Substituting for ε_i in equation (7.7) and cancelling T, we obtain

$$dS = -k[\sum (\log P_i) dP_i + \log Z \sum dP_i] \ .$$

Since $\sum P_i$ is equal to unity, $\sum dP_i$ is zero. Hence

$$dS = -k[\sum (\log P_i) \, dP_i] \ .$$

Since

$$\sum P_i \, d(\log P_i) = \sum P_i(dP_i/P_i) = \sum dP_i = 0$$

the expression for dS can be rewritten in the form

$$dS = d[-k \sum P_i \log P_i] \ .$$

Integrating, we have

$$S = -k \sum P_i \log P_i + C$$

where C is a constant of integration. At $T = 0$, the system is in its ground state, and all the P_i are zero except for P_0, which is equal to unity if the ground state is not degenerate. If $P_0 = 1$, $\log P_0$ is zero, and $\sum P_i \log P_i$ is zero at $T = 0$. For the entropy S to be zero at $T = 0$, as required by the third law of thermodynamics, C must be zero. Hence at a finite absolute temperature, the entropy of the small system in Figure 7.1 is given by

$$S = -k \sum_{(states)} P_i \log P_i \qquad (7.8)$$

This is the Boltzmann definition of entropy. It was developed in this section for the case when the Boltzmann distribution was applicable to the small system of constant volume in thermal equilibrium with the heat reservoir shown in Figure 7.1. This corresponds to the **canonical ensemble**.

Equation (7.8) can also be applied in other cases. For example, in the case of a closed system, according to the principle of equal *a priori* probabilities, when the closed system is in internal thermodynamic equilibrium all the g microstates accessible to the closed system are equally probable, so that all the P_i are equal to $1/g$. Hence there are g terms of the type

$(P_i \log P_i)$ in equation (7.8), each of which is equal to $-(1/g) \log g$, and the entropy given by equation (7.8) is equal to $k \log g$. This is in agreement with equation (3.22). This corresponds to the **microcanonical ensemble**. The case of the **grand canonical ensemble** will be considered in Problem 11.5 of Chapter 11. [The Boltzmann definition of entropy, given by equation (7.8), was compared with information theory in Section 3.10 of Chapter 3.]

7.1.3* Addition of heat to a system kept at constant volume*

To illustrate what happens to a system, when heat is added to a system which is kept at constant volume, we shall consider special cases of the apparatus shown in Figure 7.1. The volume V of the system is kept constant. With the diathermic partition in place, the system is allowed to come into thermal equilibrium with the heat reservoir, which initially has constant absolute temperature T. The probability P_i that at thermal equilibrium the system is in its ith microstate, having energy eigenvalue ε_i, is given

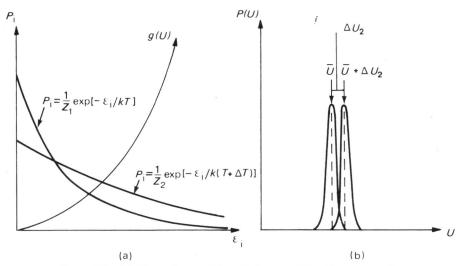

(a)   (b)

Figure 7.2—(a) The variation with ε_i of the probability P_i that a small system is in a microstate having energy eigenvalue ε_i, when it is in thermal equilibrium with heat reservoirs of absolute temperatures T and $(T + \Delta T)$. The variation with U of $g(U)$, the number of accessible microstates of the system which have energy eigenvalues U equal to ε_i, is also shown. (b) The variation with U of the probability $P(U)$ that the small system has energy U, when it is in thermal equilibrium with heat reservoirs of absolute temperatures T and $(T + \Delta T)$.

by the Boltzmann distribution, namely

$$P_i = \exp(-\varepsilon_i/kT)/Z_1 \qquad (4.8)$$

where Z_1 is the partition function at the absolute temperature T. The variation of P_i with ε_i at the absolute temperature T is sketched in Figure 7.2(a). Let $g(U)$ denote the number of microstates accessible to the system (the degeneracy) when the system has energy U equal to ε_i. The degeneracy $g(U)$ increases extremely rapidly, when U is increased. It is impossible to show how rapid this variation is on a diagram. Only a much slower variation of $g(U)$ with U can be sketched in Figure 7.2(a). The probability $P(U)$ that at thermal equilibrium the system has energy U, which is the probability that the system is in any one of the $g(U)$ microstates of the system which have energy eigenvalues equal to $U = \varepsilon_i$, is equal to the product of P_i, the probability that the system is in one particular one of these microstates, and $g(U)$, the total number of microstates having energy eigenvalues equal to $U = \varepsilon_i$. Hence, $P(U) = g(U)P_i$ where P_i is given by equation (4.8). This expression for $P(U)$ is the same as equation (4.28). The product of $g(U)$ and P_i has a very sharp maximum at an energy $\bar{U}$, which for a macroscopic system is equal to the ensemble average energy $\bar{U}$.

When the absolute temperature of the heat reservoir in Figure 7.1 is raised from T to $(T + \Delta T)$, the probability P_i that the system, of fixed volume V, is in a microstate having energy eigenvalue ε_i, when the system is in thermal equilibrium with the heat reservoir, is changed to

$$P_i = \exp[-\varepsilon_i/k(T + \Delta T)]/Z_2$$

where Z_2 is the partition function at the temperature $(T + \Delta T)$. The probability P_i is increased at high values of energy and decreased at low values of energy, as sketched in Figure 7.2(a). (A simple numerical example illustrating how P_i varies with T is given in Figure 4.7(b) of Chapter 4.) Since the volume of the system is constant, the values of the energy eigenvalues ε_i and the degeneracy $g(U)$ are unchanged. When the temperature of the heat reservoir is raised from T to $(T + \Delta T)$, the value of the product $g(U)P_i$ is increased at high energies and decreased at low energies, compared with the corresponding values of $g(U)P_i$ when the temperature of the heat reservoir was T. The maximum of the probability distribution $P(U)$, that the system has energy U, is displaced to a higher energy, when the temperature of the heat reservoir is increased from T to $(T + \Delta T)$, as sketched in Figure 7.2(b). The increase ΔU_2 in the ensemble average energy of the system is given by equation (7.6). It was shown in Section 5.2 of Chapter 5, and Section A4.1 of Appendix 4 that, at thermal equilibrium, the standard deviation in the probability distribution $P(U)$ is of the order of $\bar{U}/N^{1/2}$, where N is the number of particles in the system. It will be shown in Chapter 10, that in the classical limit, the heat capacity of an insulating solid is approximately equal to $3Nk$, and, in the classical limit, the (internal) energy of an insulating solid is given approximately by

$$\bar{U} = 3NkT \tag{7.9}$$

where k is Boltzmann's constant. According to equation (5.17), if the system in Figure 4.1(a) were an insulating solid,

$$\sigma = (3N)^{1/2}kT \ .$$

Differentiating equation (7.9), we have

$$\Delta T = \Delta U_2/3Nk \ .$$

To give an increase ΔU_2 in the internal energy of the insulating solid equal to one standard deviation σ, in the classical limit, the rise in the absolute temperature of the insulating solid would have to be

$$\Delta T = \sigma/3Nk = T/(3N)^{1/2} \ .$$

For a small macroscopic system consisting of $N = 10^{18}$ particles at 300K, ΔT would have to be of the order of 1.7×10^{-7}K for a shift of the position of the maximum of the probability distribution in Figure 7.2(b) by one standard deviation. For $N > 10^{18}$, ΔT would be even less. Hence, when a small macroscopic system is in thermal equilibrium with a heat reservoir of absolute temperature $T = 300$K, an increase in the temperature of the heat reservoir of only $\Delta T = 10^{-5}$K is sufficient to separate the two probability distributions in Figure 7.2(b) by some sixty standard deviations. For even such a small temperature rise, the increase in the ensemble average energy of a macroscopic system is a reasonably well defined quantity.

7.1.4* Graphical representation of the thermodynamic state of a system*

As an analogy, consider the case of an isolated electric charge in empty space. There is an electric field due to this charge at all other points of space. One cannot draw an electric field line through every point of space. In electrostatics, the number of electric field lines is limited such that the number of field lines crossing unit area normal to the field line is equal to, or is proportional to, the electric field intensity. The density of states D of a macroscopic thermodynamic system is such a fantastically large number that one cannot draw a line on an energy eigenvalue diagram to represent the energy eigenvalue of every microstate of the system. In principle, one could represent the variation of the density of states D with energy U on an energy eigenvalue diagram quantitatively by spacing the lines on the energy eigenvalue diagram appropriately. However, for a macroscopic system the variation of D with U is so rapid that it could not be done quantatively on a diagram. For example, it was shown in Section 2.5.2, that if one joule of energy is added to 1 kg of water at 300K, which raises its tempera-

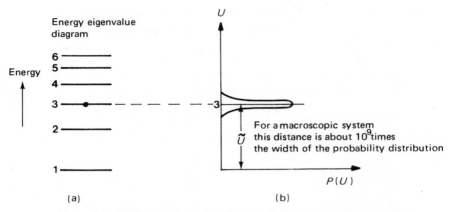

Figure 7.3—(a) Energy eigenvalue diagram. Only representative energy eigenvalues are shown. The most probable energy of the system is equal to the energy of representative energy eigenvalue 3. (b) The probability distribution $P(U)$ that, at thermal equilibrium, the system has energy U, when it is in thermal equilibrium with a heat reservoir. The most probable energy is equal to the energy of representative energy eigenvalue 3.

ture by only 2.38×10^{-4}K, the number of accessible microstates, and hence D increases by a factor of about $\exp(2.4 \times 10^{20})$. The best we can do in this Chapter is to put the lines on the energy eigenvalue diagrams closer together to represent *qualitatively* the increase in the density of states D, as the energy of the system is increased, as shown in Figure 7.3(a). These lines will be referred to as **representative energy eigenvalues.**

When the small system in Figure 4.1(a) is in thermal equilibrium with the heat reservoir, though there is a finite spread in the probability distribution $P(U)$ that the small system has energy U, it is an extremely sharp distribution. The centre of the dot on representative energy eigenvalue 3 in Figure 7.3(a) represents the most probable value of energy, which for macroscopic systems, is equal to the ensemble average energy. The reader should remember that there is a chance of finding the system within a narrow band of energies centred on the most probable value of energy, as shown in Figure 7.3(b). For macroscopic systems, the standard deviation is so small, that the probability distribution is narrower than the width of any dot we can draw. The increasing numbers on the representative energy eigenvalues are used to represent increasing values of energy.

7.1.5* A reversible adiabatic change*

Consider the hydrostatic system shown in Figure 7.1. When the system is in thermal equilibrium with the heat reservoir, which has an absolute temperature T, the diathermic partition separating the system from the heat reservoir is changed into an adiabatic one. The other walls of the

system are adiabatic walls, so that no heat can now enter or leave the
system in any expansion or compression of the system. The initial state of
the system is represented in Figure 7.4(a), in the way described in Section
7.1.4. Only representative energy eigenvalues are shown in Figure 7.4(a).
In the initial state, the most probable value of energy $\tilde{U}$, which for a
macroscopic system is equal to the ensemble average energy $\bar{U}$, is equal to
the energy of representative energy eigenvalue 4. [There is a chance of
finding the system in a microstate having an energy eigenvalue within a
very narrow range of energies centred on the energy of representative
energy eigenvalue 4.]

To obtain a reversible expansion of the system, the adiabatic piston is
released so that it is free to move. The external pressure is varied very
slowly so that it is always infinitesimally less than the pressure of the
system. In this reversible adiabatic expansion, the system does work on its
surroundings.

When the volume of the system is increased, all the energy eigenvalues
of the N particle system are decreased. For example, using the principle of
virtual work, it is shown in Appendix 5 that, if the system is in its ith

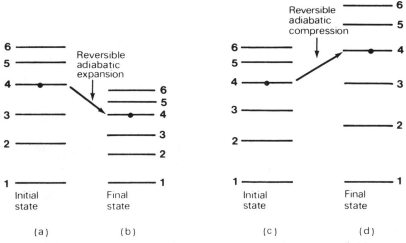

(a)   (b)   (c)   (d)

Figure 7.4—Reversible adiabatic change. In the initial state shown in (a),
the most probable energy of the system is equal to the energy of representative energy eigenvalue 4. In a reversible adiabatic expansion of
the system from (a) to (b), the energy eigenvalues of the system are all
reduced in magnitude and are closer together, but the most probable
energy remains equal to the energy of representative energy eigenvalue
4. A reversible adiabatic compression is illustrated in the change from (c)
to (d). The energy eigenvalues are further apart in (d) than in (c), but the
most probable energy remains equal to the energy of representative
energy eigenvalue 4.

microstate having energy eigenvalue $\varepsilon_i(V, N)$, the pressure of the system is

$$p_i = -(\partial\varepsilon_i/\partial V) \ . \tag{A5.2}$$

Unless the energy eigenvalue ε_i decreased as the volume V of the system was increased, the pressure p_i would not be positive. Hence, in the general case of a hydrostatic system, if the system expands the values of all the energy eigenvalues of the N particle system must decrease. This is illustrated by the changes in the positions of the representative energy eigenvalues in Figure 7.4(b) compared with Figure 7.4(a).

As a typical example of a macroscopic system, consider an ideal monatomic gas system consisting of N atoms in a cubical box of volume V. It is assumed that in an ideal monotomic gas there are no interactions between the atoms making up the system. The N monatomic gas atoms can be treated as N completely independent particles. The total energy of the N atoms is the sum of the energies of each of the N individual atoms. Each atom must be in one of the single particle quantum states of the system, which can be determined by solving Schrödinger's time independent equation for a single particle in a box in the way described in Appendix 3. According to equation (A3.13) of Appendix 3, the energy eigenvalue of each single particle state is proportional to $V^{-2/3}$. Hence, in the special case of an ideal monatomic gas, the total energy eigenvalue $\varepsilon_i(V, N)$ of a microstate of the N particle system, which is equal to the sum of the energies of the N individual atoms, must vary as $V^{-2/3}$, so that for an ideal monatomic gas

$$\varepsilon_i(V, N) = C_i V^{-2/3} \ . \tag{7.10}$$

where C_i is a constant, which will have a different numerical value for the different microstates of the N particle ideal gas system. It follows from equation (7.10) that ε_i decreases if V is increased.

When a gas is allowed to expand very slowly under adiabatic conditions, the probability P_i that the N particle system is in its ith microstate does not change. (This is the adiabatic approximation of quantum mechanics, which is sometimes called Ehrenfest's principle [1,2].) Thus in the transition from Figure 7.4(a) to 7.4(b), corresponding to a reversible adiabatic expansion of the system, the energies of all the representative energy eigenvalues are decreased and are closer together in Figure 7.4(b) than in Figure 7.4(a). The probability that the system is in a particular microstate does not change and the mean energy of the system remains equal to the energy of representative energy eigenvalue 4. The changes are represented by the positions of the dots in Figures 7.4(a) and 7.4(b). The value of the energy of representative energy eigenvalue 4 is less in Figure 7.4(b) than in Figure 7.4(a), showing that the mean energy of the system is decreased in a reversible adiabatic expansion.

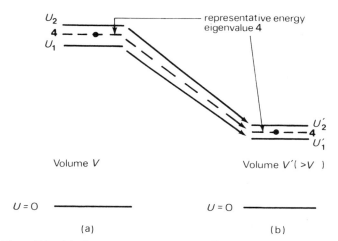

Figure 7.5—(a) The mean energies U_1 and U_2 are just on either side of the energy of representative energy eigenvalue 4. [These mean energies refer to slightly different initial temperatures.] In a reversible adiabatic expansion from the volume V in (a) to the volume V' in (b), where $V' > V$, the mean energies U_1 and U_2 are reduced to U_1' and U_2' respectively. The energy gap $(U_2' - U_1')$ is less than the energy gap $(U_2 - U_1)$.

To see how the entropy S of the system varies in a reversible adiabatic expansion, consider the Boltzmann definition of entropy, given by equation (7.8). Since according to Ehrenfrest's principle (the adiabatic approximation) all the P_i are constant in a reversible adiabatic change of volume, the entropy given by equation (7.8) is constant in a reversible adiabatic expansion.†

To see how the absolute temperature T of the system varies in a reversible adiabatic expansion, consider two mean energies U_1 and U_2 just on either side of representative energy eigenvalue 4, in the initial state illustrated in Figure 7.5(a), when the volume of the system is equal to V. Let the corresponding entropies be S_1 and S_2 respectively. Let the hydrostatic system undergo a *reversible adiabatic* expansion from a volume V to a volume V'. Assume that in this expansion, if the system starts with mean energy U_1 the mean energy is reduced to U_1', and if it starts with U_2 the mean energy is reduced to U_2', as shown in Figure 7.5(b). Since the entropy is constant

---

†Alternatively, let D and D' be the density of states in the vicinity of representative energy eigenvalue 4 in Figures 7.5(a) and 7.5(b) respectively. The total number of states between U_1 and U_2 in Figure 7.5(a) is equal to the total number of states between U_1' and U_2' in Figure 7.5(b), so that $D \times [U_2 - U_1]$ is equal to $D' \times [U_2' - U_1']$. Hence

$$S' = k \log D' = k \log D + k \log[(U_2 - U_1)/(U_2' - U_1')] \ .$$

The last term is negligible compared with $k \log D$, so that S' is equal to S.

in a reversible adiabatic change the entropies of the system, when it has mean energies U_1' and U_2', are S_1 and S_2 respectively. If the energies U_1 and U_2 are very close to each other in Figure 7.5(a), according to equation (3.23), the initial absolute temperature T of the system, when it has the energy of representative energy eigenvalue 4 in Figure 7.5(a) is

$$T = \Delta U/\Delta S = (U_2 - U_1)/(S_2 - S_1) \ . \tag{7.11}$$

The corresponding final absolute temperature T' in Figure 7.5(b) is

$$T' = (U_2' - U_1')/(S_2 - S_1) \ . \tag{7.12}$$

When a hydrostatic system expands the energy eigenvalues are all reduced so that $(U_2' - U_1')$ is less than $(U_2 - U_1)$. Hence, according to equations (7.11) and (7.12), the final absolute temperature T' is less than the initial absolute temperature T, showing that the absolute temperature decreases in a reversible adiabatic expansion.

Summarising, in a reversible adiabatic expansion of an N-particle system, the energy eigenvalues of the microstates of the N particle system decrease, but the probability that the system is in any particular N particle microstate does not change. This is illustrated in Figure 7.4(a) and 7.4(b). The entropy remains constant in a reversible adiabatic expansion, but the absolute temperature goes down.

In a reversible adiabatic compression of the system, work is down on the system, the values of the energy eigenvalues are increased, but the probability that the system is in a particular microstate does not change. This is illustrated in Figures 7.4(c) and 7.4(d). The most probable energy is increased, but remains equal to the energy of representative energy eigenvalue 4. The entropy is again constant, but in this case the absolute temperature goes up.

7.1.6* Representation of the addition of heat to a system kept at constant volume*

If the volume of a hydrostatic system is kept constant when heat is added to the system, the values of the energy eigenvalues of the microstates of the N particle system do not change. This is illustrated in Figures 7.6(a) and 7.6(b). According to equation (7.6), the addition of heat to the system changes the probability P_i that the N particle system is in its ith microstate, leading to an increase in the most probable value of energy, which for a macroscopic system is equal to the ensemble average value of energy. In the example illustrated in Figure 7.6, when heat is added to a system, which is kept at constant volume, the most probable value of energy is increased from the energy of representative energy eigenvalue 3 in Figure 7.6(a) to the energy of representative energy eigenvalue 4 in

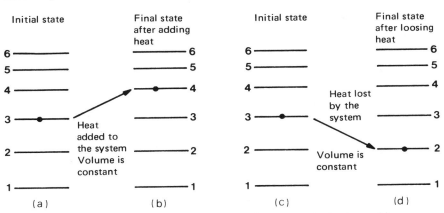

Figure 7.6—Addition of heat to a system kept at constant volume. (a) This is the initial state. When heat is added to the system, the value of the energy eigenvalues are unchanged, but the probability that the system is in a particular microstate changes such that, in the present example, the most probable energy is increased from the energy of representative energy eigenvalue 3 in (a) to the energy of representative energy eigenvalue 4 in (b). In the change from (c) to (d) heat is lost from a system kept at constant volume. The most probable energy goes down from the energy of representative energy eigenvalue 3 in (c) to the energy of representative energy eigenvalue 2 in (d).

Figure 7.6(b). When heat is added to a system kept at constant volume, the change in the entropy of the system is given by equations (7.7) and (3.47).

If heat is lost from a hydrostatic system, kept at constant volume, the energy eigenvalues ε_i of the N particle system are unchanged, but the probability P_i that the system is in its ith microstate is changed such that the most probable value of energy is decreased, as shown in Figures 7.6(c) and 7.6(d), where the most probable energy decreases from the energy of representative energy eigenvalue 3 in Figure 7.6(c) to the energy of representative energy eigenvalue 2 in Figure 7.6(d).

7.1.7* Representation of the difference between heat and work and the Kelvin–Planck statement of the second law of thermodynamics*

Comparing Figures 7.6 and 7.4, the reader can see that there is an essential difference between the effects of adding heat to a system kept at constant volume, which changes the P_i and hence the entropy S, but leaves the energy eigenvalues ε_i unchanged, and doing work on the system, which for a reversible adiabatic process leaves the P_i and hence the entropy S unchanged, but changes the energy eigenvalues ε_i.

Let a quantity of heat ΔQ be added to the system in Figure 7.7(a), which is kept at constant volume during this process. The energy eigenvalues are unchanged, but the most probable value of energy is increased

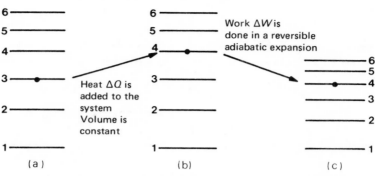

Figure 7.7—Comparison of heat and work. When heat ΔQ is added to the system, when it is in the initial state shown in (a), if the volume is kept constant, the energy eigenvalues are unchanged but the probabilities P_i are changed, such that the most probable value of energy goes up from the energy of representative energy eigenvalue 3 in (a) to the energy of representative energy eigenvalue 4 in (b). When the system then undergoes a reversible adiabatic expansion to (c) doing external work ΔW, the energy eigenvalues are all decreased, but the most probable energy remains equal to the energy of representative energy eigenvalue 4. Even if $\Delta Q = \Delta W$ the states shown in (a) and (c) are not the same.

from the energy of representative energy eigenvalue 3 in Figure 7.7(a) to the energy of representative energy eigenvalue 4 in Figure 7.7(b). Now, let the system do external work ΔW, which is numerically equal to ΔQ, in a reversible adiabatic expansion of the system. In this expansion, the most probable value of energy remains equal to the energy of representative energy eigenvalue 4, but the values of all the energy eigenvalues, including representative energy eigenvalue 4, are decreased. The internal energy U of the system is the same in Figures 7.7(a) and 7.7(c), but the system is not back in its initial state in Figure 7.7(c). Compared with the initial state shown in Figure 7.7(a), the energy eigenvalues are closer together in Figure 7.7(c) and the most probable value of energy is equal to the energy of representative energy eigenvalue 4 not representative energy eigenvalue 3. Both the ε_i and P_i are different in Figures 7.7(a) and 7.7(c). Even though all the heat given to the system in the change from Figure 7.7(a) to Figure 7.7(b) is converted into external work in the change from Figure 7.7(b) to Figure 7.7(c), the system is not back in its initial state in Figure 7.7(c).[†]

†In classical equilibrium thermodynamics, the macrostate of the system is different in Figures 7.7(a) and 7.7(c). Even though U and N are the same in both cases, the volume V is different in the two cases. This leads to differences in the other thermodynamic variables, such as entropy. For example, there is an increase in the entropy of the system when heat is added to take the system from Figure 7.7(a) to Figure 7.7(b). This increased entropy remains constant in the reversible adiabatic expansion which takes the system from Figure 7.7(b) to Figure 7.7(c).

Since the system is not back in its initial state in Figure 7.7(c), the sole result of the changes from Figure 7.7(a) to Figure 7.7(c) is not to convert all the heat added to the system into external work. This is the interpretation of the Kelvin–Planck statement of the second law of thermodynamics, according to which: 'No process is possible whose sole result is the absorption of heat from a reservoir and the conversion of all of this heat into work'. One can go around various cyclic processes, such as the Carnot cycle, which bring the system (working substance) back to its initial state, but it will be shown in Sections 7.2.1 and 7.4 that it is not possible to convert all the heat given to the system into external work in a cyclic process, which brings the working substance back to its initial state.

Figure 7.7 can be used to illustrate Carathéodory's principle that, in the immediate neighbourhood of every equilibrium state of a system, there exist states which are inaccessible (cannot be reached) in a reversible adiabatic process. Assume that the state shown in Figure 7.7(b) is the initial state. The state shown in Figure 7.7(c) is accessible from the initial state shown in Figure 7.7(b) in a reversible adiabatic expansion. The energy eigenvalues change in this process, but the mean energy of the system remains equal to the energy of representative energy eigenvalue 4. On the other hand, the state shown in Figure 7.7(a) is not accessible from the state shown in Figure 7.7(b) in a reversible adiabatic change. [The arrow between Figures 7.7(a) and 7.7(b) must now be reversed.] In the latter case, when the system starts in the state shown in Figure 7.7(b), to reach the state shown in Figure 7.7(a) the system must lose (heat) energy, at constant volume, such that the mean energy of the system is reduced from the energy of representative energy eigenvalue 4 in Figure 7.7(b) to the energy of representative energy eigenvalue 3 in Figure 7.7(a). This cannot be achieved in a reversible adiabatic change. This discussion would still be valid if ΔQ and ΔW were infinitesimal.

7.2* CONVERSION OF HEAT INTO WORK IN A CYCLIC PROCESS*

7.2.1* An ideal cycle*

It was pointed out in Section 7.1.7 that, **if all the heat added to a system is converted into external work, the system cannot be back in its initial state**. An idealised cyclic process, which brings the working substance (system) back to its initial state will now be considered. The cycle is shown on the pV indicator diagram in Figure 7.8. For purposes of discussion, state A will be chosen as the initial state. Heat is added reversibly to the hydrostatic system, by placing the system in thermal contact with a series of heat reservoirs, each of which is at an infinitesimally higher temperature than the preceding heat reservoir, until the system reaches state B in Figure 7.8.

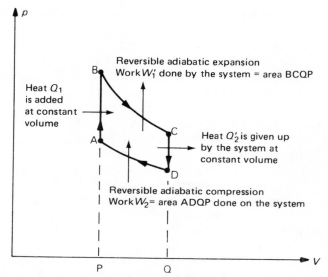

Figure 7.8—A reversible cycle. Starting at A, heat is added to the system to take it from A to B. The volume is kept constant in this change. The system then undergoes a reversible adiabatic expansion from B to C. Heat is then taken from the system, which is kept at constant volume, so that it goes from C to D. The system finally undergoes a reversible adiabatic compression to take it back from D to the initial state A.

In going from state A to state B in Figure 7.8, the volume of the system is kept constant, so that the values of the energy eigenvalues do not change. The addition of heat changes the probability P_i that the system is in its ith microstate, such that the most probable value of the energy of the system is increased, as illustrated in the change from Figure 7.9(a) to Figure 7.9(b), where the most probable energy is increased from the energy of representative energy eigenvalue 2 to the energy of representative energy eigenvalue 4. It is shown in Section 3.3 and in Section A4.2 of Appendix 4 that the heat capacity at constant volume of a normal thermodynamic system must be positive, so that when heat is added to the system to take it from state A to state B in Figure 7.8, the absolute temperature of the system is increased.

To go from state B to state C in Figure 7.8, the gas is allowed to expand reversibly and adiabatically. The work W_1' done by the gas in this expansion is equal to the area BCQP in Figure 7.8. Due to the increase in the volume of the system, the energy eigenvalues of all the microstates of the system are decreased, but the probability that the system is in any particular microstate does not change. The most probable value of energy remains

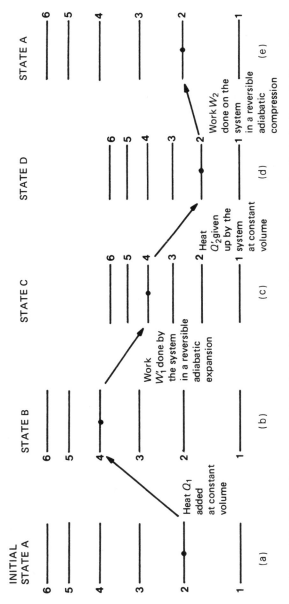

Figure 7.9—Changes in the energy eigenvalues and the most probable value of energy for the cycle shown in Figure 7.8.

equal to the energy of representative energy eigenvalue 4, as shown in Figures 7.9(b) and 7.9(c). The absolute temperature decreases in this adiabatic expansion, for the reasons given in Section 7.1.5. The system is then placed in thermal contact with a series of heat sinks, which take a total quantity of heat Q_2' from the system, as it goes from state C to state D in Figure 7.8. The volume of the system is kept constant in the change from state C to state D, so that the values of the energy eigenvalues do not change. However, the probability P_i that the system is in its ith microstate is changed, such that the most probable value of energy decreases from the energy of representative energy eigenvalue 4 to the energy of representative energy eigenvalue 2, as shown in the change from Figure 7.9(c) to Figure 7.9(d). The temperature of the system drops in the change from state C to state D.

In order to bring the system back to the initial state A in Figure 7.8, the system is compressed reversibly and adiabatically, such that it goes along the adiabatic DA. Work W_2, equal to the area ADQP in Figure 7.8, is done *on* the system. The most probable value of energy remains equal to the energy of representative energy eigenvalue 2 in the change from Figure 7.9(d) to Figure 7.9(e). The temperature of the system increases in this reversible adiabatic compression. The net work done by the system in one complete cycle is equal to the area ABCD in Figure 7.8.

Following the convention of classical equilibrium thermodynamics, the efficiency η will be defined as the ratio of the work done by the system to the total heat taken from the heat reservoirs, so that

$$\eta = (W_1' - W_2)/Q_1 .$$

Since the system is back in its initial macrostate after one complete cycle, the net change in the ensemble average anergy of the working substance in one complete cycle is zero, so that

$$Q_1 - W_1' - Q_2' + W_2 = 0 ,$$

or

$$W_1' - W_2 = Q_1 - Q_2' .$$

Hence, the efficiency is

$$\eta = (Q_1 - Q_2')/Q_1 . \tag{7.13}$$

For a given value of Q_1, to increase the efficiency of the engine, we must make Q_2' as small as possible. To achieve this the energy eigenvalues should be as close as possible in Figure 7.9(c), so that the system has to give up as little energy in the form of heat as possible, when its most probable energy is reduced from the energy of representative energy eigenvalue 4 in Figure 7.9(c) to the energy of representative energy eigenvalue 2 in Figure

7.9(d). To achieve this, the reversible adiabatic expansion from state B to state C in Figure 7.8 should be as large as possible, as shown in Figure 7.10(a). The effect of increasing the final volume of the system in state C on the spacings of the energy eigenvalues of the system is shown schematically in Figure 7.10(b). For example, it follows from equation (7.10) that, for an ideal monatomic gas, the energy eigenvalues of all the microstates of the system are proportional to $V^{-2/3}$, and all tend to zero as the volume of the gas tends to infinity. All gases approximate to an ideal gas in the limit of infinite volume (zero pressure). If V were infinite all the energy eigenvalues in state C would be zero as shown in Figure 7.10(b) and the heat Q_2' given up by the system in going from the energy of representative energy eigenvalue 4 in state C in Figure 7.9(c) to the energy of representative energy eigenvalue 2 in state D in Figure 7.9(d) would be zero. The theoretical efficiency given by equation (7.13) would be 100%.

In the limit, when the volume V of a gaseous system was infinite, all the energy eigenvalues of the system would be zero, and there would be an

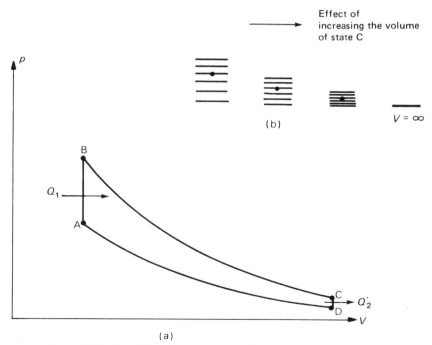

(a)

Figure 7.10—The effect of increasing the adiabatic expansion from state B to state C in Figure 7.8, (a) on the pV indicator diagram (b) on the spacing of the energy eigenvalues. In the limit, as the volume tends to infinity, all the energy eigenvalues tend to zero, as shown in (b).

infinite number of accessible microstates between $U = 0$ and an energy ΔU infinitesimally greater than zero. The ratio $\beta = (\Delta \log g / \Delta U)$ would be infinite, and the absolute temperature T would be zero, showing that if V were infinite, the system would be at the absolute zero of temperature. In practice one can never let a gas expand until its final volume is infinite, so that one cannot reach the absolute zero of temperature in a reversible adiabatic expansion. This is in agreement with the principle of the un-attainability of the absolute zero, which is one form of the third law of classical thermodynamics. Even if one could reach the state $V = \infty$, $T = 0$, one would have no practical means, such as the use of heat sinks, of controlling the change of the most probable energy of the system from the energy of representative energy eigenvalue 4 to the energy of representa-tive energy eigenvalue 2, such that the most probable energy of the system could remain equal to the energy of representative energy eigenvalue 2, during the reversible adiabatic compression from state D to state A in Figure 7.8. Summarising, since we cannot reach the state when the volume V is infinite in the reversible adiabatic expansion from state B to state C in Figure 7.8, we cannot convert all the heat, given to the system by the heat reservoirs in the change from state A to state B, into work in a cyclic process.

It is left as an exercise for the reader to go through the interpretation of the reverse cycle, in which the system goes from A to D to C to B to A in Figure 7.8. For the reverse cycle the directions of all the arrows in Figure 7.9 must be reversed. In the reverse cycle, the system acts as a refrigerator, taking in heat equal to Q_2' from the cold heat sinks and giving heat equal to Q_1 to the hot heat reservoirs each cycle. Net work is done on the system in the reverse cycle.

7.2.2* The first law of thermodynamics*

To illustrate the first law of thermodynamics, consider a slightly modi-fied form of the example shown on the pV indicator diagram in Figure 7.8. We shall compare the work that must be done on the system and the (heat) energy that must be added to the system to take the system from state D in Figure 7.8 to state B via intermediate states A and C.

The energy eigenvalue diagram for the initial equilibrium state D is shown in Figure 7.11(c). [This is the same as Figure 7.9(d).] In state D the system starts with a mean energy equal to the energy of representative energy eigenvalue 2 in Figure 7.11(c). In order to go from the initial state D to the intermediate state A, shown in Figure 7.11(d), the system must undergo a reversible adiabatic compression. The energy eigenvalues are all increased in this process, but the mean energy of the system remains equal to the energy of representative energy eigenvalue 2. The work W_2 done on the system is equal to the difference between the energies of representative energy eigenvalue 2 in Figures 7.11(c) and 7.11(d). To reach the final equilibrium state B shown in 7.11(e), heat Q_1 must now be added to the system, at constant volume, such that the mean energy of the system is raised from the energy of representative energy eigenvalue 2 in state A in Figure 7.11(d) to a mean energy equal to the energy of representative energy eigenvalue 4 in state B in Figure 7.11(e). Since the volume is constant in this process, the values of the energy eigenvalues are unchanged. The value of Q_1 is equal to the difference between the energies of representative energy eigenvalues 2 and 4 in Figures 7.11(d) and 7.11(e) respectively.

To reach the final equilibrium state B in Figure 7.8 from the initial state D, via the intermediate state C, we must start by adding heat Q_2, to the system, when it is in state D, such that the mean energy of the system is increased from the energy of representative energy eigenvalue 2 in Figure 7.11(c) to the energy of representative energy eigenvalue 4 in state C in Figure 7.11(b). Since the volume is contant, the values of the energy eigenvalues are unchanged in the process. [The heat Q_2 added to the system is numerically equal to $Q_2{'}$, the heat given up by the system in Figure 7.8.] In order to go on from intermediate state C in Figure 7.11(b) to the final equilibrium state D, shown in Figure 7.11(a), the system must undergo a reversible adiabatic compression, in which work W_1 is done on the system, to increase the energy of representative energy eigenvalue 4 in Figure 7.11(b) to its value in Figure 7.11(a). [The work W_1 done *on* the system is numerically equal to the work $W_1{'}$ done *by* the system in Figure 7.8.]

Since Figures 7.11(a) and 7.11(e) are the same, the net changes in going from initial state D in Figure 7.11(c) to the final state B in either Figure 7.11(a) or Figure 7.11(e) is the same via both the intermediate states A and C. Hence

$$W_2 + Q_1 = Q_2 + W_1 \ .$$

Going from initial state D to final state B via intermediate state A, the work W_2 is done first on the system. This is the work that must be done to raise the energy of representative energy eigenvalue 2 from its value in Figure 7.11(c) to its value in Figure 7.11(d). The heat Q_1 is the energy needed to raise the mean energy of the system from the energy of representative energy eigenvalue 2 in Figure 7.11(d) to the energy of representative energy eigenvalue 4 in Figure 7.11(e). This process is carried out when the energy eigenvalues are further apart in Figure 7.11(d), after the work W_2 is done on the system. Going via intermediate state C, heat Q_2 is added first to raise the mean energy of the system from the value of representative energy eigenvalue 2 in Figure 7.11(c) to the energy of representative energy eigenvalue 4 in Figure 7.11(b). This process is carried out when the energy eigenvalues are all closer together in figure 7.11(c) than in Figure

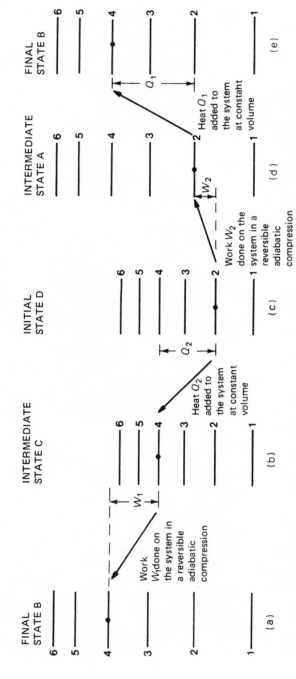

Figure 7.11—The initial state D is shown in (c). To go to the final state B via the intermediate state A, shown in (d), work W_2 is done first on the system in a reversible adiabatic compression to take the sytem to state A. Then heat Q_1 is added to the system to take it to the final state B shown in (e). An alternative way to reach state B from state D is to start by adding heat Q_2 to the system, such that the system goes to the intermediate state C shown in (b). Work W_1 is then done on the system, such that the system reaches the final state B, shown in (a). The final state B is the same as in (a) and (e). It can be seen that $(W_2 + Q_1)$ is equal to $(Q_2 + W_1)$. Less work must be done on the system, if the work W_2 is done first, before adding the heat Q_1, but in this case more heat must be added to the sytem, than if heat Q_2 were added before doing work W_1 on the system.

7.11(d), before work is done on the system. Hence Q_2 is less than Q_1. Going via intermediate state C, the work done on the system must raise the energy of representative energy eigenvalue 4 from its value in Figure 7.11(b) to its value in Figure 7.11(a), whereas, going via intermediate state A, it is only necessary to raise the energy of representative energy eigenvalue 2 during the reversible adiabatic compression. The total change of volume is the same from D to A as from C to B in Figure 7.8. As an example, consider an ideal monatomic gas system. According to equation (7.10),

$$\varepsilon_i = C_i V^{-2/3} . \tag{7.10}$$

For a given volume V, the constant C_i must be bigger for the higher energy representative energy eigenvalue 4 than for representative energy eigenvalue 2. Hence for the same decrease of volume, $\Delta \varepsilon_i$ is greater for representative energy eigenvalue 4 than representative energy eigenvalue 2. Applying equation (A5.2) of Appendix 5 to equation (7.10), it follows that the pressure of an ideal gas system is greater in state C, shown in Figure 7.11(b) than in state D, shown in Figure 7.11(c), as shown on the pV indicator diagram in Figure 7.8. The integral of $(-p \, dV)$ is greater in the reversible adiabatic compression from state C to state B than from state D to state A in Figure 7.8. This is in agreement with the conclusion that W_1 is greater than W_2 in Figure 7.11.

Summarising, it follows from Figure 7.11 that the work that must be done on the system in a reversible adiabatic compression and the heat that must be added to the system to go from state D to state B in Figure 7.8 depends on which process is performed first. It is left as an exercise for the reader to consider alternative paths from state D to state B, for example via representative energy eigenvalue 3 in Figure 7.11. By considering a series of infinitesimal processes, the reader can show that, in the general case, ΔW and ΔQ depend on how the system is taken from the initial to the final equilibrium state, but ΔU, the sum of ΔW and ΔQ is independent of the route taken, so that U is a function of the thermodynamic state of a system and dU is a perfect differential in classical equilibrium thermodynamics.

7.3* ISOTHERMAL EXPANSION*

To obtain a typical isothermal expansion, a gas is used as the system in Figure 7.1. The **diathermic partition** is used between the gaseous system and the heat reservoir. The gas is allowed to expand reversibly by releasing the piston in Figure 7.1 and adjusting the external pressure showly, such that the external pressure is always infinitesimally less than the pressure of the gas. This is a quasi-static process. The heat reservoir keeps the gas at a

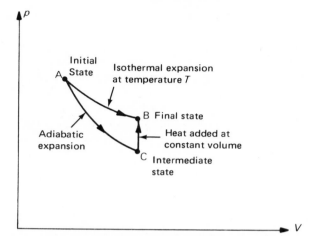

Figure 7.12—A pV indicator diagram showing an isothermal change from the initial state A to the final state B at a constant absolute temperature T. An alternative path is to go from A to the intermediate state C in a reversible adiabatic expansion; heat is then added to the system, at constant volume, such that it goes from C to B.

constant absolute temperature T, as the gas system expands isothermally from the initial equilibrium state A to the final equilibrium state B on the pV indicator diagram shown in Figure 7.12. Alternatively, one could go reversibly from the initial state A to the *intermediate* state C by using the adiabatic partition in Figure 7.1. In this reversible adiabatic expansion, the values of the energy eigenvalues of the system are decreased, but the most probable energy of the system remains equal to the energy of representative energy eigenvalue 3, as shown in the changes from Figure 7.13(a) to Figure 7.13(b). It was shown in Section 7.1.5 that the absolute temperature of the system drops in such a reversible adiabatic expansion, so that the absolute temperature of the system is less than T in the intermediate state C in Figure 7.12. The movable piston is then fixed rigidly and the adiabatic partition in Figure 7.1 is replaced by the diathermic partition. Heat flows to the gaseous system from the heat reservoir until the temperature of the system is again equal to T. In this process, the most probable value of the energy of the system increases from the energy of representative energy eigenvalue 3 to the energy of representative energy eiganvalue 4, as shown in the change from Figure 7.13(b) to Figure 7.13(c). Since the volume is constant in the second stage, the values of the energy eigenvalues are the same in Figures 7.13(b) and 7.13(c). The isothermal expansion can be represented by the direct change from Figure 7.13(a) to Figure 7.13(c). As a result of the isothermal expansion, both the P_i and the ε_i are changed in the expression $\Sigma P_i \varepsilon_i$ for the ensemble average energy of the system.

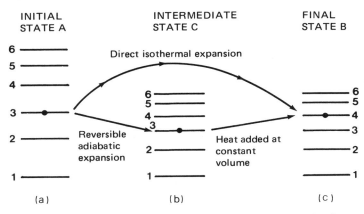

Figure 7.13—An isothermal change is illustrated in the transition from the initial state A shown in (a) to the final state B shown in (c). This isothermal change can be performed in two stages, namely a reversible adiabatic expansion to take the system from state A shown in (a) to state C, shown in (b), followed by the addition of heat, at constant volume, to take the system from state C to state B, which is shown in (c). The net result of the isothermal change from (a) to (c) is to change both the energy eigenvalues and the probability P_i that the system is in its ith microstate. In the example shown, the most probable value of energy changes from the energy of representative energy eigenvalue 3 in (a) to 4 in (c).

According to equation (5.74), in the *special* case of an ideal monatomic gas, the ensemble average energy is equal to $3NkT/2$, showing that the ensemble average energy is proportional to T, but is independent of the volume V of the ideal monatomic gas, if the temperature T is kept constant. In the *special* case of an ideal monatomic gas, the ensemble average energy represented by the position of the dot would be at the same value of energy before and after the isothermal expansion in Figures 7.13(a) and 7.13(c). For a real gas, the attractions between the molecules depend on the separations of the molecules, so that the (internal) energy of a real gas depends on its volume, even if the absolute temperature of the gas is kept constant. In the case of a real gas, the dots would not be at the same energies in Figures 7.13(a) and 7.13(c).

To determine the change of entropy in the finite isothermal change shown in Figure 7.12, consider first an infinitesimal isothermal expansion. This can be achieved by an infinitesimal adiabatic (isentropic) expansion, followed by the addition of an infinitesimal amount of heat from a heat reservoir of absolute temperature T to convert the infinitesimal adiabatic change into the equivalent of an infinitesimal isothermal change, in the way shown for a finite isothermal change in Figure 7.12. According to the thermodynamic identity, equation (3.85), which was developed from statistical mechanics in Section 3.7.5 of Chapter 3, the increase in the entropy of the system, when the infinitesimal amount of heat dQ is added to the system, at constant V and N, is

$$dS = dQ/T \ . \tag{7.14}$$

Since the entropy is constant during the infinitesimal reversible adiabatic expansion, the total change of entropy during the infinitesimal isothermal expansion is given by equation (7.14). The finite isothermal expansion between states A and B in Figure 7.12 can be divided into a series of such infinitesimal isothermal changes, all at the fixed absolute temperature T. Since T is the same for all the infinitesimal isothermal changes, integrating equation (7.14) to obtain the total change of entropy, we have

$$S_B - S_A = (1/T) \int dQ = Q/T \ . \tag{7.15}$$

where S_A and S_B are the entropies of the system in states A and B respectively in Figure 7.12, and Q is the total heat added to the system from the heat reservoir in the finite isothermal change. Since the thermodynamic identity, was developed from statistical mechanics, the temperature T in equation (7.15) is the absolute temperature defined by either equation (3.19) or equation (3.94).

7.4* THE CARNOT CYCLE*

In practice, it would be easier to use the Carnot cycle, described in detail in Appendix 2, than the ideal cycle shown in Figure 7.8. In the Carnot cycle, one only needs one hot reservoir at an absolute temperature T_1 and one cold reservoir at an absolute temperature T_2. It will be assumed that the working substance is a gas. For purposes of discussion, it will be assumed that the initial state is state A on the pV indicator diagram in Figure 7.14. The gas is allowed to expand reversibly, when it is in thermal contact with the hot reservoir, which keeps the absolute temperature of the gas equal to T_1, such that the system is taken from state A to state B in Figure 7.14. The isothermal expansion from state A to state B is illustrated by the differences in the energy eigenvalue diagrams in Figure 7.15(a) and 7.15(b). Since the volume of the gas is increased in the isothermal expansion, the values of all the energy eigenvalues are decreased. In order to keep the temperature constant, the most probable value of energy changes from the energy of representative energy eigenvalue 2 in state A in Figure 7.15(a) to the energy of representative energy eigenvalue 3 in state B in Figure 7.15(b), for the reasons given in Section 7.3. In addition to the external work done by the gas in the change from state A to state B in Figure 7.14, the gas takes in heat Q_1 from the hot reservoir. [In the special

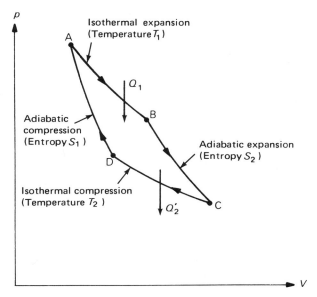

Figure 7.14—Carnot cycle pV indicator diagram.

case of an ideal monatomic gas, the most probable value for the (internal) energy of the gas would be the same in Figures 7.15(a) and 7.15(b).]

The system is then allowed to expand reversibly and *adiabatically* from state B to state C in Figure 7.14. In this reversible adiabatic expansion, the values of the energy eigenvalues are decreased further, but the probability that the system is in its ith microstate does not change, and the most probable value of energy remains equal to the energy of representative energy eigenvalue 3, as shown in the change from state B in Figure 7.15(b) to state C in Figure 7.15(c). In this reversible adiabatic expansion, the gas does external work and the absolute temperature of the gas falls from T_1 to T_2.

The system is then placed in thermal contact with the cold reservoir, which is at an absolute temperature T_2. The gas is compressed reversibly and isothermally from state C to state D in Figure 7.14. Since the volume of the gas system is decreased, the values of the energy eigenvalues are increased, but in order to keep the absolute temperature of the gas constant at T_2, the gas gives up heat Q_2' to the cold reservoir, such that in the transition from state C to state D in Figure 7.14, the most probable energy of the system changes from the energy of representative energy eigenvalue 3 in state C in Figure 7.15(c) to the energy of representative energy eigenvalue 2 in state D in Figure 7.15(d). In this isothermal compression, work is done on the system. The gas is then compressed reversibly and

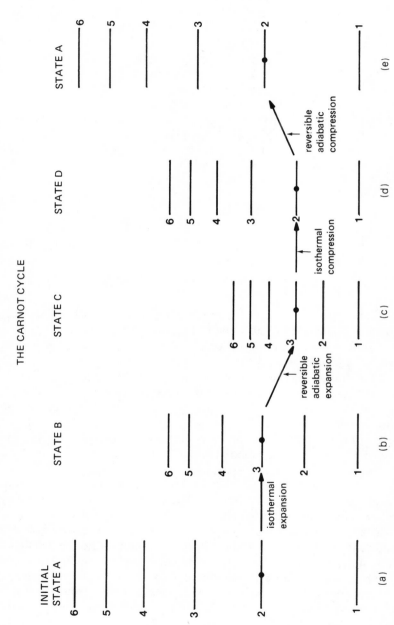

Figure 7.15—The changes in the values of the energy eigenvalues and the value of the most probable energy in a typical Carnot cycle.

adiabatically, such that it is taken from state D to state A in Figure 7.14. In this reversible adiabatic compression, the values of the energy eigenvalues are increased further, but the most probable value of energy remains equal to the energy of representative energy eigenvalue 2, as shown in the change from state D in Figure 7.15(d) to state A in Figure 7.15(e). In the transition from state D to state A in Figure 7.14 work is done on the gas, and the absolute temperature of the gas is increased from T_2 to T_1. The gas system is now back in the initial state A.

Since the net change in the ensemble average energy of the gas in a complete cycle is zero, the work done in a complete cycle is equal to $(Q_1 - Q_2')$. (Reference: equation (A2.1) of Appendix 2.) Not all the heat Q_1 taken from the hot reservoir is converted into external work. The efficiency η is defined as the ratio of the work done per cycle to the heat Q_1 taken from the hot reservoir per cycle, which is

$$\eta = (Q_1 - Q_2')/Q_1 . \tag{7.16}$$

According to equation (7.15), the change in entropy in the isothermal expansion of the gas at a constant absolute temperature T_1 between states A and B in Figure 7.14 is

$$S_2 - S_1 = Q_1/T_1 , \tag{7.17}$$

where S_1 and S_2 are the entropies of states A and B respectively in Figure 7.14. Since states A and D in Figure 7.14 are on the same adiabatic line AD, they have the same entropy S_1 for the reasons given in Section 7.1.5. Similarly, since B and C are on the same adiabatic BC, they have the same entropy S_2. Hence the difference in entropy between states D and C in Figure 7.14 is $(S_1 - S_2)$. Using equation (7.15), for the isothermal change at an absolute temperature T_2 between states C and D in Figure 7.14, we have

$$(S_1 - S_2) = -(Q_2'/T_2) . \tag{7.18}$$

Comparing equation (7.17) and (7.18), it can be seen that

$$Q_1/T_1 = Q_2'/T_2 , \quad \text{or} \quad T_1/T_2 = Q_1/Q_2' . \tag{7.19}$$

According to equation (1.9) of Section 1.7, the **thermodynamic temperature** of classical equilibrium thermodynamics is defined in terms of an ideal Carnot engine, such that the ratio of two temperatures on the thermodynamic scale is equal to the ratio of the heats absorbed and rejected by a Carnot engine working between reservoirs at these temperatures. The *absolute* temperatures in equation (7.19) were defined in Section 3.3 by the equation

$$T = (1/k)(\partial U/\partial \log g)_{V,N} \tag{3.19}$$

and in Section 3.8.2* by the equation

$$T = (1/k)(\partial U/\partial \log D)_{V,N} .\qquad(3.94)$$

It follows from equation (7.19) that the absolute temperature defined in terms of either equations (3.19) or (3.94) is the same as the thermodynamic temperature introduced into classical equilibrium thermodynamics, and defined by equation (1.9).

Substituting from equation (7.19) into equation (7.16), the efficiency of an ideal Carnot engine becomes

$$\eta = (T_1 - T_2)/T_1 .\qquad(7.20)$$

Equation (7.20) is an important restriction on the maximum efficiency of a heat engine. For example, if $T_1 = 500K$ and $T_2 = 350K$, the efficiency given by equation (7.20) is only 30%. If T_1 is raised to 1000K keeping T_2 at 350K, the efficiency is increased to 65%. In practice, it is more convenient to increase the efficiency by raising T_1 rather than by decreasing T_2 towards the absolute zero. In practice, there is always some element of irreversibility such as friction, and the efficiency given by equation (7.20) is an upper limit.

The efficiency given by equation (7.20) holds whatever the working substance. We did not have to say what the gas used was or what its properties were, when equation (7.20) was derived. This shows that the efficiency of a Carnot engine is independent of the working substance, and depends only on the absolute temperatures T_1 and T_2 of the hot and cold heat reservoirs.

7.5* AN IRREVERSBILE PROCESS*

As an example of an irreversible process, we shall consider a small solid body of mass m which falls under gravity in a viscous liquid. The small body is initially at rest at a height h above the base of a container, which is made from rigid adiabatic walls and is filled with a liquid, as shown in Figure 7.16(a). The small body of mass m is allowed to fall a vertical height h through the liquid.

Allowing for the upthrust of the liquid, the decrease in the potential energy of the falling body is $\Delta W = mgh(1 - \rho'/\rho)$, where ρ' and ρ are the densities of the liquid and small body respectively. The liquid is driven into motion by the falling body, as shown in Figure 7.16(b). After a time, the viscosity of the liquid brings the liquid to rest and the small body and liquid reach a new state of thermodynamic equilibrium, as shown in Figure 7.16(c). Experiments show that the temperature of the liquid is higher in the final equilibrium state shown in Figure 7.16(c) than in the initial state shown in Figure 7.16(a).

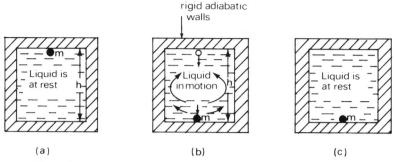

Figure 7.16—Example of an irreversible process. The mass m in (a) falls a vertical height h in the liquid. The liquid is set into motion as shown in (b). The motion of the liquid has died away in (c). The temperature of the liquid is higher in (c) than in the initial state shown in (a). The probability that the mass m will be found again at a height h in the liquid, without any external help, is negligible.

The rigid adiabatic walls keep the volume of the liquid constant, so that the energy eigenvalues of all the microstates of the liquid system are the same in the initial and the final states shown in Figure 7.16(a) and 7.16(c) respectively. The kinetic energy of the liquid in Figure 7.16(b) ends up by changing the probability P_i that the liquid system is in its ith microstate, such that the most probable energy of the liquid system, which is equal to its ensemble average energy, is increased and the temperature of the liquid is increased. In this process, the work done in setting the liquid into motion ends up with the same net result as if heat ΔQ equal to ΔW were added to the composite system made up of the small body of mass m and the liquid. If the change in the absolute temperature T of the liquid is negligible, the *increase* in the entropy of the liquid system in the change from the initial state in Figure 7.16(a) to the final equilibrium state in Figure 7.16(c) is $\Delta Q/T$, which is equal to $\Delta W/T$.

Since the entropy of the system in the final equilibrium state shown in Figure 7.16(c) is greater than the entropy in the initial state shown in Figure 7.16(a), it is extremely unlikely that the process will go spontaneously in the reverse direction, such that the mass m reaches a height h above the base of the container again. To make a rough quantitative estimate of how extremely unlikely a reverse process of this type is, it will be assumed that the amount of liquid in the container is large enough for us to treat the liquid as a heat reservoir for the falling body of mass m. According to the Boltzmann distribution, the ratio of the probability that at thermal equilibrium the mass m is at a height h in the liquid to the probability that h is zero is $\exp[-mgh(1 - \rho'/\rho)/kT]$. [Reference: equation 4.62).] If

$m = 10^{-3}$ kg, $g = 9.8$ m s^{-2}, $h = 1$ m, $\rho = 2\rho'$, $T = 300$ K, this ratio is exp(-1.18×10^{18}). This is so small that it is safe to conclude that, at thermodynamic equilibrium, one would never observe the mass m at a height of one metre, associated with a corresponding drop in the (internal) energy of the liquid. The falling of the mass m is an irreversible process.

REFERENCES

[1] Landau, L. D. and Lifshitz, E. M., *Quantum Mechanics*, Pergamon Press, 1958, page 144.
[2] Wannier, G. H., *Statistical Physics*, Wiley, 1966, page 92.

PROBLEMS

Problem 7.1

Consider the example illustrated in Figure 2.3(a) and discussed in Section 2.3. Assume that the initial energy of subsystem 1 in Figure 2.3(a) is increased from $U_1{}^0 = 5$ to $U_1{}^0 = 6$ energy units, such that $U^* = (U_1 + U_2)$ is increased from 6 to 7 energy units. Show that the total number of microstates accessible to the closed system in Figure 2.3(c) is increased from 84 to 120. Plot $P_2(U_2)$ against U_2 and P_{2i} against ε_{2i} and compare the results with the results presented in Figures 2.4(a) and 2.4(b). Show that the ensemble average energy of subsystem 2 is increased from 3 to 3.5 energy units when U^* is increased from 6 to 7 energy units. Use the equation $\Delta \bar{U}_2 = \Sigma \, \varepsilon_{2i}(\Delta P_{2i})$ to check that $\Delta \bar{U}_2$ is equal to 0.5 energy units.

Chapter 8

Negative Temperatures*

8.1* INTRODUCTION*

According to equation (3.19), in statistical mechanics the absolute temperature of a system is defined by the relation

$$T = \frac{1}{k}\left(\frac{\partial U}{\partial \log g}\right)_{V,N} = \frac{g}{k(\partial g/\partial U)_{V,N}} = \left(\frac{\partial U}{\partial S}\right)_{V,N}. \tag{8.1}$$

where $g(U, V, N)$ is the number of microstates accessible to the system, when the system has energy U, volume V and consists of N particles. The entropy S is equal to $k \log g$. Since the number of accessible microstates g must be ≥ 1, if $(\partial g/\partial U)$ is negative, that is if the number of accessible microstates decreases when the energy of the system is increased, according to equation (8.1), the absolute temperature of the system is negative. So far in this text, we have only considered normal thermodynamic systems, for which the variation of $\log g$ with U is as sketched in Figure 3.3. In these normal cases, $(\partial g/\partial U)$ is always positive, and the absolute temperature T, given by equation (8.1), is always positive. However, for systems which have an upper limit to the energy the system can have, $(\partial g/\partial U)$ can be negative for certain values of the energy U. This leads to a negative absolute temperature. This possibility will be illustrated by a simple numerical example.

8.2* NUMERICAL EXAMPLE ON NEGATIVE TEMPERATURES*

Consider two idealised systems, both of which have equally spaced non-degenerate single particle energy levels. Both systems consist of three *distinguishable* particles of spin zero. System 1 only has single particle energy levels of energy 0, 1 and 2 energy units, as shown in Figure 8.1(a). There is an upper limit of 6 energy units on the total energy system 1 can have, when all the three distinguishable particles are in the highest energy single particle state and have two energy units each. System 2 has non-

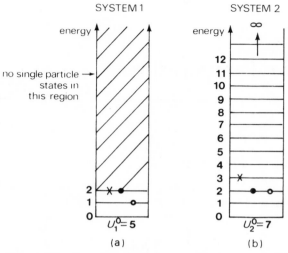

Figure 8.1—(a) Example of a system at a negative absolute temperature. There is an upper limit of 6 energy units to the energy system 1 can have, when all three distinguishable particles have two energy units each. (b) System 2 has no upper limit to the energy it can have, and is always at a positive absolute temperature.

degenerate single particle energy levels at 0, 1, 2 . . . energy units as shown in Figure 8.1(b). There is no upper limit to the total energy the three distinguishable particles in system 2 can have.

By writing out all the possibilities, the reader can check that when U_1, the total energy of system 1, is equal to 0, 1, 2, 3, 4, 5 and 6 energy units, the number of microstates $g_1(U_1)$ accessible to system 1 is 1, 3, 6, 7, 6, 3 and 1 respectively. The variation of $g_1(U_1)$ with U_1 is shown in Figure 8.2(a). It can be seen that g_1 is a maximum when U_1 is equal to three energy units. When $U_1 > 3$, the slope $(\partial g_1/\partial U_1)$ in Figure 8.2(a) is negative so that, according to equation (8.1), when $U_1 > 3$, system 1 has a negative absolute temperature. The number of microstates $g_2(U_2)$ accessible to system 2, when it has energy U_2 can be calculated using equation (2.15). These results are also plotted in Figure 8.2(a). (The variation of g_2 with U_2 is the same as the variation shown previously in Figure 2.7(a).) When there is no upper limit to the energies the particles in a system can have, the number of accessible microstates continues to increase when the energy of a system is increased, so that $(\partial g_2/\partial U_2)$ is always positive and, according to equation (8.1), the absolute temperature of system 2 is always positive.

It will be assumed that, initially, the total energy of system 1 is $U_1^0 = 5$ energy units, so that system 1 starts at a negative absolute temperature. It will be assumed that, initially, system 2 has a total energy of $U_2^0 = 7$ energy

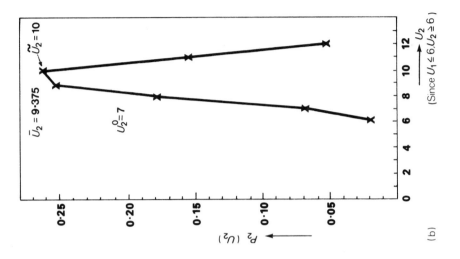

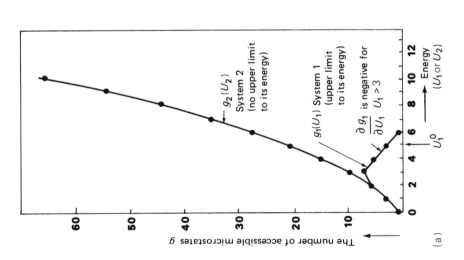

Figure 8.2—(a) The variation with energy of $g_1(U_1)$ and $g_2(U_2)$, the number of microstates accessible to systems 1 and 2 in Figure 8.1, when they have energies U_1 and U_2 respectively. System 1 has an upper limit to the energy it can have, and when $U_1 > 3$, system 1 is at a negative absolute temperature. (b) The variation with U_2 of the probability $P_2(U_2)$ that when systems 1 and 2 in Figure 8.1 are in thermal equilibrium, system 2 has energy U_2.

units. The two systems are placed in thermal contact to form two subsystems of a *closed* composite system of fixed total energy $U^* = 12$ energy units. It is being assumed that the volumes and the numbers of particles in the two subsystems are kept constant. According to the principle of equal *a priori* probabilities, which is Postulate 2 of Section 3.1, when the *closed* composite system, made up of subsystems 1 and 2, is in internal thermal equilibrium, the *closed* composite system is equally likely to be found in any one of the microstates accessible to the *composite* system, subject to the condition that

$$U_1 + U_2 = U^* = 12 \ .$$

There is an upper limit of 6 energy units to the total energy subsystem 1 can have, when each of the three distinguishable particles in subsystem 1 has 2 energy units. At thermal equilibrium U_1 can only vary from $U_1 = 0$ to $U_1 = 6$. Consequently U_2 can only vary between 6 and 12 energy units.

According to equation (2.21), the probability $P_2(U_2)$ that, at thermal equilibrium, subsystem 2 has energy U_2 is

$$P_2(U_2) = g_2(U_2)g_1(12 - U_2)/g_T \ .$$

The values of $P_2(U_2)$ are plotted in Figure 8.2(b). The total energy of subsystem 2 changes from an initial value of $U_2{}^0 = 7$ to an ensemble average value, calculated using equation (2.25), of $\bar{U}_2 = 9.375$ energy units at thermal equilibrium. The total energy of subsystem 1, which starts at a negative absolute temperature, goes down from an initial value of $U_1{}^0 = 5$ to an ensemble average of $\bar{U}_1 = 2.625$ energy units at thermal equilibrium. This result shows that, on average, energy is transferred from subsystem 1, which starts at a negative absolute temperature, to subsystem 2, which always has a positive absolute temperature, even though initially the total energy $U_1{}^0 = 5$ of subsystem 1 is less than the initial total energy $U_2{}^0 = 7$ of subsystem 2. Both subsystems 1 and 2 consist of three distinguishable particles. The separation of the non-degenerate energy levels is the same in both subsystems. The difference is that subsystem 1 has an upper limit to the energy it can have, whereas subsystem 2 does not. The reader can check that the entropy of the composite system increases from an initial value of $k \log 108$ to a value of $k \log 1512$ at thermal equilibrium.

8.3* A MACROSCOPIC EXAMPLE OF NEGATIVE TEMPERATURES—THE IDEAL SPIN SYSTEM*

Consider the macroscopic idealised spin system of N spatially separated atomic magnetic dipole moments due to N spatially separated unpaired electrons, shown previously in Figure 4.3(a). Since the potential energy of

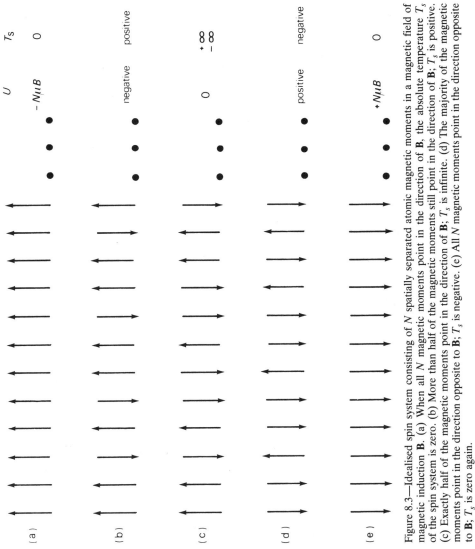

Figure 8.3—Idealised spin system consisting of N spatially separated atomic magnetic moments in a magnetic field of magnetic induction **B**. (a) When all N magnetic moments point in the direction of **B**, the absolute temperature T_s of the spin system is zero. (b) More than half of the magnetic moments point in the direction of **B**; T_s is positive. (c) Exactly half of the magnetic moments point in the direction of **B**; T_s is infinite. (d) The majority of the magnetic moments point in the direction opposite to **B**; T_s is negative. (e) All N magnetic moments point in the direction opposite to **B**; T_s is zero again.

a magnetic moment μ pointing in the direction of a magnetic field of magnetic induction **B** is $(-\mu B)$, the total potential energy of the N particle system in the case shown in Figure 8.3(a), when all the magnetic moments are pointing in the direction of the magnetic field, is $N(-\mu B)$. For this value of energy, there is only one accessible microstate, namely the microstate in which all N magnetic moments point in a direction parallel to the magnetic field, as illustrated in Figure 8.3(a). For this value of energy, since $g = 1$, $S = k \log g$ is zero. Since $(\partial \log g/\partial U)$ is infinite when $U = (-N\mu B)$, according to equation (8.1), the absolute temperature T_s of the spin system is zero in this state. To increase the energy U of the system, the orientations of some of the magnetic moments must be changed, such that they point in the direction opposite to the direction of the magnetic field, as shown in Figure 8.3(b). If n is the number of magnetic moments still pointing parallel to the magnetic field, the number pointing antiparallel to the magnetic field is $(N - n)$ and the total energy of the N particle spin system is

$$U = (N - n)\mu B + n(-\mu B) = (N - 2n)\mu B \ . \qquad (8.2)$$

The number of accessible microstates g, when n magnetic moments are pointing parallel to the magnetic field, is equal to the number of ways n magnetic moments can be selected from a total of N magnetic moments. According to equation (A1.3) of Appendix 1

$$g = N!/[n! \ (N - n)!] \ . \qquad (8.3)$$

The variation with U of $\log g$, which is equal to S/k, for the special case when N is equal to 50, is shown in Figure 8.4. The entropy S increases from zero, when $n = N$ and U is equal to $(-N\mu B)$, to reach a maximum when $n = N/2$, which is when exactly half of the magnetic moments are pointing in the direction of the magnetic field. When U is between $(-N\mu B)$ and zero, that is when $n > N/2$, as shown in Figure 8.3(b), $(\partial \log g/\partial U)$ and $(\partial S/\partial U)$ are positive in Figure 8.4, and, according to equation (8.1), the absolute temperature T_s of the spin system is positive. When $n = N/2$, as shown in Figure 8.3(c), $U = 0$ and $(\partial \log g/\partial U)$ is zero in Figure 8.4. According to equation (8.1) the absolute temperature T_s of the spin system is infinite in this state. When $n < N/2$, that is when over half the magnetic moments are pointing in a direction opposite to the magnetic field, as shown in Figure 8.3(d), $(\partial \log g/\partial U)$ is negative in Figure 8.4 and according to equation (8.1), the absolute temperature T_s of the spin system is negative. As U increases from zero to $+N\mu B$, more and more of the magnetic moments must point in the direction opposite to the direction of the magnetic field, $(\partial \log g/\partial U)$ becomes more and more negative, and according to equation (8.1) the absolute temperature T_s of the spin system varies through negative values from $-\infty$ towards zero. When all the magnetic

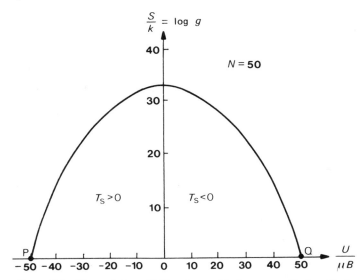

Figure 8.4—The variation of $S/k = \log g$ with the energy U of an ideal-ised spin system consisting of $N = 50$ atomic magnetic moments in a magentic field **B**. When n, the number of magnetic moments pointing parallel to **B**, is equal to 50, $U = (-50\,\mu B)$, $g = 1$ and $S = 0$. As n decreases from 50 to 25, U, gets less negative, S increases and T_s is positive. When $n < 25$, U is positive, S decreases with increasing U and T_s is negative.

moments are pointing in the direction opposite to the magnetic field, as shown in Figure 8.3(e), $U = N\mu B$, and there is only one microstate, so that $g = 1$ and the entropy $S = k \log g$ is zero.

Summarising, as U is increased in Figures 8.4 from $(-N\mu B)$ through zero to $(+N\mu B)$, the absolute temperature T_s of the spin system varies from $0, \ldots +\infty, -\infty \ldots 0$. The temperature parameter β_s, which is given by the slope of the curve in Figure 8.4, varies from $+\infty \ldots 0 \ldots -\infty$, as U is increased from $(-N\mu B) \ldots 0 \ldots (+N\mu B)$.

8.4* TRANSFER OF HEAT*

Consider the example of thermal interaction illustrated in Figures 8.5(a) and 8.5(b). Subsystem 1 has an upper limit to the energy it can have. It will be assumed that in the initial state shown in Figure 8.5(a), the energy of subsystem 1 is such that subsystem 1 is initially at a negative absolute temperature, so that, according to equation (8.1), $(\partial U_1/\partial S_1)$, and its reciprocal $(\partial S_1/\partial U_1)$ are both negative. Subsystem 2 is a normal thermodynamic system, which is always at a positive absolute temper-ature so that $(\partial U_2/\partial S_2)$ and its reciprocal $(\partial S_2/\partial U_2)$ are always positive.

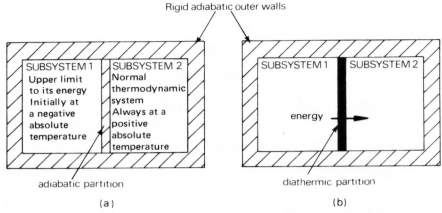

Figure 8.5—(a) Subsystem 1 has an upper limit to the energy it can have, whereas subsystem 2 is a normal thermodynamic system, which has no upper limit to the energy it can have and is always at a positive absolute temperature. In the initial state, with the adiabatic partition in place, subsystem 1 starts at a negative absolute temperature. (b) The adiabatic partition is replaced by a diathermic partition. In the approach to thermal equilibrium, heat is transferred, on average, from sybsystem 1, which is initially at a negative absolute temperature, to subsystem 2 which is always at a positive absolute temperature.

The adiabatic partition in Figure 8.5(a) is changed to a diathermic partition, as shown in Figure 8.5(b). According to equation (3.36), as the *closed* composite system shown in Figure 8.5(b) approaches thermal equilibrium, on average $dS_1 + dS_2 > 0$. Expanding dS_1 and dS_2 using equation (A1.32) of Appendix 1, for constant V_1, V_2, N_1 and N_2, and substituting in the above inequality, we have

$$\left(\frac{\partial S_1}{\partial U_1}\right)_{V_1,N_1} dU_1 + \left(\frac{\partial S_2}{\partial U_2}\right)_{V_2,N_2} dU_2 > 0 .$$

Since $(U_1 + U_2)$ is constant dU_1 is equal to $(-dU_2)$.

Hence $$\left[\left(\frac{\partial S_2}{\partial U_2}\right)_{V_2,N_2} - \left(\frac{\partial S_1}{\partial U_1}\right)_{V_1,N_1}\right] dU_2 > 0 . \qquad (8.4)$$

According to equation (8.4), since $(\partial S_2/\partial U_2)$ is always positive, when $(\partial S_1/\partial U_1)$ is negative dU_2 must be positive and, on average, energy will be transferred from subsystem 1, which is initially at a negative absolute temperature, to subsystem 2 which is always at a positive absolute temperature.

In the example of the ideal spin system shown in Figure 8.3(d), when more than half the magnetic moments point in the direction opposite to the magnetic field direction, the spin system has a negative absolute temperature. If the spin system shown in Figure 8.3(d) is to reach thermal equilibrium with the rest of the crystal lattice, which is always at a positive absolute temperature, some of the magnetic moments must change their orientations from antiparallel to parallel to the magnetic field such that, at thermal equilibrium, when the Boltzmann distribution, equation (4.21), is applicable

$$N_1/N_0 = \exp(-2\mu B/kT) \ . \tag{8.5}$$

In equation (8.5), N_1 is the number of magnetic moments which point in the direction opposite to the magnetic field and have potential energy $(+\mu B)$ each, and N_0 is the number of magnetic moments which point in the direction of the magnetic field and have potential energy $(-\mu B)$ each. The temperature T in equation (8.5) is the *final* equilibrium temperature of the crystal lattice and spin subsystem. Since there is no upper limit to the energy the crystal lattice can have, it will absorb sufficient energy from the spin subsystem for them both to reach thermal equilibrium at a positive absolute temperature T, when, according to equation (8.5), N_1 will be less than N_0. When some of the magnetic moments change their orientations from antiparallel to parallel to the magnetic field during the approach to thermal equilibrium, each change of orientation gives an energy $2\mu B$ to the rest of the crystal lattice. This illustrates how, on average, during the approach to thermal equilibrium, heat is transferred from the system which is initially at a negative absolute temperature, namely the spin system, to the system at a positive absolute temperature, namely the crystal lattice.

In classical equilibrium thermodynamics, it is assumed that when two systems are placed in thermal contact, heat flows from the hot system to the cold system. If this criterion were adopted, one would say that a system at a negative absolute temperature was hotter than a system at a positive absolute temperature.

If in Figure 8.5(b), both subsystems 1 and 2 started at negative absolute temperatures, for dU_2 to be positive in equation (8.4), so that on average heat flowed from subsystem 1 to subsystem 2, $(\partial S_1/\partial U_1)$ would have to be more negative than $(\partial S_2/\partial U_2)$. Using equation (8.1) this requires that $|\ 1/T_1\ | > |\ 1/T_2\ |$, or $|\ T_1\ | < |\ T_2\ |$. Hence, for heat to flow on average from subsystem 1 to subsystem 2, when they both started at negative absolute temperatures, T_1 would have to be less negative than T_2. For example, if T_2 were $(-300K)$ and T_1 were $(-100K)$, heat would flow, on average, from subsystem 1 to subsystem 2. The scale of increasing hotness is from $T = 0$ through increasing positive absolute temperatures to $+\infty$, then immediately to $-\infty$ and then through less and less negative absolute temperatures towards zero.

8.5* POPULATION INVERSION*

Consider an isolated spin system consisting of a large number N of spatially separated atomic magnetic moments in a magnetic field of magnetic induction **B**. Let N_1 and N_0 be the number of magnetic moments in the single particle states of energy $+\mu B$ and $(-\mu B)$ respectively. If N_1 is greater than N_0, we have what is known as a *population inversion*. The action of a laser depends on a population inversion. It is sometimes said that, when there is a population inversion, the spin system has a negative absolute temperature. It will be assumed that $(N - 1)$ of the magnetic moments in the spin system act as a heat reservoir of absolute temperature T_s for the other magnetic moment. Applying the Boltzmann distribution, equation (4.6), to this one magnetic moment, we find that the ratio of the probabilities P_1 and P_0 that the chosen magnetic moment is in the states of potential energy $+\mu B$ and $(-\mu B)$ respectively is

$$P_1/P_0 = \exp(-2\mu B/kT_s) \qquad (= N_1/N_0) \ . \qquad (8.6)$$

The average value of the ratio N_1/N_0 at thermal equilibrium is equal to P_1/P_0. If $N_1 > N_0$, that is if over half the spins are in the higher energy single particle state, which is true for the case shown in Figure 8.3(d), the spin absolute temperature T_s in equation (8.6) must be negative.

In practice, a spin system is never completely isolated, but forms part of a paramagnetic crystal. For N_1 to remain greater than N_0, the magnetic moments in the spin system must interact between themselves, but not with the rest of the crystal lattice, which is always at a positive absolute temperature. In practice, there is always some interaction between the magnetic moments of the spin subsystem and the rest of the crystal lattice, and N_1 will decrease gradually (unless it is sustained by some external means such as optical pumping in the case of a ruby laser), until the spin subsystem and crystal lattice reach thermal equilibrium at a positive absolute temperature T. Equation (8.5) is then applicable and N_1 will be less than N_0.

8.6* EXPERIMENTAL DEMONSTRATION OF NEGATIVE TEMPERATURES*

In a paramagnetic salt, the interactions between the atomic magnetic moments (the unpaired electrons) and the crystal lattice are too strong for a state of negative temperature in the spin subsystem to last long enough to be detected experimentally. In the experiment performed by Purcell and Pound [1] in 1951, the *nuclei* of the lithium atoms in lithium fluoride were aligned in an external applied magnetic field. The initial state, when most of the nuclear magnetic moments pointed in the direction of the magnetic field, was close to the point P in Figure 8.4. When the direction of the applied magnetic field was reversed, the nuclear spin subsystem was close to the point Q on the entropy–energy curve in Figure 8.4. This corresponded to a region where $(\partial \log g / \partial U)$ was negative, corresponding to a negative absolute absolute temperature for the nuclear spin subsystem. The interactions between the nuclear magnetic moments and the lattice were weak enough in the case of lithium fluoride for the state of negative temperatures in the nuclear spin subsystem to persist for several minutes.

Purcell and Pound [1] used nuclear magnetic resonance techniques (NMR) to establish the existence of the negative absolute temperature of the nuclear spin subsystem. At positive absolute temperatures, the $^7_3 \text{Li}$ nuclei in the nuclear spin subsystem absorbed energy from an incident electromagnetic signal of frequency ε/h, where ε was the energy gap between the states in which a nuclear magnetic moment pointed parallel and antiparallel to the magnetic field. Purcell and Pound [1] found that for a lithium nuclear spin subsystem at a negative absolute temperature, corresponding to a population inversion, the amplitude of the resonant electromagnetic signal was increased by stimulated emission from the upper levels. For fuller details the reader is referred to the original paper of Purcell and Pound [1] and to Andrew [2]. A survey of negative absolute temperature is given by Proctor [3] and Zemansky [4].

REFERENCES

[1] Purcell, E. M. and Pound, R. V., *Phys. Rev.*, (1951), Vol. **81**, page 279.
[2] Andrew, E. R., *Nuclear Magnetic Resonance*, Cambridge University Press, 1955, page 179.
[3] Proctor, W. G., *Scientific American*, August 1978, page 78.
[4] Zemansky, M. W., *Heat and Thermodynamics*, 5th Edition, McGraw–Hill, 1957, page 487.

Chapter 9

Planck's Radiation Law

9.1 INTRODUCTION

The quantum theory of radiation was first introduced by Planck in 1900 to interpret the experimentally determined spectrum of black body radiation. (An account of the historical development of the theory is given by ter Haar [1].) The wave model of radiation will be used to develop Planck's law in Section 9.2. An introduction to the photon model will be given in Section 9.4. The photon will be treated as a boson of spin 1 and the photon gas treated as a special case of the ideal boson gas later in Section 12.5.4* of Chapter 12. This last approach is generally the best approach for the advanced reader to adopt.

9.2 PLANCK'S LAW (WAVE MODEL)

Consider the electromagnetic (heat) radiation inside the cavity inside the heat reservoir of constant absolute temperature T, shown in Figure 9.1. The radiation inside such a cavity is sometimes called black body radiation. The radiation will be treated as a wave motion, which gives standing wave oscillations inside the cavity. These are the normal modes. According to equation (A6.13) of Appendix 6, the number of independent standing electromagnetic wave solutions (normal modes) per unit volume, which have angular frequencies in the range ω to $(\omega + d\omega)$ is

$$D(\omega)\, d\omega = \omega^2\, d\omega / \pi^2 c^3 \ . \tag{9.1}$$

In the wave model, it is assumed that each of these independent standing wave oscillations (normal modes) can be treated as an independent harmonic oscillator of natural angular frequency ω in thermal equilibrium with a heat reservoir of absolute temperature T. According to equation (4.37), the mean energy of such a harmonic oscillator in thermal equilibrium with a heat reservoir is

$$\bar{\varepsilon} = \tfrac{1}{2}\hbar\omega + \hbar\omega / [\exp(\beta\hbar\omega) - 1] \tag{9.2}$$

In order to obtain the formula given by Planck, it is conventional to ignore the temperature independent zero point energy term $\frac{1}{2}\hbar\omega$.

The radiant energy per unit volume in the angular frequency range ω to $(\omega + d\omega)$, denoted $u_\omega\, d\omega$, is equal to the product of the number of harmonic oscillators (standing wave solutions) in this angular frequency range and the mean energy of each harmonic oscillator. Using equations (9.1) and (9.2) and ignoring the zero point energy we have

$$u_\omega\, d\omega = \frac{\hbar\omega^3\, d\omega}{\pi^2 c^3[\exp(\hbar\omega/kT) - 1]}\,. \tag{9.3}$$

Alternative forms of equation (9.3) are

$$u_\nu\, d\nu = \frac{8\pi h\nu^3\, d\nu}{c^3[\exp(h\nu/kT) - 1]}\,. \tag{9.4}$$

where $u_\nu\, d\nu$ is the radiant energy per unit volume in the frequency range ν to $(\nu + d\nu)$,

and

$$u_\lambda\, d\lambda = \frac{8\pi\, d\lambda}{\lambda^5}\left\{\frac{hc}{[\exp(hc/\lambda kT) - 1]}\right\}\,, \tag{9.5}$$

where $u_\lambda\, d\lambda$ is the radiant energy per unit volume in the wavelength range λ

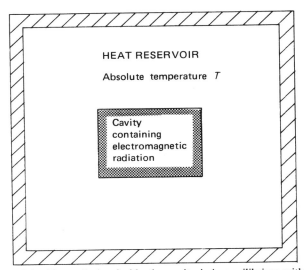

HEAT RESERVOIR

Absolute temperature T

Cavity containing electromagnetic radiation

Figure 9.1—The radiation inside the cavity is in equilibrium with the heat reservoir.

to $(\lambda + d\lambda)$. The absolute temperature T in equations (9.3), (9.4) and (9.5) is the absolute temperature of the heat reservoir surrounding the cavity in Figure 9.1, and $h = 2\pi\hbar$ is Planck's constant.

Equations (9.3), (9.4) and (9.5) are in agreement with the experimental results. In optical type experiments, it is the wavelength of the radiation which is generally measured, and it is convenient to use the variation of u_λ with λ. In radioastronomy, one generally measures the energy of the radiation in a particular frequency bandwidth, and it is convenient to compare the results with either equation (9.3) or equation (9.4).

9.3 DISCUSSION OF PLANCK'S RADIATION LAW

The variation of u_λ with λ is shown in Figure 9.2(a), and the variation of u_ν with ν is shown in Figure 9.2(b).

As the wavelength λ tends to infinity, corresponding to the extreme infra-red, $(hc/\lambda kT)$ tends to zero and

$$\exp(hc/\lambda kT) - 1 \simeq 1 + (hc/\lambda kT) - 1 \simeq (hc/\lambda kT) \ .$$

Hence, as λ tends to infinity, equation (9.5) becomes

$$u_\lambda \, d\lambda \simeq (8\pi \, d\lambda/\lambda^4)kT \ . \tag{9.6}$$

This is the Rayleigh–Jeans equation. In the original Rayleigh–Jeans theory it was assumed that the mean energy of each of the harmonic oscillators was equal to the classical value of kT, given by the equipartition theorem developed in Section 4.10.4.

As the wavelength λ tends to zero, corresponding to the extreme ultraviolet, $(hc/\lambda kT)$ tends to infinity, $\exp(hc/\lambda kT) \gg 1$, and equation (9.5) becomes

$$u_\lambda \, d\lambda \simeq (8\pi hc/\lambda^5)\exp(-hc/\lambda kT)d\lambda \ . \tag{9.7}$$

This is Wien's law. As λ tends to zero, the exponential term dominates the right hand side of equation (9.7) and u_λ tends to zero, as shown in Figure 9.2(a).

The total energy per unit volume u in the cavity in Figure 9.1 can be obtained by integrating equation (9.4) over all frequencies. We have

$$u = \int_0^\infty u_\nu \, d\nu = \frac{8\pi h}{c^3} \int_0^\infty \frac{\nu^3 \, d\nu}{[\exp(h\nu/kT) - 1]} \ .$$

Let $\qquad x = h\nu/kT, \quad$ so that $\quad \nu^3 \, d\nu = (kT/h)^4 x^3 \, dx \ .$

At an absolute temperature T, we have

$$u = \frac{8\pi k^4 T^4}{h^3 c^3} \int_0^\infty \frac{x^3 \, dx}{[\exp(x) - 1]} \ . \tag{9.8}$$

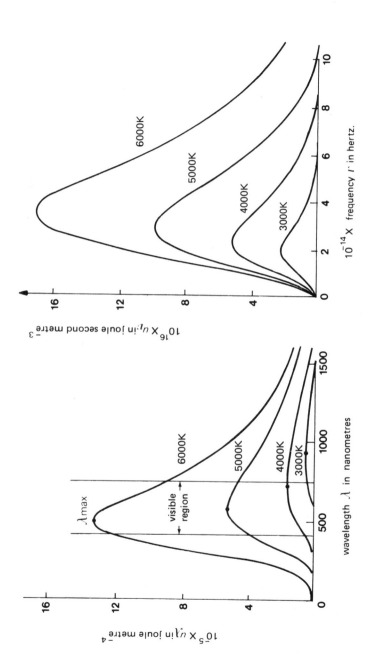

(b)

(a)

Figure 9.2—Planck's radiation law. (a) The variation of u_λ with λ is shown for various values of the absolute temperature T of the heat reservoir in Figure 9.1; u_λ tends to zero as λ tends to zero and as λ tends to infinity. The area under the curves is proportional to T^4. The values of the wavelength λ_{max} at which u_λ is a maximum lie on the rectangular hyperbola $\lambda_{max} T$ is a constant. (b) The variation of u_ν with frequency ν. (Adapted with permission from *Thermal Physics* by P. C. Riedi; published by The Macmillan Press Limited.)

Since the definite integral in equation (9.8) is a number, it can be seen immediately that the total energy u per unit volume is proportional to T^4. The definite integral is a standard integral, which has the numerical value of $\pi^4/15$. (Reference: Zemansky [2a].) Hence

$$u = (8\pi^5 k^4/15h^3c^3)T^4 = aT^4 \ . \tag{9.9}$$

Substituting numerical values for k, h and c, we find that

$$u = 7.56 \times 10^{-16}T^4 \, \text{J m}^{-3} \ .$$

Equation (9.9) gives the total radiant energy per unit volume inside the cavity, which is surrounded by a heat reservoir of constant absolute temperature T in Figure 9.1. The total radiant energy is equal to the area under the curve of u_λ against λ in Figure 9.2(a). The higher the temperature of the enclosure the larger the area under the curve of u_λ against λ, as shown in Figure 9.2(a). According to Stefan's law, the energy e emitted per unit area per second by a black body is equal to σT^4, where σ is Stefan's constant. Since the radiation inside the cavity in Figure 9.1 is continually absorbed and emitted by the inner surface of the heat reservoir surrounding the cavity, it is possible to relate the energy of radiation per unit volume, which is always moving at the speed of light c, to the energy emitted per unit area of the surface per unit time. Using a straightforward geometrical calculation it can be shown that σ is equal to $(ca/4)$. (Reference: Roberts and Miller [3].) Hence *Stefan's law* is

$$e = \sigma T^4 = (2\pi^5 k^4/15h^3c^2)T^4 \ . \tag{9.10}$$

Substituting numerical values for k, h and c, we find that σ is equal to $5.67 \times 10^{-8} \, \text{W m}^{-2} \, \text{K}^{-4}$. This is in agreement with the experimental results.

The wavelength λ_{max} at which u_λ has its maximum value is that wavelength at which the denominator in equation (9.5) is a minimum. This is when

$$(\text{d}/\text{d}\lambda)[\lambda^5\{\exp(hc/\lambda kT) - 1\}] = 0 \ .$$

This reduces to $$[1 - \exp(-y)] = y/5 \ , \tag{9.11}$$

where $$y = hc/\lambda kT \ .$$

Equation (9.11) is a transcendental equation, which can be solved graphically. The numerical solution is $y = 4.96$ (Reference: Zemansky [2(b)].) The reader can check this by substituting $y = 4.96$ into equation (9.11). Hence

$$\lambda_{\text{max}}T = hc/4.96k \tag{9.12}$$

This is *Wien's displacement law*. Substituting for *h*, *c* and *k* we find that

$$\lambda_{max}T = 2.9 \times 10^{-3} \text{ m K} \ .$$

Alternatively, the experimental values of Stefan's constant σ and $\lambda_{max}T$ can be used to estimate the values of *h* and *k* if *c* is known.

According to equation (9.12), as the temperature of the heat reservoir in Figure 9.1 is increased, the wavelength at which the curve of u_λ against λ is a maximum is displaced to lower wavelengths, as shown in Figure 9.2(a). The variation of λ_{max} with *T* is a rectangular hyperbola. Summarising, as the temperature of the heat reservoir in Figure 9.1 is increased, the total radiant energy per unit volume inside the cavity is increased in pro-

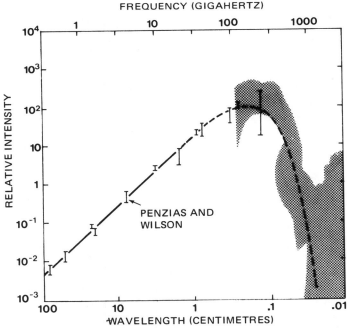

Figure 9.3—Spectrum of the cosmic background radiation. The observed energy spectrum follows the energy spectrum of the radiation in equilibrium with a black body of temperature 3K. Notice the non linear scales on the abscissa. The measurement of Penzias and Wilson at 0.075 m wavelength (corresponding to 4×10^9 Hz) is shown. Most of the subsequent measurements, which were also done at single wavelengths, are indicated by vertical bars. Recently P. L. Richards and his co-workers at the University of California at Berkeley have measured the higher frequency portion of the curve with a wide-band technique, obtaining the results indicated by the shaded area. (Reproduced with permission from an article by R. A. Muller in *Scientific American*. © 1978 Scientific American Inc. All rights reserved.)

portion to T^4 and the wavelength of the peak of the curve relating u_λ to λ decreases as $1/T$.

If T is equal to 6000K, corresponding to the temperature of the photosphere of the Sun, λ_{max} is equal to 483 nm, which is in the visible region of the electromagnetic spectrum. If T is equal to 300K, corresponding to the surface temperature of the Earth, λ_{max} is approximately equal to 10 µm, which is in the infra-red region of the spectrum.

In 1964 Penzias and Wilson made the first observation of the cosmic background (electromagnetic) radiation, when they were investigating radio signals from earth satellites at 0.075 m wavelength. This cosmic background radiation has since been confirmed at other wavelengths. The experimental results are shown in Figure 9.3. It can be seen that the results fit in with the spectrum of black body radiation, predicted by Planck's law for the radiation inside a cavity in a heat reservoir of absolute temperature of about 3K. The cosmic background radiation fills the whole of space. Since the spectrum of the cosmic background radiation fits in with Planck's radiation law, the radiation must have been in equilibrium with matter at some time in the past. It is believed that the temperature of the cosmic background radiation has decreased to approximately 3K during the expansion of the Universe. The existence of the cosmic background radiation is now taken to be evidence in favour of the big bang theory of the origin of the Universe. A reader interested in the cosmological significance of the cosmic background radiation is referred to Riedi [4], Webster [5] and to Weinberg [6].

9.4 PLANCK'S LAW (INTRODUCTION TO THE PHOTON MODEL)

In this section a brief introductory survey of the photon model is developed, largely by analogy with the wave model given in Section 9.2. These new ideas will be confirmed, when the photon gas is treated as a special case of the ideal boson gas in Section 12.5.4* of Chapter 12.

Consider again the radiation inside the cavity, which is inside the heat reservoir shown in Figure 9.1. In Section 9.2, each normal mode (standing wave solution) of angular frequency ω was treated as a harmonic oscillator which could have the energy eigenvalues $n\hbar\omega$ where $n = 0, 1, 2, \ldots$. (The zero point energy $\frac{1}{2}\hbar\omega$ is being ignored.) Just as in the case of a particle in a box, in this section we shall treat each independent stationary wave solution of the wave equation inside a box as a single particle (photon) state. [Reference: Section A6.1.5 of Appendix 6.] If the angular frequency of a stationary wave solution is ω, the energy of a photon in that single particle state is $\hbar\omega$. Since the photon is a boson, there is no restriction on the number of photons that can be in each single particle state. In the photon model, it is reasonable to assume that, if the total energy associated with a single particle state of angular frequency ω is $n\hbar\omega$, there are n photons each of energy $\hbar\omega$ in that single particle state. When equation (9.2) was derived in Section 4.7 of Chapter 4, it was assumed that the probability that the energy of the harmonic oscillator was $n\hbar\omega$ was equal to $\exp(-n\hbar\omega/kT)/Z$, where T was the absolute temperature of the heat reservoir and Z was the partition function.† By analogy, it is reasonable to assume that the probability that there are n photons, each of energy $\hbar\omega$ in a single particle state of (single particle) energy eigenvalue $\hbar\omega$, giving a total energy of $n\hbar\omega$ for the single particle state, is proportional to $\exp(-n\hbar\omega/kT)$. Photons are continually absorbed and emitted by the surface of the walls of the cavity in the heat reservoir in Figure 9.1. According to equation (A.1.7), the mean number of photons in the single particle (photon) state of (single particle) energy eigenvalue $\hbar\omega$ is given by

$$\bar{n} = \Sigma\, n\, \exp(-\beta n\hbar\omega)/\Sigma\, \exp(-\beta n\hbar\omega) = \Sigma\, nx^n/\Sigma\, x^n \ . \qquad (9.13)$$

where x is equal to $\exp(-\beta\hbar\omega)$ and the summation is over all values of n from 0 to ∞. The reader can check by expanding using the binomial theorem, that

$$\Sigma\, x^n = (1 - x)^{-1} \ .$$

We also have $x(d/dx)\, \Sigma\, x^n = x\, \Sigma\, nx^{(n-1)} = \Sigma\, nx^n \ .$

Hence $\Sigma\, nx^n = x(d/dx)[1/(1 - x)] = x/(1 - x)^2 \ . \qquad (9.14)$

Substituting in equation (9.13), we find that

$$\bar{n} = 1/[\exp(\beta\hbar\omega) - 1]. \qquad (9.15)$$

The total energy per unit volume of all the photons having energies between $\hbar\omega$ and $\hbar(\omega + d\omega)$ is equal to the number of single particle states (normal modes) per unit volume in the angular frequency range ω to $(\omega + d\omega)$, times the mean number of photons in each single particle state, times $\hbar\omega$, the energy of each photon. According to equation (A6.19) of Appendix 6, the number of single particle (photon) states, per unit volume, in which the photon has angular frequencies in the range ω to $(\omega + d\omega)$, is

$$D(\omega)\, d\omega = \omega^2\, d\omega/\pi^2 c^3 \ .$$

Hence $u_\omega\, d\omega = \dfrac{\omega^2\, d\omega}{\pi^2 c^3} \times \left(\dfrac{1}{e^{\beta\hbar\omega} - 1}\right) \times \hbar\omega = \dfrac{\hbar\omega^3\, d\omega}{\pi^2 c^3(e^{\beta\hbar\omega} - 1)} \qquad (9.16)$

This is the same as equation (9.3), and is Planck's law.

The mean number of photons per unit volume having angular frequen-

---

†This is the **Boltzmann distribution**. In general, when the number of particles in a small system in thermodynamic equilibrium with a heat and particle reservoir varies, one should use the grand canonical distribution, given by equation (11.8) of Chapter 11. It will be shown in Section 12.5* of Chapter 12 that the grand canonical distribution reduces to the Boltzmann distribution in the case of a photon gas.

cies between ω and $(\omega + d\omega)$, which will be denoted by $N(\omega)\,d\omega$, is equal to the product of the number of single particle (photon) states in the angular frequency range ω to $(\omega + d\omega)$ and the mean number of photons in each of these states. Using equation (A6.19) and (9.15), we have

$$N(\omega)\,d\omega = \left(\frac{\omega^2}{\pi^2 c^3}\right)\left(\frac{1}{[\exp(\beta\hbar\omega) - 1]}\right) d\omega .$$

The mean total number of photons per unit volume is

$$\bar{N} = \int_0^\infty N(\omega)\,d\omega = \frac{1}{\pi^2 c^3}\int_0^\infty \frac{\omega^2\,d\omega}{[e^{\beta\hbar\omega} - 1]} .$$

Using $x = \beta\hbar\omega$, we have

$$\bar{N} = \frac{1}{\pi^2 c^3}\left(\frac{kT}{\hbar}\right)^3 \int_0^\infty \frac{x^2\,dx}{(e^x - 1)} .$$

The definite integral is a standard integral which has the numerical value of 2.404 (Reference: Landau and Lifshitz [7].) Hence,

$$\bar{N} = \frac{2.404}{\pi^2 c^3}\left(\frac{kT}{\hbar}\right)^3 = \frac{2.404 \times 8\pi}{c^3}\left(\frac{kT}{h}\right)^3 = 2.02 \times 10^7 T^3\,\text{m}^{-3} . \qquad (9.17)$$

For $T = 273K$, the mean number of photons per unit volume is $\bar{N} \simeq 4 \times 10^{14}$ photons per metre^3. At the same temperature, an ideal gas would contain about 3×10^{25} molecules per metre^3 at atmospheric pressure. According to equation (9.17) the mean number of photons per metre^3 is proportional to T^3. Dividing equation (9.9) by (9.17), we find that the mean energy of the photons in the cavity in Figure 9.1 is

$$\langle \hbar\omega \rangle = u/\bar{N} = \left(\frac{\pi^4}{15 \times 2.404}\right)kT = 2.7kT . \qquad (9.18)$$

According to equation (9.18) the mean energy of the photons is very roughly of the order of kT.

9.5* THE PHOTON GAS*

It will be assumed that the photons are inside a cavity in a heat reservoir of absolute temperature T, as shown in Figure 9.1. Photons are continually absorbed and emitted by the walls of the cavity inside the heat reservoir in Figure 9.1, so that total number of photons in the photon gas is not constant. The photons do not interact with each other.† The zero point energy will again be ignored. Since the photons are bosons, there is no restriction on the number of photons that can be in any single particle

---

†The extremely small quantum electrodynamical effect which leads to a small photon-photon interaction can be neglected. (Reference: Jauch and Rohrlich [8].)

(photon) state, so that the number of photons in any single particle (photon) state can vary from zero to infinity. The partition function of the photon gas is

$$Z = \Sigma \exp[-\beta(n_1\hbar\omega_1 + n_2\hbar\omega_2 + \ldots)] \tag{9.19}$$

where $n_1, n_2, \ldots$ are the numbers of photons in single particle (photon) states, which have (single particle) energy eigenvalues $\hbar\omega_1, \hbar\omega_2, \ldots$ respectively. (The use of the partition function approach will be justified in Section 12.5.4* of Chapter 12.) All the numbers $n_1, n_2, \ldots$ in equation (9.19) can vary independently from 0 to ∞. Hence the partition function, given by equation (9.19), can be factorised to give

$$Z = \sum_{n_1=0}^{\infty} \exp(-\beta n_1 \hbar\omega_1)[\Sigma \exp\{-\beta(n_2\hbar\omega_2 + n_3\hbar\omega_3 + \ldots)\}] \ .$$

Summing over $n_2, n_3, \ldots$ in turn, we have

$$Z = \left(\sum_{n_1=0}^{\infty} e^{-\beta n_1 \hbar\omega_1} \right)\left(\sum_{n_2=0}^{\infty} e^{-\beta n_2 \hbar\omega_2} \right) \ldots \tag{9.20}$$

The reader can check, by using the binomial theorem, that Σx^n is equal to $(1 - x)^{-1}$. Putting $x = \exp(-\beta\hbar\omega_i)$ we find that

$$\sum_{n_i=0}^{\infty} \exp(-\beta n_i \hbar\omega_i) = [1 - \exp(-\beta\hbar\omega_i)]^{-1} \ .$$

Substituting into equation (9.20) and taking the natural logarithm of both sides, we obtain

$$\log Z = -\Sigma \log[1 - \exp(-\beta\hbar\omega_i)] \ . \tag{9.21}$$

The summation is over all the single particle (photon) states. According to equation (A6.19) of Appendix 6, the number of single particle (photon) states $D(\omega)d\omega$ in a volume V having angular frequencies in the range ω to $(\omega + d\omega)$ is

$$D(\omega) \, d\omega = V\omega^2 \, d\omega/\pi^2 c^3.$$

Replacing the sum by an integral in equation (9.21), we obtain

$$\log Z = -\frac{V}{\pi^2 c^3}\int_0^{\infty} \omega^2 \log[1 - \exp(-\beta\hbar\omega)] \, d\omega$$

Integrating by parts by letting du be equal to $\omega^2 \, d\omega$ and v be equal to $\log[1 - \exp(-\beta\hbar\omega)]$ in the equation $\int v \, du = uv - \int u \, dv$, and using the standard integral, (reference: Zemansky [2a]),

$$\int_0^{\infty} \frac{x^3 \, dx}{(e^x - 1)} = \frac{\pi^4}{15} \ ,$$

the reader can show that

$$\log Z = \frac{\pi^2}{45} \left(\frac{kT}{\hbar c}\right)^3 V = \frac{8\pi^5}{45} \left(\frac{kT}{hc}\right)^3 V \ . \tag{9.22}$$

In equation (9.22), V is the volume of the cavity in the heat reservoir of absolute temperature T in Figure 9.1.

Using equation (5.41), we have

$$\mu = - kT(\partial \log Z/\partial N)_{V,T} = 0 \ , \tag{9.23}$$

showing that, according to equation (9.22), the chemical potential of the photon gas is zero. (This point will be discussed in more detail in Section 12.5.4* of Chapter 12.) Using equations (5.37) and (9.22) we obtain

$$U = kT^2(\partial \log Z/\partial T)_{V,N} = 8\pi^5 k^4 T^4 V/15 h^3 c^3 \ . \tag{9.24}$$

Since the energy density u is equal to U/V, equations (9.9) and (9.24) are the same. Using equation (5.38) and (9.22) we have

$$p = kT(\partial \log Z/\partial V)_{T,N} = 8\pi^5 k^4 T^4/45 h^3 c^3 \ . \tag{9.25}$$

Combining equations (9.24) and (9.25) we have

$$pV = \tfrac{1}{3}U \ . \tag{9.26}$$

Hence $$p = \tfrac{1}{3}(U/V) = u/3 \tag{9.27}$$

Equation (9.27) can be developed from electromagnetic theory using Maxwell's stress tensor. (Reference: Panofsky and Phillips [9].) According to equation (5.76) of Chapter 5, in the case of an ideal classical monatomic gas in the non-relativistic limit, we have

$$pV = \tfrac{2}{3}U \ . \tag{5.76}$$

The difference between equations (9.26) and (5.76) arises from the fact that the photon gas is a relativistic gas in which the photons travel at the speed of light. Equation (5.76) holds for particles of finite rest mass in the non-relativistic limit. Using equations (5.39), (9.22) and (9.24) we find that the entropy of the photon gas is

$$S = 32\pi^5 k^4 T^3 V/45 h^3 c^3 \tag{9.28}$$

For an isentropic expansion of a photon gas, S is constant, so that according to equation (9.28),

$$T^3 V = \text{a constant} \qquad (S \text{ is constant}) \ . \tag{9.29}$$

Using equation (9.24) to eliminate T from equation (9.29), the reader can show that for an isentropic change,

$$uV^{4/3} = \text{a constant} \qquad (S \text{ is constant}) \qquad (9.30)$$

where $u = U/V$ is the energy density.

REFERENCES

[1] ter Haar, D., *The Old Quantum Theory*, Pergamon Press 1967.
[2] Zemansky, M. W., *Heat and Thermodynamics*, Fifth Edition, McGraw-Hill, 1968, (a) pages 430 and 636 (b) page 429.
[3] Roberts, J. K. and Miller, A. R., *Heat and Thermodynamics*, Fifth Edition Blackie, 1960, page 498.
[4] Riedi, P. C., *Thermal Physics*, The Macmillan Press 1976, page 224.
[5] Webster, A., *Scientific American*, August 1974.
[6] Weinberg, S., *The First Three Minutes, A Modern View of the Origin of the Universe*, Fontana/Collins, 1978.
[7] Landau, L. D. and Lifshitz, E. M., *Statistical Physics*, Pergamon Press, 1969 page 155.
[8] Jauch, J. M. and Rohrlich, F., *The Theory of Photons and Electrons*, Addison-Wesley, 1955, page 287.
[9] Panofsky, W. K. H. and Phillips, M., *Classical Electricity and Magnetism*, 2nd Edition, Addison-Wesley, 1962, page 191.

PROBLEMS

Problem 9.1
Compare the total energy emitted by a black body when its absolute temperature is (a) 3K, (b) 300K, (c) 30 000K.

Problem 9.2
Consider the black body radiation inside the cavity shown in Figure 9.1. Calculate the mean radiant energy per metre^3, the mean number of photons per metre^3 and the pressure of the photon gas, if the absolute temperature of the heat reservoir is (a) 3K, (b) 300K, (c) 30 000K.

Problem 9.3
Treat the Sun as a black body of absolute temperature 5800K. Use Stefan's law to show that the total radiant energy emitted by the Sun per second is 3.95×10^{26} J. Show that the rate at which energy is reaching the top of the Earth's atmosphere is 1.40 kW m^{-2}. (This is equivalent to about one electric fire per square metre.)

Stefan's constant σ	$= 5.67 \times 10^{-8}$ W m^{-2} K^{-4}
Radius of the Sun	$= 7.00 \times 10^8$ m
Distance from the Earth to the Sun	$= 1.5 \times 10^{11}$ m

Problem 9.4

Calculate the value of the wavelength λ_{max} at which the energy density per unit wavelength range (u_λ) of the radiation inside the cavity in Figure 9.1 is a maximum, when the temperature of the heat reservoir is 1000K.

Problem 9.5

The Universe is full of the cosmic background radiation. Find the frequency at which the energy per unit volume per unit frequency range (u_ν) given by equation (9.4) is a maximum. Assume that the temperature of the radiation is 3K.

[The solution of $3\{1 - \exp(-x)\} = x$ is $x = 2.82$.]

Chapter 10

The Heat Capacity of an Insulating Solid

10.1 INTRODUCTION

If the increase in the temperature of a substance, kept at constant volume, is dT when a quantity of heat dQ is added to the substance, the ratio

$$C_V = (dQ/dT)_V \tag{10.1}$$

is defined as the *heat capacity at constant volume*. From the thermodynamic identity, equation (3.85), if dV and dN are both zero, $dQ = dU = T\,dS$. Hence, equation (10.1) can be rewritten in the forms

$$C_V = (\partial U/\partial T)_V = T(\partial S/\partial T)_V \ . \tag{10.2}$$

If the heat capacity refers to unit mass of the substance, it is called the specific heat capacity and is denoted c_V. If the heat capacity refers to a mole of the substance it is called the molar heat capacity and is denoted $C_{V,m}$.

The heat capacity determined experimentally is generally the heat capacity at constant pressure, which is denoted by C_p. Using classical equilibrium thermodynamics, it can be shown that $C_p - C_V = KTV\alpha^2$, where T is the absolute temperature, V is the volume, α is the isobaric cubic (volume) expansivity and K is the isothermal bulk modulus. (Reference: Zemansky [1(a)].) This relation can be used to estimate C_V from the measured value of C_p. For insulating solids at room temperatures, the difference between C_p and C_V is generally a few per cent. At very low temperatures, the difference is very small.

Experiments have shown that, at room temperatures, the heat capacities at constant volume of many solids are approximately equal to $3Nk$, where N is the total number of atoms in the solid and k is Boltzmann's constant. This result is known as Dulong and Petit's law. For example, at room temperatures the molar heat capacities of the elements are generally approximately equal to $3N_Ak$, which is equal to $3R = 24.93$ J mole^{-1} K^{-1}, where N_A is Avogadro's constant.

Experiments at low temperatures have shown that C_V tends to zero as the absolute temperature T tends to zero. This is consistent with the laws of classical equilibrium thermodynamics. [Reference: Section 1.11.]

10.2 EINSTEIN'S THEORY FOR THE HEAT CAPACITY OF AN INSULATING SOLID

The atoms in a crystal are held together by the attractive forces of all the other atoms in the crystal. In Einstein's theory, it is assumed that any particular atom oscillates about its mean position (lattice site) in a steady potential well due to all the other atoms in the crystal. When a particle in a steady potential well is displaced a *small* amount from its equilibrium position and released, the particle will oscillate with simple harmonic motion. The equation of motion is

$$\ddot{\mathbf{r}} = -\omega^2\mathbf{r} \ , \tag{10.3}$$

where $\mathbf{r}$ is the displacement vector of the particle from its equilibrium position. If $\mathbf{r}$ is equal to $(\mathbf{i}x + \mathbf{j}y + \mathbf{k}z)$, where $\mathbf{i}, \mathbf{j}$ and $\mathbf{k}$ are unit vectors in the x, y and z directions respectively, substituting in equation (10.3), we obtain

$$\mathbf{i}\ddot{x} + \mathbf{j}\ddot{y} + \mathbf{k}\ddot{z} = -\omega^2\mathbf{i}x - \omega^2\mathbf{j}y - \omega^2\mathbf{k}z \ . \tag{10.4}$$

Equation (10.4) separates into three equations of the form $\ddot{x} = -\omega^2 x$. The motion of the particle can be treated as three independent linear simple harmonic oscillations. In Einstein's theory of heat capacities, a solid consisting of N atoms is treated as $3N$ independent harmonic oscillators. Surface effects are ignored.

In Einstein's theory it is assumed that each of the $3N$ oscillators has the *same* angular frequency ω_E, as shown in Figure 10.1(a). It will be assumed that as far as one particular linear harmonic oscillator is concerned, the rest of the crystal lattice acts as a heat reservoir of constant absolute temperature T. The mean energy $\bar{\varepsilon}$ of each of the harmonic oscillators is given by equation (4.37), which was developed in Section 4.7 using the Boltzmann distribution. The mean energy of the $3N$ oscillators is

$$U = 3N\bar{\varepsilon} = 3N\left\{\tfrac{1}{2}\hbar\omega_E + \frac{\hbar\omega_E}{[\exp(\hbar\omega_E/kT) - 1]}\right\} \ .$$

Using equation (10.2), we have

$$C_V = \left(\frac{\partial U}{\partial T}\right)_V = 3Nk\left(\frac{\theta_E}{T}\right)^2 \frac{\exp(\theta_E/T)}{[\exp(\theta_E/T) - 1]^2} \ . \tag{10.5}$$

where the Einstein characteristic temperature θ_E is given by

$$\theta_E = \hbar\omega_E/k \ . \tag{10.6}$$

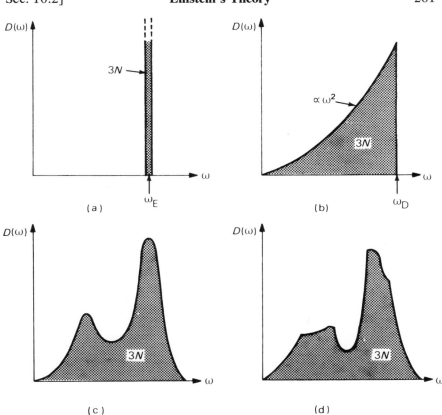

Figure 10.1—Angular frequency spectrum of the lattice vibrations in a solid composed of N atoms. (a) In the Einstein model, all the $3N$ harmonic oscillators have the same angular frequency ω_E. (b) In the Debye model, the oscillators are the normal modes of a continuous solid. $D(\omega)$ is proportional to ω^2 right up to the cut off angular frequency ω_D. (c) A typical spectrum calculated by Blackman for a cubic crystal. (d) A more rigorous calculation allowing for atomic structure. (Reproduced with permission from: *Heat and Thermodynamics* by Mark W. Zemansky. © McGraw-Hill, 1957.)

The variation of the molar heat capacity $C_{V,m}$, predicted by equation (10.5), with (T/θ_E) is shown in Figure 10.2.

At high temperatures, when $T \gg \theta_E$ and $\theta_E/T \ll 1$, $\exp(\theta_E/T) \simeq 1 + (\theta_E/T)$. Hence at high temperatures, equation (10.5) becomes

$$C_V \simeq 3Nk(\theta_E/T)^2(1 + \theta_E/T)/[1+ \theta_E/T - 1]^2 \simeq 3Nk \ . \tag{10.7}$$

For a mole of atoms, N is equal to Avogadro's constant N_A so that the molar heat capacity $C_{V,m}$ is approximately equal to $3N_A k = 3R$ at high temperatures, as shown in Figure 10.2. This is Dulong and Petit's law. It is the classical limit of equation (10.5). This result was to be expected, since according to the equipartition theorem developed in Section 4.10.4*, the mean energy of a linear harmonic oscillator is equal to kT in the classical limit. If there are $3N_A$ oscillators each having energy kT, the total energy U is equal to $3N_A kT$ giving a value of $3N_A k$ for $C_{V,m}$ in the classical limit.

At low temperatures, when $T \ll \theta_E$, $\exp(\theta_E/T) \gg 1$ and equation (10.5) becomes

$$C_V = 3Nk(\theta_E/T)^2 \exp[-(\theta_E/T)] \ . \tag{10.8}$$

As T tends to zero, the exponential term $\exp(-\theta_E/T)$ tends to zero much faster than the increase in the $(\theta_E/T)^2$ term, so that C_V tends to zero exponentially as shown in Figure 10.2.

In the expression for C_V given by equation (10.5), the only unknown is the Einstein temperature θ_E, corresponding to the unknown value of the Einstein angular frequency ω_E. The mathematical form of equation (10.5) is the same for all substances. If the heat capacity $C_{V,m}$ of a mole of atoms is plotted against (T/θ_E), the values of $C_{V,m}$ for all insulating solids should lie on the same curve, which is plotted in Figure 10.2. A value for θ_E can be determined from the best fit of equation (10.5) to the experimental data.

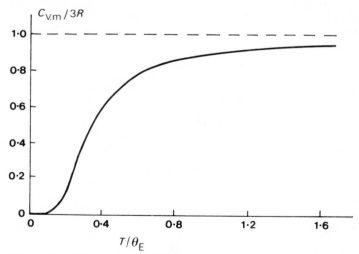

Figure 10.2—Temperature variation of the molar heat capacity of a solid according to the Einstein model. At low temperatures, the predicted values of the molar heat capacity of an insulating solid tend to zero, as T tends to zero, faster than the experimental results, which follow a T^3 law. (Reproduced with permission from: *Statistical Mechanics and Properties of Matter*, by E. S. R. Gopal.)

For many inorganic substances, θ_E is of the order of 200 to 300K, corresponding to an Einstein angular frequency ω_E of the order of 2.6×10^{13} to 3.9×10^{13} rad s^{-1} and a vibrational frequency $\nu_E = \omega_E/2\pi$ of the order of 4.2×10^{12} to 6.3×10^{12} Hz. For diamond, θ_E is approximately 1320K. The agreement with the experimental data is reasonably good down to a temperature of $0.2\theta_E$, which is about 40 to 60K for most substances. At lower temperatures, the experimental values of C_V are significantly larger than the values predicted by equation (10.5). According to equation (10.8), C_V should tend to zero exponentially as T tends to zero. However, the experimental results for insulating solids show that, at very low temperatures, C_V is proportional to T^3.

10.3 DEBYE'S THEORY FOR THE HEAT CAPACITY OF AN INSULATING SOLID

In Einstein's theory of heat capacities, it was assumed that all the atoms oscillated with simple harmonic motion in a steady potential well with the same angular frequency ω_E, as shown in Figure 10.1(a). In practice, if one of the atoms in a crystal is displaced, this affects the restoring forces acting on the other atoms in its vicinity, so that the collective motions of the atoms in a solid can be important as, for example, in sound waves in a solid. In Debye's theory, the solid is treated as a continuous elastic medium. (In practice, solids have an atomic structure which can lead to important effects.)

According to equation (A6.14) of Appendix 6, the number of standing wave solutions (normal modes) for elastic waves in a continuous solid, which have angular frequencies in the range ω to $(\omega + d\omega)$, is

$$D(\omega)\,d\omega = \frac{V\omega^2\,d\omega}{2\pi^2}\left[\frac{1}{c_L{}^3} + \frac{2}{c_T{}^3}\right] \tag{10.9}$$

where c_L and c_T are the velocities of longitudinal and transverse elastic (sound) waves in the solid, and V is the volume of the solid. Debye assumed that equation (10.9) held right up to a maximum angular frequency ω_D, known as the Debye angular frequency.† Debye assumed that the total number of normal modes was equal to $3N$, where N was the total number of atoms. Hence, according to Debye's theory,

$$\int_0^{\omega_D} D(\omega)\,d\omega = 3N \; . \tag{10.10}$$

---

†In an atomic theory of solids, there is an upper cut off frequency when the wavelength of the standing wave is equal to twice the separation of neighbouring atoms in the solid.

Substituting from equation (10.9) into equation (10.10) and integrating, we have

$$3N = \frac{V}{2\pi^2} \left[\frac{1}{c_L^3} + \frac{2}{c_T^3} \right] \left(\frac{\omega_D^3}{3} \right)$$

Rearranging, we have

$$\omega_D^3 = \frac{18N\pi^2}{V[(1/c_L^3) + (2/c_T^3)]} . \tag{10.11}$$

Equation (10.11) expresses the Debye cut off angular frequency ω_D in terms of the velocities of longitudinal and transverse elastic (sound) waves in the solid. Substituting for $V(1/c_L^3 + 2/c_T^3)$ from equation (10.11) into equation (10.9) we obtain

$$D(\omega) \, d\omega = 9N\omega^2 \, d\omega/\omega_D^3 . \tag{10.12}$$

The variation of $D(\omega)$, the number of normal modes per unit angular frequency range, with ω is shown in Figure 10.1(b). According to equation (10.12), $D(\omega)$ is proportional to ω^2 right up to the Debye cut off angular frequency ω_D. In Debye's theory, each normal mode (standing wave solution), is treated as an independent linear harmonic oscillator in thermal equilibrium with the rest of the crystal, which acts as a heat reservoir of constant absolute temperature T for each normal mode. According to equation (4.37) the mean energy of a harmonic oscillator of natural angular frequency ω is

$$\bar{\varepsilon} = \tfrac{1}{2}\hbar\omega + \hbar\omega/[\exp(\hbar\omega/kT) - 1] . \tag{10.13}$$

The total average energy of the solid is

$$U = \int_0^{\omega_D} \bar{\varepsilon} D(\omega) \, d\omega . \tag{10.14}$$

Substituting for $\bar{\varepsilon}$ from equation (10.13) and for $D(\omega)$ from equation (10.12) and integrating, we obtain

$$U = \frac{9N\hbar\omega_D}{8} + \frac{9N\hbar}{\omega_D^3} \int_0^{\omega_D} \frac{\omega^3 \, d\omega}{[\exp(\hbar\omega/kT) - 1]} . \tag{10.15}$$

The first term on the right hand side of equation (10.15) is independent of temperature. Using equation (10.2) we find that the heat capacity at constant volume is

$$C_V = \left(\frac{\partial U}{\partial T} \right)_V = \frac{9N\hbar}{\omega_D^3} \frac{\partial}{\partial T} \left[\int_0^{\omega_D} \frac{\omega^3 \, d\omega}{[\exp(\hbar\omega/kT) - 1]} \right] .$$

Differentiating under the intergral sign with respect to T, we have

$$C_V = \frac{9Nk}{\omega_D^3} \left(\frac{\hbar}{kT} \right)^2 \int_0^{\omega_D} \frac{\omega^4 \exp(\hbar\omega/kT) \, d\omega}{[\exp(\hbar\omega/kT) - 1]^2} . \tag{10.16}$$

Let
$$x = \hbar\omega/kT \tag{10.17}$$

so that
$$\omega = (kT/\hbar)x; \qquad d\omega = (kT/\hbar)\, dx \ .$$

Let
$$x_D = \hbar\omega_D/kT = \theta_D/T \tag{10.18}$$

where $\theta_D = \hbar\omega_D/k$ is the Debye temperature. Substituting in equation (10.16), we obtain

$$C_V = \frac{9Nk}{x_D{}^3} \int_0^{x_D} \frac{x^4 e^x\, dx}{(e^x - 1)^2} \ . \tag{10.19}$$

The definite integral in equation (10.19) is a function of the upper limit x_D. The integral has to be evaluated numerically. It is generally denoted by either $F(x_D)$ or $F(\theta_D/T)$. Numerical values are given by Zemansky [1(b)]. We have

$$C_V = (9Nk/x_D{}^3)F(x_D) = 9Nk(T/\theta_D)^3\, F(\theta_D/T) \ . \tag{10.20}$$

The variation of the molar heat capacity $C_{V,m}$ with (T/θ_D), predicted by equation (10.20), is shown in Figure 10.3(a). The experimental results for copper are also shown. The appropriate value of $\theta_D = 308$K for copper is determined from the best fit of the data to equation (10.20). Above a few degrees Kelvin the agreement is reasonably good for copper.

At high temperatures, $x = \theta_D/T$ is very much less than unity, so that $e^x \simeq (1 + x)$. Hence as T tends to infinity

$$F(x_D) = \int_0^{x_D} \frac{x^4 e^x\, dx}{(e^x - 1)^2} \simeq \int_0^{x_D} \frac{x^4(1 + x)\, dx}{x^2} \simeq \int_0^{x_D} x^2\, dx = \frac{x_D{}^3}{3} \ .$$

Hence at *high* temperatures, for a mole of atoms, equation (10.20) becomes

$$C_{V,m} \simeq (9N_A k/x_D{}^3)(x_D{}^3/3) = 3N_A k = 3R \ .$$

as shown in Figure 10.3(a). This is Dulong and Petit's law. This result was to be expected since, according to equation (10.10), for a mole of atoms there are $3N_A$ independent harmonic oscillators, each of which has an energy kT in the classical limit, so that U is equal to $3N_A kT$ and $C_{V,m}$ should be equal to $3R$ at high temperatures.

At very low temperatures, the upper limit $x_D = (\hbar\omega/kT)$ in equation (10.19) becomes very large. The value of the variable $x = (\hbar\omega/kT)$ is also very large in the vicinity of the upper limit x_D. At very low temperatures, when $x \gg 1$ and $e^x \gg 1$, we have

$$x^4 e^x/(e^x - 1)^2 \simeq x^4 e^x/(e^x)^2 = x^4 e^{-x} \ .$$

When $x \gg 1$, the e^{-x} term dominates the $(x^4 e^{-x})$ term, which becomes very small. Hence at very low temperatures, the integrand in equation (10.19) is very small near the upper limit x_D, whereas x_D itself is very large. No great error is introduced at very low temperatures, if the upper limit x_D in the

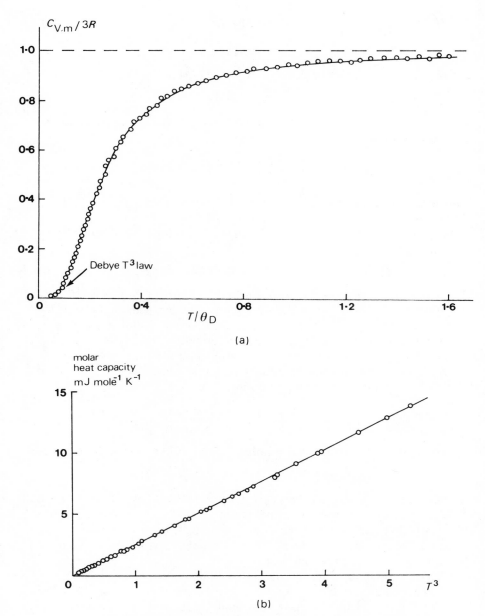

(a)

(b)

Figure 10.3—(a) Temperature variation of the molar heat capacity of a solid according to the Debye model. The observed heat capacity of copper is compared with the predictions for a Debye temperature $\theta_D = 308K$. (Below 5K, the free electrons make a significant contribution to the heat capacity of copper). (b) The experimental heat capacity data for solid argon below 2K are plotted against T^3. The linear relationship confirms the Debye T^3 law in this temperature range. ((a) and (b) are reproduced with permission from *Statistical Mechanics and Properties of Matter* by E. S. R. Gopal and *Introductory Statistical Mechanics* by R. E. Turner and D. S. Betts respectively.)

integral in equation (10.19) is replaced by infinity. The definite integral is then a number whose numerical value is $4\pi^4/15$. (Reference: Zemansky [1(c)].) Hence at very low temperatures the molar heat capacity, given by equation (10.19), becomes

$$C_{V,m} = \frac{9N_A k}{x_D^3}\left(\frac{4\pi^4}{15}\right) = \frac{12R\pi^4}{5}\left(\frac{T}{\theta_D}\right)^3 = 1.94 \times 10^3 \left(\frac{T}{\theta_D}\right)^3 \text{ J mole}^{-1}\text{ K}^{-1} .$$

(10.21)

This is the Debye T^3 law. It is valid experimentally for insulating solids at temperatures $T \ll \theta_D$. Values for θ_D can be obtained from experimental determinations of the variation of C_V with T. For example, $\theta_D = 93\text{K}$ for solid argon.

The low temperature results for solid argon, an insulating solid, are shown in Figure 10.3(b), where $C_{V,m}$ is plotted against T^3 at temperatures below 2K. The linear relation in Figure 10.3(b) confirms the Debye T^3 law in this temperature range. [In the case of a metal, such as copper, the free electrons make a significant contribution to the total heat capacity at temperatures below 5K. This point will be elaborated in Section 12.2.2 and illustrated in Figure 12.7.]

At low temperatures Debye's theory is superior to Einstein's theory. Debye's model of standing waves in a continuous solid is expected to be reasonably good at low temperatures, since at low temperatures it is only the low frequency, that is the long wavelength normal modes which are excited to any significant extent. It is a reasonable approximation to treat the solid as a continuous body, when the wavelengths of the standing waves (normal modes) are much bigger than the separations of the atoms in the solid. At high temperatures, both Einstein's theory and Debye's theory give reasonably good agreement with the experimental results and in both theories the molar heat capacity tends to the classical value of $3R$, when the energy of each of the $3N_A$ harmonic oscillators tends to kT. In the intermediate temperature range, the experimental results differ significantly from Debye's theory. To obtain better agreement with the experimental results in the intermediate temperature range, an atomic model must be used for the solid. This leads to a different expression for $D(\omega)$ to equation (10.12). It is beyond the scope of this text book to give an account of the Born-Karman theory of a linear diatomic chain and a general discussion of the calculation of $D(\omega)$ for various crystal structures. A reader interested in an introductory discussion is referred to Gopal [2] or to a text book on solid state physics, such as Kittel [3]. Some calculated variations of $D(\omega)$ with ω based on atomic models are shown in Figures 10.1(c) and 10.1(d). The variation shown in Figure 10.1(c) was obtained by Blackman for a simple cubic lattice. A more rigorous calculation is shown in Figure 10.1(d). These predictions are generally in qualitative agreement with the variation of $D(\omega)$ with ω determined by slow neutron scattering experiments.

Values of the Debye temperature θ_D can be estimated from the measured values of the elastic constants. It follows from equations (10.11) and the definition of θ_D as $\hbar\omega_D/k$ by equation (10.18) that

$$\theta_D{}^3 = \frac{18N\pi^2\hbar^3}{V[(1/c_L{}^3) + (2/c_T{}^3)]k^3} \, .$$

According to the theory of elastic waves in solids

$$c_L = \left(\frac{K + 4G/3}{\rho}\right)^{1/2}, \quad \text{and} \quad c_T = \left(\frac{G}{\rho}\right)^{1/2},$$

where K is the bulk modulus, G is the rigidity modulus and ρ is the density. (The velocities c_L and c_T correspond to the P and S waves of seismology respectively.) The agreement of the measured and calculated values of θ_D is generally reasonably good, but not exact.

10.4 PHONONS

In Chapter 9, two models of radiation were used to develop the Planck distribution law, namely the wave model in Section 9.2 and the photon model in Section 9.4. In Section 10.3, when Debye's theory of heat capacities was developed, a wave model was used. There is great similarity between the Nineteenth Century wave models of electromagnetic radiation based on the hypothetical ether and Debye's theory of heat capacities. In the Nineteenth Century it was assumed that the ether filled the whole of space. It was assumed that the ether had elastic properties. It was assumed that light waves were vibrations of the ether and it was possible to interpret many of the properties of light using this elastic model. According to Debye's theory of heat capacities based on standing elastic waves in a solid, from equations (10.9) and (10.13), putting the volume V equal to unity and ignoring the zero point energy $\frac{1}{2}\hbar\omega$, we have

$$u_\omega \, d\omega = \frac{\omega^2 \, d\omega}{2\pi^2} \left[\frac{1}{c_L{}^3} + \frac{2}{c_T{}^3}\right] \frac{\hbar\omega}{[\exp(\beta\hbar\omega) - 1]} \, . \qquad (10.22)$$

where $u_\omega \, d\omega$ is the energy, per unit volume, of the normal modes that have angular frequencies between ω and $(\omega + d\omega)$. No longitudinal light waves have been observed. If the term in c_L in equation (10.22) is omitted, equation (10.22) reduces to equation (9.3), which is Planck's law.

Since both the wave and the particle (photon) models give the same energy distribution in the case of electromagnetic radiation, it is reasonable in solid state physics to try using a particle type model for the internal excitation energy of a crystal lattice. This approach is useful, for example, in the interpretation of the thermal conductivity of an insulating solid. The energy levels of a harmonic oscillator of angular frequency ω are $\varepsilon_n = (n + \frac{1}{2})\hbar\omega$. If the energy of one of the normal modes of the lattice, which is treated as a harmonic oscillator, changes, the energy of that normal mode must change by an integral multiple of $\hbar\omega$. By analogy with the case of electromagnetic radiation, considered in Section 9.4, in the *phonon* model of lattice vibrations it is assumed that, when the energy of a normal mode increases (decreases) by $\hbar\omega$, the number of phonons in that normal mode increases (decreases) by one, and by analogy with equation (9.15), the mean number of phonons of energy $\hbar\omega$ in that normal mode is

$$\bar{n} = 1/[\exp(\hbar\omega/kT) - 1] \ . \tag{10.23}$$

The phonon can be thought of as a quantum of energy of the lattice vibrations. In the case of optical interactions in solids and slow neutron scattering in solids, the photons and neutrons do not interact with just one of the atoms in a solid, but with a group of atoms. The energy transferred to the group of atoms is quantised. In the phonon model, it is assumed that phonons are created or destroyed in such interactions. Energy is conserved in these processes. (A reader interested in the subtleties of momentum conservation in such processes is referred to a text book on solid state physics, such as Kittel [3].)

The phonons are sometimes treated as an ideal boson gas which has zero chemical potential. However, the phonon is not a true particle, since one cannot have free phonons in empty space without a crystal lattice. For this reason, the phonon is sometimes called a quasi-particle.

The phonon model is used extensively in solid state physics. For fuller details, the reader is referred to a text book on solid state physics, such as Kittel [3].

10.5 DISCUSSION

Our discussion has been confined to insulating solids. At very low temperatures, the experimentally determined variation of the heat capacity of a *metal* with temperature is given, to a good approximation, by an equation of the form

$$C_V = AT^3 + \gamma T, \tag{10.24}$$

where A and γ are constants. It will be shown in Section 12.2.2 of Chapter 12 that the γT term in equation (10.24) represents the contribution of the conduction (free) electrons to the heat capacity of the metal. This contribution must be added to the contribution of the lattice vibrations, given by the Debye T^3 law.

There can be other contributions to the total heat capacity of a solid. For example, it was shown in Section 5.8.3 that the paramagnetic ions in a solid give a large contribution to the heat capacity of a solid in the temperature range where $(kT/\mu B)$ is approximately equal to 0.8. There are also significant heat capacity anomalies associated with other phase changes, such as for example, near the Curie temperature, when there is a transition from ferromagnetic to paramagnetic behaviour.

REFERENCES

[1] Zemansky, M. W. *Heat and Thermodynamics*, 5th Edition, McGraw-Hill, 1957, (a) page 294, (b) page 316, (c) page 314.
[2] Gopal, E. S. R. *Statistical Mechanics and Properts of Matter*, Ellis Horwood, Chichester, 1974, page 147.
[3] Kittel, C. *Introduction to Solid State Physics*, 5th Edition, Wiley, 1976.

Chapter 11

The Grand Canonical Distribution

11.1 INTRODUCTION

When the Boltzmann (canonical) distribution was developed in Section 4.2 of Chapter 4, it was assumed that there was no interchange of particles between the small system and the heat reservoir in Figure 4.1(a), so that the number of particles in the small system was fixed. We shall now consider the more general case shown in Figure 11.1, in which there can be interchange of particles between the reservoir and the small system. This will lead up to the grand canonical distribution and the grand partition function. The grand canonical distribution will be used to develop the Fermi–Dirac and the Bose–Einstein distribution functions in Sections 11.3 and 11.4 respectively. A brief introduction to the role of the grand partition function in statistical thermodynamics will be given in Section 11.5*, for the benefit of advanced readers.

11.2 THE GRAND CANONICAL DISTRIBUTION

Consider the case of a small system, which is separated from a reservoir of constant absolute temperature T by a rigid, diathermic partition with holes in it as shown in Figure 11.1. Both particles and heat can go from the reservoir to the small system and from the small system to the reservoir. The (heat and particle) reservoir and the small system are inside a rigid outer adiabatic enclosure, and form a *closed* isolated composite system of constant total volume V^*, constant total energy U^* and constant total number of particles N^*, all of which are of the same type.

Let the *small* system consist of N particles and be in the N particle quantum state of energy eigenvalue ε_i, where ε_i is a function of both the number of particles N in the small system and the volume V of the small system. The energy of the (heat and particle) reservoir is

$$U_R = U^* - \varepsilon_i \ . \tag{11.1}$$

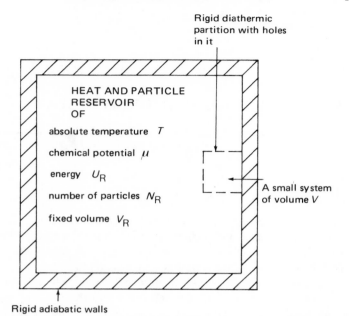

Figure 11.1—The grand canonical distribution. The small system and the heat and particle reservoir are inside rigid, adiabatic, outer walls and form a closed system of constant total energy, constant total volume and constant total number of particles. The small system and the heat and particle reservoir are separated by a rigid, diathermic partition with holes in it to allow the exchange of both energy and particles.

The number of particles in the (heat and particle) reservoir is

$$N_R = N^* - N \ . \tag{11.2}$$

The volume V_R of the (heat and particle) reservoir is kept constant by the rigid outer adiabatic walls and the rigid diathermic partition with holes in it, which separates the reservoir from the small system.

According to the principle of equal *a priori* probabilities, which is Postulate 2 of Section 3.1, when the closed composite system, made up of the (heat and particle) reservoir and small system in Figure 11.1, has had time to reach internal thermodynamic equilibrium, the closed composite system is equally likely to be found in any one of the microstates accessible to the closed composite system, which are consistent with the values U^* for the total energy, N^* for the total number of particles and V_R and V for the constant volumes of the (heat and particle) reservoir and small system respectively. By analogy with equations (2.26) and (4.2), it follows that the probability $P_{N,i}$ that, at thermodynamic equilibrium, the small system consists of N particles and is in a microstate which has energy eigenvalue

$\varepsilon_i(V, N)$, is proportional to the number of microstates $g_R(U_R, N_R, V_R)$ accessible to the (heat and particle) reservoir when the (heat and particle) reservoir has energy U_R equal to $(U^* - \varepsilon_i)$, consists of $N_R = (N^* - N)$ particles and has a fixed volume V_R. Hence

$$P_{N,i} = C'g_R(U_R, N_R, V_R) = C'g_R(U^* - \varepsilon_i, N^* - N, V_R) \qquad (11.3)$$

where $C' = 1/g_T$ is a constant.

If a function $f(x + a, y + b)$ of the variables $(x + a)$ and $(y + b)$ is expanded in a Taylor series, according to equation (A1.30) of Appendix 1, provided $a \ll x$ and $b \ll y$, we have

$$f(x + a, y + b) \simeq f(x,y) + a(\partial f/\partial x) + b(\partial f/\partial y) . \qquad (11.4)$$

The partial differential coefficients are evaluated at the point (x, y).

Using equation (11.4) to expand $\log g_R(U^* - \varepsilon_i, N^* - N)$, if $\varepsilon_i \ll U^*$ and $N \ll N^*$, for constant V_R we have

$$\log g_R(U^* - \varepsilon_i, N^* - N) = \log g_R(U^*, N^*) - \varepsilon_i(\partial \log g_R/\partial U_R)$$
$$- N(\partial \log g_R/\partial N_R) . \qquad (11.5)$$

The differential coefficients are evaluated at $U_R = U^*$ and $N_R = N^*$. According to equations (3.14) and (3.19)

$$(\partial \log g_R/\partial U) = \beta = 1/kT$$

where T is the constant absolute temperature of the heat reservoir. According to equation (3.78)

$$(\partial \log g_R/\partial N) = -(\mu/kT) = -\beta\mu$$

where μ is the constant chemical potential of the particle reservoir. Substituting in equation (11.5) and taking exponentials we have

$$g_R(U^* - \varepsilon_i, N^* - N) = g_R(U^*, N^*) \exp[\beta(\mu N - \varepsilon_i)] . \qquad (11.6)$$

The quantity $g_R(U^*, N^*)$ is the number of microstates accessible to the (heat and particle) reservoir when the (heat and particle) reservoir has all the energy and all the particles, and is a constant. Substituting from equation (11.6) into equation (11.3) we have

$$P_{N,i} = C \exp[\beta(\mu N - \varepsilon_i)] = C \exp[(\mu N - \varepsilon_i)/kT] \qquad (11.7)$$

where $C = C'g_R(U^*, N^*)$ is a constant. In equation (11.7), T is the absolute temperature of the heat reservoir, μ is the chemical potential of the particle reservoir, N is the number of particles in the ith N-particle state of the small system and ε_i is the total energy (kinetic plus potential) of the N particles in the ith N-particle state of the small system. Equation (11.7) is the *grand canonical distribution*. It is left as an exercise for the reader to prove equation (11.7) for the case of separate heat and particle reservoirs.

The summation $\sum_N \sum_i P_{N,i}$, which is over all the states of the small system, for all values of N, must be equal to unity. Hence

$$C \sum_N \sum_i \exp[\beta(\mu N - \varepsilon_i)] = 1 \ .$$

Substituting for C, we have

$$P_{N,i} = \exp[\beta(\mu N - \varepsilon_i)]/\Xi \tag{11.8}$$

where Ξ, (the Greek capital xi), is given by

$$\Xi = \sum_N \sum_i \exp[\beta(\mu N - \varepsilon_i(V, N))] \tag{11.9}$$

and is called the *grand partition function*. One way, *inter alia*, to evaluate Ξ is to sum first over all the states of the N particle small system for a fixed value of N to give

$$\Xi = \sum_N \exp(\beta\mu N)\left[\sum_i \exp(-\beta\varepsilon_i)\right] = \sum_N \exp(\beta\mu N)Z(N) \tag{11.10}$$

where $Z(N)$ is the partition function for a fixed value of N, and is given by equation (4.9). To determine Ξ, we must then sum $\exp(\beta\mu N)Z(N)$ for all values of N. The grand partition function Ξ is a function of the absolute temperature T of the heat reservoir, of the chemical potential μ of the particle reservoir and of the volume V of the small system in Figure 11.1.

If, when the small system in Figure 11.1 is in a microstate in which it consists of N particles and has total energy ε_i, the value of any variable α is $\alpha_{N,i}$, according to equation (A1.7) of Appendix 1, the ensemble average of the variable α is

$$\bar{\alpha} = \sum_N \sum_i P_{N,i}\alpha_{N,i} = \frac{\sum_N \sum_i \alpha_{N,i} \exp[\beta(\mu N - \varepsilon_i)]}{\Xi} \ . \tag{11.11}$$

The grand canonical distribution will now be used to develop the Fermi–Dirac and Bose–Einstein distribution functions. The role of the grand partition function in statistical thermodynamics will be outlined in Section 11.5* for the benefit of advanced readers.

11.3 THE FERMI–DIRAC DISTRIBUTION FUNCTION

Particles of half integral spin $(\frac{1}{2}, \frac{3}{2}, \frac{5}{2}, \ldots)$ obey Fermi–Dirac statistics and are called fermions. For example, an electron is a fermion of spin $\frac{1}{2}$. Only one fermion of a group of identical fermions can occupy any single particle quantum state (orbital) with a given set of quantum numbers. For example, only one of the electrons in an atom can have a given set of values of the quantum numbers n, l, m and s. The electrons must differ in at least one of these quantum numbers. [Reference: Section 2.1.2.]

Consider N identical fermions inside a box made of a rigid, diathermic

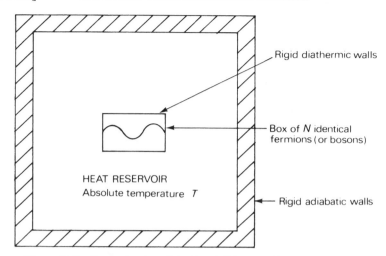

Figure 11.2—Determination of the Fermi–Dirac and Bose–Einstein distribution functions. The small system chosen for the application of the grand canonical distribution is one of the single particle states for a fermion (boson) in the box. This is represented symbolically by the sine curve drawn inside the box. In the case of identical fermions, there can only be 0 or 1 fermions in this single particle state. (In the case of identical bosons, there can be any number of bosons in the chosen single particle state.) The fermions (bosons) in the other single particle states in the box act as the particle reservoir for the chosen single particle state.

material. The box is in thermal equilibrium with a heat reservoir of constant absolute temperature T, as shown in Figure 11.2. One of the *single* particle states (orbitals) for a fermion in the box is chosen as *the* small system. This single particle state is represented symbolically by the sine curve inside the box in Figure 11.2. [The solutions given by equation (A6.6) of Appendix 6 are of this form.] The chosen *single* particle state has (single particle) energy eigenvalue ε. The fermions in the other single particle states act as the particle reservoir for the chosen single particle state. In the case of an *ideal* fermion gas, in which there is no interaction between the fermions, a fermion goes from one single particle state to another by exchanging energy with the heat reservoir.[†] According to the Pauli exclusion principle, there can only be zero or one fermion in the chosen single particle state. Since the number of fermions in our small

[†]In practice, there may be some interaction between the fermions. In this case the fermions in the other single particle states form part of the heat reservoir. In the idealised case of a fermion gas in an isolated box composed of perfect adiabatic walls, it is the fermions in the other single particle states, which act as the heat as well as the particle reservoir for the chosen single particle state.

system (the chosen single particle state) can vary, the grand canonical distribution must be used.

If there are no fermions in the chosen single particle state (our small system), $N = 0$ and the total energy of the fermions in the chosen single particle state is zero. When there is one fermion in the chosen single particle state (our small system), $N = 1$ and the total energy of the $N = 1$ fermions in the single particle state is equal to ε. Summarising, there are only two possible states for our small system, namely, $N = 0$, energy $= 0$ and $N = 1$, energy $= \varepsilon$. The grand partition function, given by equation (11.9), is

$$\Xi = \exp(0) + \exp[\beta(\mu - \varepsilon)] = 1 + \exp[\beta(\mu - \varepsilon)] \ . \qquad (11.12)$$

Putting $\alpha_{N,i}$ equal to N in equation (11.11), we find that $f(\varepsilon)$, the mean (ensemble average) number of fermions in the chosen single particle state, which has (single particle) energy eigenvalue ε, is

$$f(\varepsilon) = \bar{N} = (1/\Xi) \sum_N \sum_i N \exp[\beta(\mu N - \varepsilon_i)]$$
$$f(\varepsilon) = (1/\Xi)\{0 \exp[0] + 1 \times \exp[\beta(\mu - \varepsilon)]\} = (1/\Xi)\exp[\beta(\mu - \varepsilon)] \ .$$

Substituting for Ξ from equation (11.12), we have

$$f(\varepsilon) = \exp[\beta(\mu - \varepsilon)]/\{1 + \exp[\beta(\mu - \varepsilon)]\} \ . \qquad (11.13)$$

Multiplying the numerator and the denominator of the right hand side by $\exp[\beta(\varepsilon - \mu)]$, we have

$$f(\varepsilon) = \frac{1}{\{\exp[\beta(\varepsilon - \mu)] + 1\}} = \frac{1}{\{\exp[(\varepsilon - \mu)/kT] + 1\}} \qquad (11.14)$$

In equation (11.14), T is the absolute temperature of the heat reservoir, and μ is the chemical potential of the particle reservoir. In solid state physics, the chemical potential μ is often called the *Fermi level*. The mean fermion occupancy $f(\varepsilon)$ of the chosen single particle state is called the *Fermi–Dirac distribution function*. It will be used extensively in Chapter 12.

11.4 THE BOSE–EINSTEIN DISTRIBUTION FUNCTION

Particles of integral spin (0, 1, 2, . . .) obey Bose-Einstein statistics and are called bosons. There is no limit to the number of indentical bosons that can be in any particular single particle state. [Reference: Section 2.1.5.]

Assume that the box in Figure 11.2 contains N indentical bosons. We shall again treat one of the *single* particle states (orbitals) as *the* small system for the application of the grand canonical distribution. The (single particle) energy eigenvalue of the chosen single particle state is ε.

The bosons in the other single particle states act as the particle reservoir for the chosen single particle state. In the case of indentical bosons, the total number of bosons in the chosen single particle state (our small system) can have any of the values 0, 1, 2 . . . . The corresponding total energies of the bosons in the chosen single particle state are 0, ε, 2ε, ... Hence the possible states of our small system are $N = 0$, energy $= 0$; $N = 1$, energy $= \varepsilon$; $N = 2$, energy $= 2\varepsilon$, $N = 3$, energy $= 3\varepsilon$ etc. According to equation (11.9), the grand partition function is

$$\Xi = \exp(0) + \exp[\beta(\mu - \varepsilon)] + \exp[\beta(2\mu - 2\varepsilon)] + \ldots . \quad (11.15)$$

Putting

$$\exp[\beta(\mu - \varepsilon)] = x \quad (11.16)$$

we have

$$\Xi = \sum_{N=0}^{\infty} x^N = (1 - x)^{-1} . \quad (11.17)$$

The reader can check equation (11.17) by expanding $(1 - x)^{-1}$ using the binomial theorem, which is equation (A1.25) of Appendix 1.

Putting $\alpha_{N,i}$ equal to N in equation (11.11), we find that $b(\varepsilon)$, the mean (ensemble average) number of bosons in the chosen single particle state, is

$$b(\varepsilon) = \bar{N} = (1/\Xi) \sum_{N} \sum_{i} N \exp[\beta(\mu N - \varepsilon_i)]$$

$$= (1/\Xi)\{0 \exp[0] + 1 \exp[\beta(\mu - \varepsilon)] + 2 \exp[\beta(2\mu - 2\varepsilon)] + \ldots\} .$$

Using equation (11.16), we have

$$b(\varepsilon) = \sum N x^N / \Xi . \quad (11.18)$$

According to equation (9.14)

$$\sum_{N=0}^{\infty} N x^N = \frac{x}{(1 - x)^2} . \quad (11.19)$$

Substituting from equations (11.17) and (11.19) into equation (11.18) we find that the mean (ensemble average) number of bosons in the chosen single particle state, which has (single particle) energy eigenvalue ε, is

$$b(\varepsilon) = x/(1 - x) = \exp[\beta(\mu - \varepsilon)]/\{1 - \exp[\beta(\mu - \varepsilon)]\} .$$

The mean boson occupancy $b(\varepsilon)$ is called the *Bose–Einstein distribution function*. Multiplying the numerator and the denominator of the extreme right hand side by $\exp[\beta(\varepsilon - \mu)]$, we have

$$b(\varepsilon) = \frac{1}{\{\exp[\beta(\varepsilon - \mu)] - 1\}} = \frac{1}{\{\exp[(\varepsilon - \mu)/kT] - 1\}} \quad (11.20)$$

In equation (11.20), T is the absolute temperature of the heat reservoir and μ is the chemical potential of the particle reservoir. Notice that there is a (-1) in the denominator of equation (11.20) compared with a $(+1)$ in the denominator of equation (11.14), which is the corresponding equation for fermions.

11.5* THE GRAND PARTITION FUNCTION AND THERMODYNAMICS*

For the benefit of advanced readers, an outline is given in this Section of the approach to statistical thermodynamics based on the grand partition function Ξ, which is given by equation (11.9) and is a function of T, the absolute temperature of the heat reservoir, μ, the chemical potential of the particle reservoir and V, the volume of the small system in Figure 11.1.

In classical equilibrium thermodynamics the grand potential Ω is defined by the equation

$$\Omega = U - TS - \mu N \ . \tag{11.21}$$

The grand potential is a function of the thermodynamic state of the system. It is an extensive variable. For infinitesimal changes, we have

$$d\Omega = dU - T\,dS - S\,dT - \mu\,dN - N\,d\mu \ . \tag{11.22}$$

According to the thermodynamic identity, which is equation (3.85) of Chapter 3,

$$dU = T\,dS - p\,dV + \mu\,dN \ .$$

Substituting in equation (11.22), we have

$$d\Omega = -S\,dT - N\,d\mu - p\,dV \ . \tag{11.23}$$

The natural variables for the grand potential Ω are T, μ and V. Using equation (A1.31) of Appendix 1, we have

$$d\Omega = (\partial\Omega/\partial T)_{\mu,V}\,dT + (\partial\Omega/\partial\mu)_{T,V}\,d\mu + (\partial\Omega/\partial V)_{T,\mu}\,dV \ . \tag{11.24}$$

Comparing equations (11.23) and (11.24), it can be seen that

$$S = -(\partial\Omega/\partial T)_{\mu,V}; \qquad N = -(\partial\Omega/\partial\mu)_{T,V}; \qquad p = -(\partial\Omega/\partial V)_{T,\mu} \ . \tag{11.25}$$

For a one component system, according to equations (6.41) and (6.31)

$$G = \mu N = U - TS + pV \ . \tag{11.26}$$

Hence, equation (11.21) can be rewritten in the form

$$\Omega = -pV \ . \tag{11.27}$$

In statistical mechanics, the grand potential Ω is defined in terms of the grand partition function Ξ by the equation

$$\Omega = -kT \log \Xi \tag{11.28}$$

where $\Xi(T, \mu, V)$ is given by equation (11.9) and is a function of V, the volume of the system, T, the absolute temperature of the heat reservoir and μ, the chemical potential of the particle reservoir. By analogy with equations (11.25), one would expect that

$$p = \bar{p} = -(\partial\Omega/\partial V)_{T,\mu} = kT(\partial \log \Xi/\partial V)_{T,\mu} \tag{11.29}$$

$$N = \bar{N} = -(\partial\Omega/\partial\mu)_{V,T} = kT(\partial \log \Xi/\partial\mu)_{V,T} \tag{11.30}$$

$$S = -(\partial\Omega/\partial T)_{\mu,V} = (\partial(kT \log \Xi)/\partial T)_{\mu,V} . \tag{11.31}$$

By analogy with equation (11.27) one would expect that

$$pV = kT \log \Xi = -\Omega . \tag{11.32}$$

Equations (11.29), (11.30), (11.31) and (11.32) are developed from statistical mechanics in Problems (11.4), (11.3), (11.6) and (11.7) respectively. According to equation (11.26), U is equal to $(TS + \mu N - pV)$. Substituting for S, N and pV from equations (11.31), (11.30) and (11.32), it follows that

$$U = kT\mu(\partial \log \Xi/\partial\mu)_{T,V} + kT^2(\partial \log \Xi/\partial T)_{\mu,V} . \tag{11.33}$$

Equations (11.33), (11.29), (11.31) and (11.30) give the thermodynamic variables U, p, S and N, if the grand partition function Ξ of the small system in Figure 11.1 is known as a function of T, the absolute temperature of the heat reservoir, μ the chemical potential of the particle reservoir and V the volume of the small system. The approach to statistical thermodynamics based on the grand partition function will be applied to the ideal fermion gas in Section 12.1.8* and to the ideal boson gas in Section 12.5.3* of Chapter 12. A reader interested in a fuller discussion is referred to *Introductory Statistical Mechanics* by R. E. Turner and D. S. Betts, (Sussex University Press, 1974).

PROBLEMS

Problem 11.1*
Using a method based on classical equilibrium thermodynamics, similar to the methods used in Sections 6.2.2 and 6.4.2, show that, when the small system in Figure 11.1 is in thermodynamic equilibrium with the (heat and particle) reservoir, the grand potential Ω of the small system is a minimum.

Problem 11.2*

Using a method based on statistical mechanics, similar to the methods used in Sections 6.2.3 and 6.4.3, show that, when the small system in Figure 11.1 is in equilibrium with the heat and particle reservoir, the grand potential Ω of the small system is a minimum.

(*Hint*: Substitute the expression for g_R, given by equation (11.6), into the condition that $g_R g$ is a maximum at equilibrium.)

Problem 11.3*

Substitute for the grand partition function Ξ from equation (11.9) into the expression

$$kT(\partial \log \Xi/\partial \mu)_{T,V} = (kT/\Xi)(\partial \Xi/\partial \mu)_{T,V}$$

and carry out the partial differentiation to show that

$$\bar{N} = kT(\partial \log \Xi/\partial \mu)_{T,V}$$

where according to equation (11.11)

$$\bar{N} = (1/\Xi) \sum_N \sum_i N \exp[\beta(\mu N - \varepsilon_i)] \ .$$

Problem 11.4*

Substitute for the grand partition function Ξ from equation (11.9) into the expression

$$kT(\partial \log \Xi/\partial V)_{\mu,T} = (kT/\Xi)(\partial \Xi/\partial V)_{\mu,T}$$

and carry out the partial differentiation to show that

$$\bar{p} = kT(\partial \log \Xi/\partial V)_{\mu,T}$$

where according to equations (11.11) and (A5.2) of Appendix 5

$$\bar{p} = (1/\Xi) \sum_N \sum_i [-(\partial \varepsilon_i/\partial V)] \exp[\beta(\mu N - \varepsilon_i)] \ .$$

Problem 11.5*

Follow the method of Section 7.1.2, using the expression for $P_{N,i}$ given by equation (11.8), to show that, for the grand canonical ensemble (constant T and μ), the entropy is

$$S = -k \sum_N \sum_i P_{N,i} \log P_{N,i} \ .$$

This is the Boltzmann definition of entropy.

(*Hint*: In this case, comparing equations (7.4) and (7.5) with the thermodynamic identity, equation (3.85), we have

$$\sum_N \sum_i \varepsilon_i \, dP_{N,i} = T \, dS + \mu \, d\bar{N}.)$$

Problem 11.6*

Substitute the value for the probability $P_{N,i}$, given by equation (11.8), in the Boltzmann expression for the entropy, namely

$$S = -k \sum_N \sum_i P_{N,i} \log P_{N,i}$$

to show that for the grand canonical ensemble

$$S = -k \sum_N \sum_i P_{N,i}[-(\varepsilon_i/kT) + (\mu N/kT) - \log \Xi] \ .$$

Use equation (11.11) to show that, for fixed T and μ,

$$-k \sum_N \sum_i P_{N,i}[-(\varepsilon_i/kT)] = U/T; \qquad -k \sum_N \sum_i P_{N,i}[\mu N/kT] = -(\mu \bar{N}/T).$$

Since Ξ is a constant, if T, V and μ are constant, we have

$$k \sum_N \sum_i P_{N,i} \log \Xi(T, V, \mu) = k \log \Xi \sum_N \sum_i P_{N,i} = k \log \Xi \ .$$

Hence show that

$$S = (U/T) - (\mu \bar{N}/T) + k \log \Xi \ . \tag{11.34}$$

Consider the relation

$$(\partial(kT \log \Xi)/\partial T)_{\mu,V} = k \log \Xi + (kT/\Xi)(\partial \Xi/\partial T)_{\mu,V} \ .$$

Substitute for Ξ from equation (11.9) into the $(\partial \Xi/\partial T)$ term, carry out the partial differentiation, and hence show that

$$S = (\partial(kT \log \Xi)/\partial T)_{\mu,V} \ . \tag{11.31}$$

Problem 11.7*

It follows from equation (11.34) of Problem 11.6 that

$$\Omega = -kT \log \Xi = U - TS - \mu N \ .$$

Show that $[-kT \log \Xi \ (T, \mu, V)]$ is an extensive variable, so that if the intensive variables T and μ are kept constant, $(-kT \log \Xi)$ is proportional to the volume V. Hence show that

$$(\partial \log \Xi/\partial V)_{T,\mu} = (\log \Xi)/V \ . \tag{11.35}$$

Show that equation (11.29) for the pressure p can be rewritten in the form

$$pV = kT \log \Xi = -\Omega \ . \tag{11.32}$$

Chapter 12

Some Applications of the Fermi–Dirac and the Bose–Einstein Distributions

12.1 THE IDEAL FERMION GAS

12.1.1 Introduction

Fermions are particles of half integral spin which obey Fermi–Dirac statistics. Consider an ideal fermion gas consisting of N identical fermions, each of mass m, in a box of volume V, kept at a constant absolute temperature T, as shown in Figure 11.2 of Chapter 11. It will be assumed that there are no interactions between the fermions, and that each fermion occupies a single particle state. (In this context, a single particle state is sometimes referred to as an orbital.) These single particle states (orbitals) can be determined by solving Schrödinger's time independent equation for a single particle in a box. This is done for the case of a cubical box in Appendix 3. According to the Pauli exclusion principle there can only be one or zero fermions in any single particle state with a given set of quantum numbers. [References: Section 2.1.5 of Chapter 2 and Section 11.3 of Chapter 11.] According to equations (A3.18) of Appendix 3, in the non-relativistic limit, the number of *single* particle states (orbitals) for a free particle in a cubical box, which differ in at least one of the quantum numbers n_x, n_y and n_z and spin quantum number s, and which have energy eigenvalues in the energy range ε to $(\varepsilon + d\varepsilon)$ is

$$D(\varepsilon)\, d\varepsilon = \frac{gV}{4\pi^2} \left(\frac{2m}{\hbar^2}\right)^{3/2} \varepsilon^{1/2}\, d\varepsilon \ . \tag{12.1}$$

In equation (12.1), $g = (2s + 1)$ is the spin degeneracy. For fermions of spin $\frac{1}{2}$, such as electrons, g is equal to 2. It can be shown that equation (12.1) is the correct expression for the density of states $D(\varepsilon)$ whatever the shape of the box. Equation (12.1) holds for both fermions and bosons. The variation of D with ε is shown in Figure 12.1(a). According to equation (12.1), D is proportional to $\varepsilon^{1/2}$ at a fixed volume V.

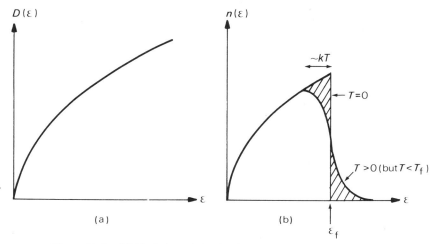

Figure 12.1—(a) Variation of the density of single particle states $D(\varepsilon)$ for a fermion in a box with the energy eigenvalue ε of the single particle state. The volume V of the box is kept constant. (b) Energy distribution of N fermions in a box at $T = 0$ and at a temperature T less than the Fermi temperature T_f, where T_f is given by equation (12.10).

12.1.2 The Fermi–Dirac distribution function

According to equation (11.14), the ensemble average number of fermions in a particular single particle state, which has (single particle) energy eigenvalue ε, is given by the Fermi–Dirac distribution function:

$$f(\varepsilon) = \frac{1}{[\exp\{(\varepsilon - \mu)/kT\} + 1]} = \frac{1}{[\exp\{\beta(\varepsilon - \mu)\} + 1]} \qquad (12.2)$$

where T is the absolute temperature of the heat reservoir and μ is the chemical potential of the particle reservoir, which in the present case is the fermions in the other single particle states inside the box. The determination of μ in terms of N, V and T will be discussed in Section 12.1.4. In this Section, we shall quote the appropriate values for μ. The variation of $f(\varepsilon)$ with ε, at various absolute temperatures T, is shown in Figure 12.2.

If $T = 0$ and if ε is less than $\mu(0)$, the value of the chemical potential at the absolute zero of temperature, $f(\varepsilon) = (e^{-\infty} + 1)^{-1} = 1$. If $T = 0$ and $\varepsilon > \mu(0)$, then $f(\varepsilon) = (e^{+\infty} + 1)^{-1} = 0$. Hence, according to equation (12.2), at the absolute zero of temperature all the single particle states which have energy eigenvalues up to the value of the chemical potential $\mu(0)$ at the absolute zero are occupied. All the single particle states having energy eigenvalues greater than $\mu(0)$ are empty at the absolute zero, as shown in Figure 12.2(a). This result follows directly from the Pauli exclusion principle, according to which there can only be a maximum of one

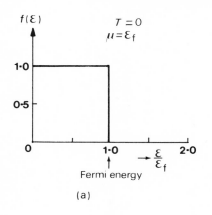

(a)

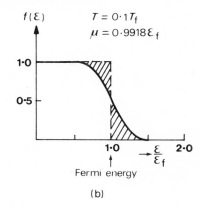

(b)

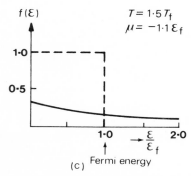

(c)

Figure 12.2—The Fermi–Dirac distribution function $f(\varepsilon)$, given by equation (12.2), is plotted against $\varepsilon/\varepsilon_f$, where ε is the energy eigenvalue of the single particle state and $\varepsilon_f = kT_f$ is the Fermi energy given by equation (12.10), for absolute temperatures of (a) $T = 0$ (b) $T = 0.1T_f$ (c) $T = 1.5T_f$. At $T = 0$, all the single particle states up to the Fermi energy ε_f are filled, but the states above ε_f, are empty. When $T = 0.1T_f$, $f(\varepsilon)$ is only modified significantly in a temperature range of a few kT centred on ε_f. When $T = 1.5T_f$, $f(\varepsilon)$ is less than 0.5 for all the single particle states including the ground state, and the chemical potential $\mu(T)$ is negative. [This example illustrates how changing the temperature T of the heat reservoir changes $f(\varepsilon)$. The values of ε_f and T_f, given by equation (12.10), do not vary with temperature if N and V are kept constant.]

fermion in each single particle state. All the N fermions in the system cannot all go into the single particle ground state at the absolute zero, but they occupy the N single particle states having the N lowest energy eigenvalues. These range in energy from zero up to the value $\mu(0)$ of the chemical potential at $T = 0$, as shown for the case of a simple system composed of eight identical fermions of spin $\frac{1}{2}$ in Figure 12.3(a). For fermions of spin $\frac{1}{2}$ in a cubical box, there can be two fermions having spin quantum numbers $+\frac{1}{2}$ and $-\frac{1}{2}$ respectively, associated with each set of the quantum numbers n_x, n_y and n_z. The two different spin states with spin quantum numbers $+\frac{1}{2}$ and $-\frac{1}{2}$ are represented by arrows pointing in opposite directions in Figure 12.3(a). The value $\mu(0)$ of the chemical potential at the absolute zero of temperature is called the *Fermi energy* and is denoted ε_f. The *Fermi temperature T_f* is defined by the equation

$$\varepsilon_f = kT_f \ . \tag{12.3}$$

Summarising, at $T = 0$ the N fermions occupy all the N single particle states having energy eigenvalues up to the Fermi energy. It will be shown in Section 12.1.6 that the Fermi energy ε_f and the Fermi temperature T_f are given in terms of m, N, V and the spin degeneracy g by equation (12.10), which is

$$\varepsilon_f = \frac{\hbar^2}{2m} \left(\frac{6\pi^2 N}{gV} \right)^{2/3} = kT_f \ .$$

For example, it is shown in Problem 12.7 that the Fermi temperature of a gas of $^3_2$He atoms at STP is 0.07K, which is much less than room temperatures. On the other hand, it will be shown in Section 12.2.1 that the Fermi temperature of the free electrons in copper is 8×10^4K, which is much greater than room temperature.

The Fermi–Dirac distribution function for an absolute temperature $T \ll T_f$ is shown in Figure 12.2(b). The example chosen is $T = 0.1T_f$. The numerical values for this case are developed in Problem 12.1. According to equation (12.6), which will be introduced in Section 12.1.4, when $T = 0.1T_f$ the value of the chemical potential $\mu(T)$ is $0.9918\varepsilon_f$, which is only just below the Fermi energy ε_f. It can be seen that at low values of ε, much less than ε_f, $f(\varepsilon)$ is very close to unity. For $T = 0.1T_f$, the large changes in $f(\varepsilon)$ are confined to an energy range of about kT on either side of the Fermi energy ε_f. In this energy range some of the fermions, which at $T = 0$ would have occupied single particle states having energy eigenvalues just below the Fermi energy, will, at a temperature $T \ll T_f$, be found in single particle states having energy eigenvalues just above the Fermi energy, as shown for the case of eight identical fermions of spin $\frac{1}{2}$ in Figure 12.3(b). The Fermi–Dirac distribution function $f(\varepsilon)$ is equal to 0.5 when $\varepsilon = \mu(T) = 0.9918\varepsilon_f$ in Figure 12.2(b).

According to equation (12.2), for a single particle state having an energy eigenvalue equal to $(\mu(T) + x)$, which is x above $\mu(T)$, the value of the chemical potential at the temperature T, we have

$$f(\mu + x) = [\exp(x/kT) + 1]^{-1} \ .$$

For a single particle state having energy eigenvalue $(\mu - x)$, which is x

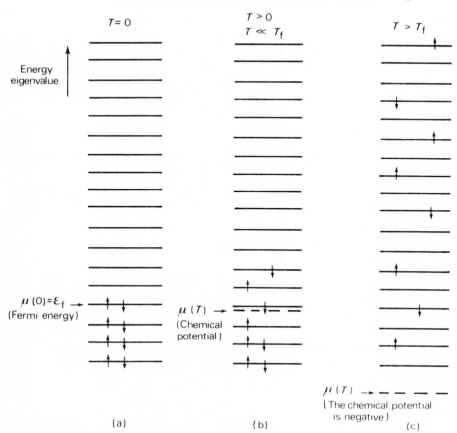

Figure 12.3—Pictorial representation of the variation of the Fermi–Dirac distribution function $f(\varepsilon)$ with the single particle energy eigenvalue ε for a system consisting of 8 identical fermions of spin $\frac{1}{2}$. The two different directions of the arrows represent the two states with spin quantum numbers $+\frac{1}{2}$ and $-\frac{1}{2}$ respectively. (a) At $T = 0$, all the single particle states up to the Fermi energy ε_f are filled. All the single particle states above ε_f are empty. (b) When $T \ll T_f$, some of the fermions, which at $T = 0$ would be in single particle states just below ε_f are in single particle states just above ε_f. In this case, the chemical potential $\mu(T)$ is just below the Fermi energy ε_f. (c) When $T > T_f$, $f(\varepsilon)$ is less than 0.5 for the single particle ground state, and the chemical potential $\mu(T)$ is negative.

below the chemical potential $\mu(T)$,

$$f(\mu - x) = [\exp(-x/kT) + 1]^{-1} .$$

The reader can check by direct substitution that

$$f(\mu + x) = 1 - f(\mu - x) .$$

The quantity $[1 - f(\mu - x)]$ is the probability that there is no fermion in the single particle state having energy eigenvalue $(\mu - x)$. Hence the probability $f(\mu + x)$ that there is a fermion in a single particle state having an energy eigenvalue x above the chemical potential $\mu(T)$ is equal to the probability that there is a vacancy (hole) in the single particle state having an energy eigenvalue x below the chemical potential. This leads to the similar shapes of the shaded areas above and below the chemical potential $\mu(T)$ in Figure 12.2(b).

The variation of the Fermi–Dirac distribution function $f(\varepsilon)$ with ε for the case when $T = 1.5T_f$ is shown in Figure 12.2(c). In this case $f(\varepsilon)$ is less than 0.5 for the single particle ground state, and $f(\varepsilon)$ decreases continuously with increasing ε. The case when $T > T_f$ is illustrated in Figure 12.3(c) for a system consisting of eight identical fermions of spin $\frac{1}{2}$. It will be shown in Section 12.1.7 that, in the high temperature classical limit when $T \gg T_f$, $f(\varepsilon)$ is proportional to $\exp(-\varepsilon/kT)$ and $f(\varepsilon)$ is very much less than unity at all energies.

For a given value of ε, the expression for $f(\varepsilon)$, given by equation (12.2), depends on the variables μ and T. The volume V and the number of particles N do not appear explicitly in equation (12.2). What, then, happens to the average number of fermions $f(\varepsilon)$ in a single particle state of energy eigenvalue ε, if, for example, N, the total number of fermions, is doubled at fixed values of V and T? The answer is that if N is changed, the value of the chemical potential μ is changed. [Reference: equation (12.5) of Section 12.1.4.] A change in μ at constant temperature T leads to a change in $f(\varepsilon)$, the average number of fermions in the single particle state of energy eigenvalue ε. [An illustrative example is developed in Problem 12.4.] Similarly, a change in the volume V of the box at constant N and T, changes μ and hence $f(\varepsilon)$. [A change in V also changes the energy eigenvalues of the single particle states.] The dependence of $f(\varepsilon)$ on the absolute temperature T, at fixed V and N, comes via changes in both μ and T in equation (12.2). This is illustrated by the differences between Figures 12.2(a), 12.2(b) and 12.2(c). [For fixed V and N, the energy eigenvalues ε are unchanged.]

12.1.3 The energy spectrum of an ideal fermion gas

The average number of fermions $n(\varepsilon)\,d\varepsilon$ in single particle states which have energy eigenvalues between ε and $(\varepsilon + d\varepsilon)$ is given by the product of $D(\varepsilon)\,d\varepsilon$, the number of single particle states (orbitals) having energy eigenvalues in the energy range ε to $(\varepsilon + d\varepsilon)$ and $f(\varepsilon)$, the mean number of fermions in each of these single particle states. Using equations (12.1) and (12.2), we have

$$n(\varepsilon)\,d\varepsilon = \frac{gV}{4\pi^2}\left(\frac{2m}{\hbar^2}\right)^{3/2}\frac{\varepsilon^{1/2}\,d\varepsilon}{[\exp\{(\varepsilon - \mu)/kT\} + 1]}. \tag{12.4}$$

At the absolute zero of temperature, $T = 0$, $\mu = \varepsilon_f$ and $n(\varepsilon)$ is proportional to $\varepsilon^{1/2}$ for fermion energies up to the Fermi energy ε_f, and $n(\varepsilon)$ is zero for $\varepsilon > \varepsilon_f$, as shown in Figure 12.1(b). If the temperature is increased a little above the absolute zero, but with $T \ll T_f$, some of the fermions which at $T = 0$ would have been in single particle states having energy eigenvalues just below the Fermi energy, now occupy some of the single particle states having energy eigenvalues just above the Fermi energy. When $T \ll T_f$, that is when $kT \ll \varepsilon_f$, the energy distribution of the fermions is only changed significantly in the vicinity of the Fermi energy, as shown in Figure 12.1(b). At much higher temperatures when $T \gg T_f$, that is when $kT \gg \varepsilon_f$, a large proportion of the fermions will be in single particle states having energy eigenvalues well above the Fermi energy. It will be shown in Section 12.1.7 that, in the classical limit, when $T \gg T_f$, the Fermi–Dirac energy distribution goes over to the Maxwell–Boltzmann distribution, which was developed in Section 4.9 of Chapter 4.

12.1.4 The chemical potential of an ideal fermion gas

The value of the chemical potential $\mu(T, V, N)$ in equations (12.2) and (12.4) has not yet been determined. It is being assumed that there is no interchange of fermions between the ideal fermion gas and the container. In the case of the ideal fermion gas, it is the fermions in the other single particle states which act as the particle reservoir for the single particle state chosen as the small system for the application of equation (12.2), which was derived from the grand canonical distribution in Section 11.3 of Chapter 11. The total number N of fermions in the ideal fermion gas is equal to the integral of equation (12.4) from $\varepsilon = 0$ to $\varepsilon = \infty$. We have

$$N = \int_0^\infty n(\varepsilon)\, d\varepsilon = \frac{gV}{4\pi^2}\left(\frac{2m}{\hbar^2}\right)^{3/2}\int_0^\infty \frac{\varepsilon^{1/2}\, d\varepsilon}{[\exp\{(\varepsilon - \mu)/kT\} + 1]} \; . \quad (12.5)$$

This is an integral equation which can be solved numerically to determine the chemical potential μ in terms of T, V and N. If N, V or T is changed, the chemical potential μ is also changed.

The variation of the chemical potential μ of an ideal fermion gas with absolute temperature T is shown in Figure 12.4. (Reference: Turner and Betts [1a].) At $T = 0$, the chemical potential $\mu(0)$ is equal to the Fermi energy ε_f, where ε_f is given by equation (12.10). When T is very much less than the Fermi temperature T_f, the chemical potential μ is only just below the Fermi energy. As the temperature increases further the chemical potential decreases significantly, and is zero at a temperature just below the Fermi temperature T_f. As T increases above the Fermi temperature, the chemical potential gets more and more negative. [This variation in the value of the chemical potential μ with T is illustrated by the position of μ in Figures 12.3(a), 12.3(b) and 12.3(c), for the simple case of eight identical

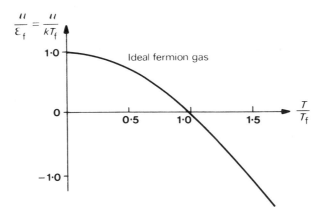

Figure 12.4—Variation of the chemical potential μ of an ideal fermion gas with absolute temperature T. The Fermi energy $\varepsilon_f = kT_f$ is given by equation (12.10). At $T = 0$, $\mu = \varepsilon_f$. The chemical potential μ is zero when T is just below T_f. At temperatures $T > T_f$, μ is negative. [Adapted with permission from *Introductory Statistical Mechanics*, by R. E. Turner and D. S. Betts.]

fermions of spin $\frac{1}{2}$.] Changes in V and N at a fixed value of T lead to a change in the value of μ through their effects on the values of ε_f and T_f to be used in Figure 12.4.

It is possible to interpret the main features of Figure 12.4 in terms of the Fermi–Dirac distribution function $f(\varepsilon)$, given by equation (12.2). According to equation (12.2), when ε is equal to μ, $f(\varepsilon)$ is equal to 0.5. Hence for an ideal fermion gas, the chemical potential μ is equal to that value of energy at which $f(\varepsilon)$, the mean number of fermions in the single particle state, is equal to 0.5.† At the absolute zero of temperature, μ is equal to the Fermi energy ε_f. When the temperature is just above the absolute zero, only fermions which at $T = 0$ would have been in single particle states having energy eigenvalues just below the Fermi energy ε_f are in excited single particle states having energy eigenvalues just above ε_f. The energy at which the mean occupancy $f(\varepsilon)$ is equal to 0.5 remains close to, but just below, the Fermi energy ε_f. As the temperature is increased further, more and more fermions occupy excited single particle states above the Fermi energy, and the value of the energy at which $f(\varepsilon)$ is equal to 0.5, moves to lower and lower energies until eventually $f(\varepsilon)$ is equal to 0.5 for the single particle ground state. At this temperature μ is equal to the energy of the single particle ground state, which for fermions in a cubical

†For semiconductors, the density of states $D(\varepsilon)$ is not given by equation (12.1) and the chemical potential generally comes in the gap between the valence and conduction bands.

box is given by equation (2.5) of Section 2.1.4. At even higher temperatures, more and more of the fermions are in excited states above the Fermi energy and the mean occupancy of the single particle ground state is less than 0.5. Putting $\varepsilon = 0$ and $f < 0.5$ in equation (12.2), we have

$$(e^{-\mu/kT} + 1)^{-1} < 0.5 \ .$$

Rearranging, we have

$$e^{-\mu/kT} > 1 \ .$$

Taking the natural logarithm of both sides, we have

$$-\mu/kT > \log 1 = 0 \ .$$

This shows that the chemical potential μ is negative in the high temperature limit.

It can be shown analytically that, in the low temperature limit when $T \ll T_f$, the chemical potential of an ideal fermion gas is given, to a good approximation, by

$$\mu \simeq \varepsilon_f \left[1 - \frac{\pi^2}{12} \left(\frac{T}{T_f} \right)^2 \right] \qquad (T \ll T_f) \ . \qquad (12.6)$$

(Reference: Turner and Betts [1b].) For example, according to equation (12.6), if $T = 0.01 T_f$, μ is equal to $0.999918 \varepsilon_f$, whereas, if $T = 0.1 T_f$, μ is equal to $0.9918 \varepsilon_f$. Hence in the low temperature limit, when $T \ll T_f$, the chemical potential μ is only just fractionally less than the Fermi energy ε_f, which is the value of the chemical potential at the absolute zero.

In the high temperature limit, when $T \gg T_f$, the ideal fermion gas approximates to an ideal classical gas. It was shown in Section 5.9.8 that, for an ideal monatomic gas in the classical limit,

$$\exp[-\mu/kT] = (gV/N)(mkT/2\pi\hbar^2)^{3/2} \ . \qquad (5.84)$$

(A derivation of equation (5.84) as the classical limit of the ideal fermion gas is given by Turner and Betts [1c].) As an example of a fermion gas in the classical limit, consider a mole of $_2^3$He gas atoms at STP. It is shown in Problem 12.7 that in this example, $\exp(-\mu/kT)$ is equal to 3.4×10^5, so that $\mu/kT = -12.7$. This illustrates that, in the high temperature limit, when $T \gg T_f$, the chemical potential μ is negative, and $\exp(-\mu/kT) \gg 1$. To reach this classical limit, the curve in Figure 12.4 would have to be extrapolated to larger values of T/T_f.

12.1.5 Energy, pressure and heat capacity of an ideal fermion gas

Using equation (12.4), we find that the ensemble average energy of the ideal fermion gas is

$$U = \int_0^\infty \varepsilon n(\varepsilon) \, d\varepsilon = \frac{gV}{4\pi^2} \left(\frac{2m}{\hbar^2}\right)^{3/2} \int_0^\infty \frac{\varepsilon^{3/2} \, d\varepsilon}{[\exp\{\beta(\varepsilon - \mu)\} + 1]} \quad . \quad (12.7)$$

The variation of U with the absolute temperature T, at constant volume V, is shown in Figure 12.5(a). According to equation (5.74) of Section 5.9.4, U is equal to $3NkT/2$ for an ideal monatomic classical gas. This relation is shown by the dotted line in Figure 12.5(a). It can be seen from Figure 12.5(a) that the energy of the ideal fermion gas is always greater than the energy an ideal monatomic classical gas would have if the volume V, the temperature T and the number of particles N were the same. When T is greater than the Fermi temperature T_f, the behaviour of the ideal fermion gas begins to approximate to the ideal monatomic classical gas laws developed in Section 5.9 of Chapter 5. When $T < T_f$, the ideal fermion gas departs markedly from the ideal classical gas laws. The ideal fermion gas is said to be *degenerate* in this temperature range. It will be shown in Section 12.1.6 that, when $T = 0$, the mean energy of the fermion gas is equal to $\frac{3}{5}N\varepsilon_f$.

The heat capacity at constant volume is given by $C_V = (\partial U/\partial T)_V$, where U is given by equation (12.7). The variation of the molar heat capacity $C_{V,m}$ of the ideal fermion gas with temperature is shown in Figure 12.5(b). At low temperatures when $T \ll T_f$, the variation of $C_{V,m}$ with T is linear. When $T > T_f$ the molar heat capacity approaches the value of $\frac{3}{2}R$, which is valid for an ideal monatomic gas in the classical limit.

According to equation (A5.6) of Appendix 5, the mean pressure of the ideal fermion gas is

$$p = \sum P_i \left(-\frac{\partial E_i(V, N)}{\partial V}\right) , \qquad (A5.6)$$

where $E_i(V, N)$ is the energy eigenvalue of the ith N particle state of the N particle fermion system, and P_i is the probability that the N particle system is in that N particle state. According to equation (2.5) of Section 2.1.4, for a cubical box, the energy eigenvalues of the *single* particle states are given by

$$\varepsilon = (\hbar^2\pi^2/2mV^{2/3})(n_x^2 + n_y^2 + n_z^2) .$$

In an ideal fermion gas, in which there are no interactions between the fermions, the energy of the N fermion system is equal to the sum of the energies of the N individual fermions. Hence the energy eigenvalue $E_i(V, N)$ of the ith microstate of the N fermion system is

$$E_i(V, N) = (\hbar^2\pi^2/2mV^{2/3})\sum_j [(n_x)_j^2 + (n_y)_j^2 + (n_z)_j^2] = K_i/V^{2/3} .$$

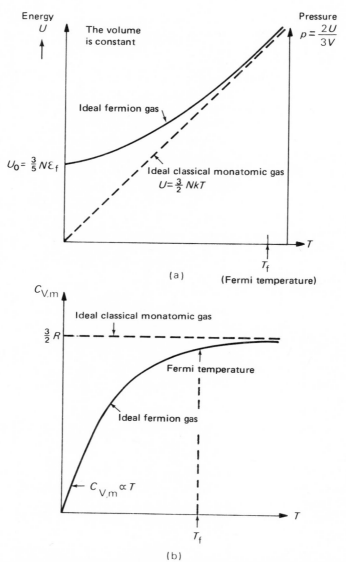

Figure 12.5—Behaviour of an ideal fermion gas. (a) The variations of the energy U and pressure p with absolute temperature T are shown. (b) The variation of the molar heat capacity $C_{V,m}$ with absolute temperature T is shown. When $T \ll T_f$, $C_{V,m}$ is proportional to T. At high temperatures $C_{V,m}$ tends to the classical value of $3R/2$. (Adapted with permission from *Statistical Mechanics and Properties of Matter* by E. S. R. Gopal, Ellis Horwood Ltd., Chichester.)

The summation is over the N single particle states occupied by the N fermions, when the system is in its ith N particle state. Notice that E_i is proportional to $V^{-2/3}$, and K_i is a constant for the ith N particle state. Substituting for E_i in the expression for p given by equation (A5.6) of Appendix 5 and carrying out the partial differentiation, at a particular value of volume V, we have

$$p = \Sigma P_i(2K_i/3V^{5/3}) = \tfrac{2}{3}\Sigma P_i(E_i/V) = 2U/3V \ . \qquad (12.8)$$

where U is the ensemble average energy of the N fermion system. Equation (12.8) is valid for all values of U for an *ideal* fermion gas, in which there is no interaction between the fermions.[†] The same argument can be applied to derive equation (12.8) for an ideal boson gas.

Since, according to equation (12.8), the pressure p of an ideal fermion gas is proportional to U, if V is constant, the variation of p with T at constant volume has the same form as the variation of U with T shown in Figure 12.5(a). It can be seen from Figure 12.5(a), that the pressure of an ideal fermion gas is always greater than the pressure an ideal classical monatomic gas would have, if V, T and N were the same. [According to equation (5.76), the pressure of an ideal monatomic gas is equal to NkT/V in the classical limit.]

Summarising, the differences between the properties of the ideal fermion gas and the ideal monatomic classical gas (developed in Section 5.9 using the Boltzmann distribution and the partition function) are most significant when the temperature of the gas is below the Fermi temperature. The ideal fermion gas is said to be **degenerate** in this temperature range.

12.1.6 The completely degenerate ideal fermion gas

At the absolute zero, when $T = 0$, there are no fermions having energies above the Fermi energy ε_f, and the upper limit of infinity in equation

---

[†]At a particular value of volume V, equation (12.8) can be rewritten in the form

$$pV^{5/3} = \tfrac{2}{3}\Sigma P_i K_i$$

According to the adiabatic approximation of quantum mechanics (Ehrenfest's principle), the probabilities P_i are constant in a reversible adiabatic (isentropic) change of volume of the ideal fermion gas. [Reference: Section 7.1.5* of Chapter 7.] The K_i are all constant, if the occupancies of all the single particle states making up the ith N particle state are unchanged. (It is being assumed that the fermions do not have any internal degrees of freedom which can be excited thermally such as the rotational energy of a diatomic molecule.) If the P_i and K_i are all constant in a reversible adiabatic change of volume of the ideal fermion gas, we have $pV^{5/3} = $ a constant (S is constant). This equation is valid at all temperatures for an *ideal* fermion gas, even if it is degenerate. It is also true for an ideal boson gas. (Reference: Landau and Lifshitz [2a].)

(12.5) can be replaced by ε_f. Hence at $T = 0$, equation (12.5) becomes

$$N = \frac{gV}{4\pi^2}\left(\frac{2m}{\hbar^2}\right)^{3/2}\int_0^{\varepsilon_f}\varepsilon^{1/2}\,d\varepsilon = \frac{gV}{4\pi^2}\left(\frac{2m}{\hbar^2}\right)^{3/2}\frac{2\varepsilon_f^{3/2}}{3}\,. \tag{12.9}$$

Rearranging, we find that the Fermi energy ε_f, which according to equation (12.3) is also equal to kT_f, is given by

$$\varepsilon_f = kT_f = \frac{\hbar^2}{2m}\left(\frac{6\pi^2 N}{gV}\right)^{2/3}\,. \tag{12.10}$$

The Fermi energy ε_f and Fermi temperature T_f are both proportional to $(N/V)^{2/3}$, so that the Fermi energy and Fermi temperature both increase, if the concentration of the fermions (N/V) is increased. The Fermi energy ε_f and Fermi temperature T_f are inversely proportional to the mass of the fermions so that, for example, for the same concentration (N/V), the Fermi energy and Fermi temperature are bigger for electrons than for a gas of $_2^3\text{He}$ atoms.

The ensemble average energy of a fermion in an ideal fermion gas at the absolute zero of temperature is

$$\bar{\varepsilon} = \frac{1}{N}\int_0^{\varepsilon_f}\varepsilon\,n(\varepsilon)\,d\varepsilon = \frac{gV}{4\pi^2 N}\left(\frac{2m}{\hbar^2}\right)^{3/2}\int_0^{\varepsilon_f}\varepsilon^{3/2}\,d\varepsilon$$

$$\bar{\varepsilon} = \frac{gV}{10\pi^2 N}\left(\frac{2m}{\hbar^2}\right)^{3/2}\varepsilon_f^{5/2}\,.$$

Using equation (12.10), we find that at $T = 0$ the average fermion energy is

$$\bar{\varepsilon} = \tfrac{3}{5}\varepsilon_f\,. \tag{12.11}$$

The total energy of an ideal fermion gas consisting of N non-interacting fermions is the product of N and the mean energy of each fermion. Hence, at $T = 0$, the total energy of an ideal fermion gas is

$$U_0 = \tfrac{3}{5}N\varepsilon_f\,, \tag{12.12}$$

as shown in Figure 12.5(a).

Using equation (12.12) in equation (12.8), we find that the pressure of an ideal fermion gas at the absolute zero of temperature is

$$p_0 = \tfrac{2}{5}(N/V)\varepsilon_f = \frac{\hbar^2}{5m}\left(\frac{6\pi^2}{g}\right)^{2/3}\left(\frac{N}{V}\right)^{5/3}\,. \tag{12.13}$$

This pressure can have enormous values. [Reference: Problem 12.5.] It is important in the theory of white dwarf and neutron stars. [Reference: Section 12.4.]

At $T = 0$, all the single particle states up to the Fermi energy are occupied. Since the fermions are identical, this corresponds to only one state of the N particle system. Hence at $T = 0$, the entropy given by equation (3.22) is equal to $k \log 1$, which is zero, showing that the entropy of an ideal fermion gas is zero at the absolute zero, in agreement with the third law of thermodynamics.

12.1.7 The classical limit

It can be seen from Figures 12.5(a) and 12.5(b) that when $T \gg T_f$, the behaviour of an ideal fermion gas approaches the ideal monatomic classical gas laws. It is shown in Problem 12.7 that the Fermi temperature of a gas made up of $^3_2$He atoms at STP is 0.07K, showing that at STP, $T \gg T_f$.

According to equation (5.84) of Section 5.9.8, for an ideal monatomic gas in the classical limit, we have

$$\exp[-\mu/kT] = (gV/N)(mkT/2\pi\hbar^2)^{3/2} \gg 1 \ . \tag{12.14}$$

A derivation of this equation, as the classical limit of the ideal fermion gas, is given by Turner and Betts [1c]. It is shown in Problem 12.7 that, for a mole of $^3_2$He gas atoms at STP, $\exp(-\mu/kT)$ is equal to 3.4×10^5. According to equation (5.74) of Section 5.9.4, the mean kinetic energy of a monatomic gas atom in the classical limit is $3kT/2$. Choosing $\varepsilon = 3kT/2$ to represent a typical fermion in the classical limit, and using a value of 3.4×10^5 for $\exp(-\mu/kT)$, we find that according to equation (12.2), at STP the value of $f(\varepsilon = 3kT/2)$ is 0.66×10^{-6}. This result shows that the mean occupancy of any particular single particle state is very small in the classical limit.

Since the energy ε is always positive, $\exp(\varepsilon/kT)$ is always greater than unity. Hence in the classical limit, when $\exp(-\mu/kT) \gg 1$, then

$$e^{\varepsilon/kT}e^{-\mu/kT} \gg 1$$

for all values of ε and equation (12.2) approximates to

$$f(\varepsilon) = e^{\mu/kT}e^{-\varepsilon/kT} \ .$$

The $\exp(\mu/kT)$ term is called the **absolute activity**, and is generally denoted by the symbol λ. It can be seen that, in the classical limit, the Fermi–Dirac distribution function $f(\varepsilon)$, giving the probability that there is a fermion in a single particle state of energy eigenvalue ε, is equal to $\lambda \exp(-\varepsilon/kT)$. This is the classical *Boltzmann distribution function*[†], which was used in our discussion of the ideal monatomic gas, in the classical limit, in Section 5.9 of Chapter 5.

[†]The classical Boltzmann distribution function can also be applied to calculcate $f(\varepsilon)$ at high values of ε, even if $T < T_f$, provided $\exp[(\varepsilon - \mu)/kT] \gg 1$. In this region $f(\varepsilon)$ is again very small.

In the classical limit, when $\exp[(\varepsilon - \mu)/kT]$ is much greater than unity, equation (12.4) reduces to

$$n(\varepsilon)\,d\varepsilon = (gV/4\pi^2)(2m/\hbar^2)^{3/2}e^{\mu/kT}e^{-\varepsilon/kT}\varepsilon^{1/2}\,d\varepsilon \ \ .$$

Substituting for $e^{\mu/kT}$ from equation (12.14), we obtain

$$n(\varepsilon)\,d\varepsilon = (2N/\pi^{1/2})(1/kT)^{3/2}\varepsilon^{1/2}e^{-\varepsilon/kT}\,d\varepsilon \ \ .$$

Using $\varepsilon = \tfrac{1}{2}mv^2$ so that $d\varepsilon = mv\,dv$, we have

$$n(v)\,dv = (4N/\pi^{1/2})(m/2kT)^{3/2}\exp(-mv^2/2kT)v^2\,dv \ \ . \qquad (12.15)$$

Equation (12.15) is the Maxwell–Boltzmann distribution. It is the same as equation (4.55) which was developed in Section 4.9. Notice that $\hbar$ has cancelled out in equation (12.15) in the classical limit. Even though N does not appear explicitly in the energy distribution, given by equation (12.4), it appears in the Maxwell–Boltzmann distribution given by equation (12.15). The factor N in equation (12.15) arose when we substituted for $\exp(-\mu/kT)$ using equation (12.14). This illustrates how the dependence of $n(\varepsilon)\,d\varepsilon$ on N in equation (12.4) comes in via the chemical potential μ. Since all atomic particles are either fermions or bosons, it is preferable to think of the Maxwell–Boltzmann distribution as the classical limit of either the Fermi–Dirac or Bose–Einstein velocity distributions as appropriate, rather than use the method of Section 4.9, which was based on the Boltzmann distribution.

According to equation (12.14), in the classical limit

$$V/N \gg (1/g)(2\pi\hbar^2/mkT)^{3/2} \ \ .$$

The ratio V/N is the mean volume per fermion, which is equal to d^3, where d is the mean separation of the fermions. Hence in the classical limit

$$d \gg (2\pi\hbar^2/g^{2/3}mkT)^{1/2} \ \ . \qquad (12.16)$$

According to equation (5.74), the mean kinetic energy of an atom of an ideal monatomic gas is equal to $3kT/2$ in the classical limit. If p is the *momentum* corresponding to this value of energy

$$\tfrac{1}{2}mv^2 = p^2/2m = \tfrac{3}{2}kT \ \ .$$

Hence, $$mkT = \tfrac{1}{3}\,p^2 \ \ .$$

Substituting in equation (12.16), in the classical limit we have

$$d \gg (6\pi\hbar^2/g^{2/3}p^2)^{1/2} = (h/p)(3/2\pi g^{2/3})^{1/2} \ \ .$$

For a fermion of spin $\tfrac{1}{2}$, the spin degeneracy $g = 2$ and $(3/2\pi g^{2/3})^{1/2}$ is equal to 0.55, which is of order unity. Hence in the classical limit

$$d \gg h/p \ \ . \qquad (12.17)$$

The expression h/p is the de Broglie wavelength of a fermion which has a kinetic energy of $3kT/2$. Equation (12.17) illustrates how, in the classical limit, the mean separation of the fermions is much greater than the de Broglie wavelengths of the fermions.

To illustrate the other extreme, assume that a fermion, in a fermion gas as $T = 0$, has the mean fermion energy $\bar{\varepsilon}$, which according to equation (12.12) is equal to $0.6\varepsilon_f$. The corresponding momentum p is given by

$$p = (2m\bar{\varepsilon})^{1/2} = (1.2m\varepsilon_f)^{1/2} \ .$$

Substituting for ε_f from equation (12.10), for a fermion of spin $\frac{1}{2}$, we have

$$p = [1.2m(\hbar^2/2m)(6\pi^2N/gV)^{2/3}]^{1/2} = 0.38h(N/V)^{1/3} \ .$$

Rearranging we have,

$$(V/N)^{1/3} = 0.38h/p = 0.38\lambda \ ,$$

where λ is the de Broglie wavelength of a particle of energy $0.6\varepsilon_f$. The expression $(V/N)^{1/3}$ is equal to the mean separation d the fermions would have, if they could be treated as classical point particles. In this limit, d is less than the de Broglie wavelength, so that the idea of spatially separated classical point particles breaks down when $T < T_f$. The full quantum theory must be used in this temperature range.

It can be shown that the pressure of a slightly degenerate ideal fermion gas is given, to a good approximation, by

$$pV = NkT\{1 + (N/2gV)(\pi\hbar^2/mkT)^{3/2}\} \ . \tag{12.18}$$

(Reference: Landau and Lifshitz [2b].) For a mole of $^3_2$He gas atoms at STP, the second term inside the bracket on the right hand side of equation (12.18) is approximately 5×10^{-7}, and equation (12.18) for the ideal fermion gas reduces to

$$pV \simeq NkT \tag{12.19}$$

which is the equation of state for an ideal classical gas. In the case of *real* gases at STP, the effect of the second term inside the bracket on the right hand side of equation (12.18) is much less than the effect of the attractions between the gas atoms and molecules. The latter effect can lead to significant departures from the ideal classical gas laws at room temperatures.

12.1.8* The thermodynamics of an ideal fermion gas*

Consider the case of an N particle system, composed of N non-interacting identical fermions inside a box. Using equation (12.8) and equation (11.32), which was developed from statistical mechanics in Problem 11.7* of Chapter 11, we have

$$pV = \tfrac{2}{3}U = -\Omega = kT \log \Xi$$

where Ω is the grand potential and Ξ is the grand partition function. Substituting the value of U for an ideal fermion gas in a box, given by equation (12.7), we obtain equation (12.24), which is given later. For the benefit of advanced readers this result will now be developed in the more traditional way by calculating the grand partition function.

According to equation (11.9), for fixed N, the grand partition function of the N fermion system is

$$\Xi = \sum_i \exp[\beta\{\mu N - E_i(V, N)\}] \qquad (12.20)$$

where $E_i(V, N)$ is the energy eigenvalue of the ith N particle state of the N fermion system, V is the volume of the system, β is the temperature parameter of the heat reservoir and μ is the chemical potential of the particle reservoir. The summation in equation (12.20) is over all the possible N particle states of the N fermion system, for fixed N. [The value of μ is given in terms of V, T and N by equation (12.5).] Expressing N and $E_i(V, N)$ in terms of the properties of the single particle states of the system, we have for the ith N particle state

$$N = n_1 + n_2 + \ldots + n_j + \ldots$$
$$E_i = n_1\varepsilon_1 + n_2\varepsilon_2 + \ldots + n_j\varepsilon_j + \ldots$$

where n_j is the number of fermions in the jth *single* particle state, which has (single particle) energy eigenvalue ε_j, when the N particle system is in its ith N particle state. Equation (12.20) can be rewritten in the form

$$\Xi = \sum \exp[\beta\mu(n_1 + n_2 + \ldots)]\exp[-\beta(n_1\varepsilon_1 + n_2\varepsilon_2 + \ldots)] \; .$$

Consider *single* particle state number 1. In the case of fermions, n_1 can only have the values 0 or 1. To sum over all the N particle states, both the values $n_1 = 0$ and $n_1 = 1$ must be combined with every possible combination of the numbers $n_2, n_3 \ldots$. Hence the grand partition function Ξ can be factorised to give

$$\Xi = \{1 + \exp[\beta(\mu - \varepsilon_1)]\} \sum \exp[\beta\mu(n_2 + n_3 + \ldots)]\exp[-\beta(n_2\varepsilon_2$$
$$+ n_3\varepsilon_3 + \ldots)] \; .$$

The same argument can be applied to each *single* particle state in turn to give

$$\Xi = \{1 + \exp[\beta(\mu - \varepsilon_1)]\}\{1 + \exp[\beta(\mu - \varepsilon_2)]\} \ldots \; . \qquad (12.21)$$

Using equation (11.28), the grand potential is

$$\Omega = -kT \log \Xi = -kT \sum_j \log\{1 + \exp[\beta(\mu - \varepsilon_j)]\} \; . \qquad (12.22)$$

The summation in equation (12.22) is over all the *single* particle states of the system. Using equation (12.2), equation (12.22) can be rewritten in the form

$$\Omega = kT \sum_j \log[1 - f(\varepsilon_j)] \ . \tag{12.23}$$

In the case of an ideal fermion gas in a box, the density of single particle states $D(\varepsilon)$ is given by equation (12.1). Replacing the summation in equation (12.22) by an integration over the single particle states, we have for an ideal fermion gas in a box

$$\Omega = -kT \left(\frac{gV}{4\pi^2}\right) \left(\frac{2m}{\hbar^2}\right)^{3/2} \int_0^\infty \varepsilon^{1/2} \log[1 + e^{\beta(\mu-\varepsilon)}] d\varepsilon \ .$$

Integrating by parts and rearranging, we obtain

$$\Omega = -\frac{2}{3} \left(\frac{gV}{4\pi^2}\right) \left(\frac{2m}{\hbar^2}\right)^{3/2} \int_0^\infty \frac{\varepsilon^{3/2} d\varepsilon}{[\exp\{\beta(\varepsilon - \mu)\} + 1]} = -\frac{2U}{3} \ . \tag{12.24}$$

The chemical potential μ is given by equation (12.5). The other thermodynamic variables can be developed from equation (12.24) using equations (11.29), (11.30), (11.31) and (11.33).

The entropy S can be calculated by applying equation (11.31) to equation (12.22). It is left as an exercise for the reader to show that

$$S = -k \sum_j [f(\varepsilon_j)\log f(\varepsilon_j) + \{1 - f(\varepsilon_j)\}\log\{1 - f(\varepsilon_j)\}] \tag{12.25}$$

where $f(\varepsilon_j)$ is the mean occupancy of the *single* particle state, of energy eigenvalue ε_j. The summation is over all the *single* particle states of the system. The reader may find it useful to use some results that follow from equation (12.2), namely

$$\exp[(\mu - \varepsilon_j)/kT] = f(\varepsilon_j)/[1 - f(\varepsilon_j)] \ .$$

Taking the natural logarithm of both sides, we have

$$(\mu - \varepsilon_j)/kT = \log\{f(\varepsilon_j)/[1 - f(\varepsilon_j)]\} \ .$$

A reader interested in a fuller discussion of the thermodynamics of an ideal fermion gas is referred to Huang [3a], Landau and Lifshitz [2c] and Turner and Betts [1d].

12.2 THE FREE ELECTRON THEORY OF METALS

12.2.1 Introduction
In metals, some of the electrons in the outer shells of the atoms may become detached from individual atoms. These are the conduction electrons. In solid metals, the positive ions are confined to the near vicinity of

their mean lattice positions, whereas the conduction electrons are able to move anywhere in the metal. Experiments, for example on the photoelectric effect, have shown that work must be done to remove an electron from a metal. In Sommerfeld's free electron theory of metals, it is assumed that the conduction electrons move in a uniform potential well of the type shown in Figure 12.6(a). (A more realistic model is shown in Figure 12.6(b), where the potential varies in the vicinity of the positive ions with the periodicity of the crystal lattice.) In Sommerfeld's theory the conduction electrons are treated as an ideal fermion gas of *free* electrons. It was assumed in Appendix 3, when the energy eigenvalues of a single particle in a cubical box were derived, that the potential was infinite on the walls of the box. The density of states for a particle in a deep potential well approximates to the density of states $D(\varepsilon)$ for a particle in a box, given by equation (12.1), provided the energy ε is measured from the bottom of the potential well in Figure 12.6(a).

The number of free electrons per metre³ (N/V) in a lump of copper at room temperature is $8.50 \times 10^{28}\ \mathrm{m}^{-3}$. Using the values $m = 9.11 \times 10^{-31}\ \mathrm{kg}, g = 2$ and $\hbar = 1.055 \times 10^{-34}\ \mathrm{J\,s}$ in equation (12.10), we find that, if the free electrons are treated as an ideal fermion gas, the value of the Fermi energy for the free electrons in copper at room temperature is

$$\varepsilon_f = (\hbar^2/2m)(6\pi^2 N/gV)^{2/3} = 1.13 \times 10^{-18}\ \mathrm{J} \equiv 7.0\ \mathrm{eV}\ .$$

The corresponding Fermi temperature is

$$T_f = \varepsilon_f/k = 8.2 \times 10^4 \mathrm{K}\ .$$

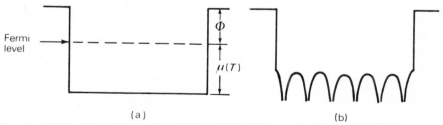

Fermi level
Φ
$\mu(T)$
(a)   (b)

Figure 12.6—(a) The potential well used in the free electron theory of metals. It is assumed that the conduction electrons are free to move inside the uniform potential well. (b) This is a sketch to impress on the reader that, in practice, the potential varies in the vicinity of the positive ions in a crystal, with the periodicity of the lattice. This periodicity leads to a band structure in the variation of the density of single particle states with energy. In a metal there is a gap between the valence band and the conduction band. The electrons in the conduction band are fairly free to move in the metal.

The ratio (T/T_f) for a specimen of copper at 300K is 0.0037, showing that the free electron gas is degenerate.

According to equation (12.6), the chemical potential of the free electrons in copper at an absolute temperature of 300K is given, to a very good approximation, by

$$\mu = \varepsilon_f\{1 - (\pi^2/12)(0.0037)^2\} = 0.999989\varepsilon_f \ . \qquad (12.26)$$

This is only fractionally below the Fermi energy ε_f, which is the value the chemical potential would have at $T = 0$, if the concentration (N/V) of the free electrons remained constant. As a good approximation, it is often assumed in the theory of metals that the chemical potential at room temperature is equal to the Fermi energy ε_f. This can lead to some confusion in the theory of metals and semiconductors, where the chemical potential $\mu(T)$ at an absolute temperature T is sometimes called the **Fermi level**. The reader should be clear about the distinction between the Fermi level and the Fermi energy. The Fermi level, which is the chemical potential $\mu(T)$ at any absolute temperature T, is *not* equal to the Fermi energy, which is the value of the chemical potential $\mu(0)$ at the absolute zero, except at the absolute zero. The two are different, if the fermion gas is at any temperature $T > 0$. It is only an *approximation*, valid when $T \ll T_f$, to use the value of the Fermi energy as an *approximation* for the chemical potential (Fermi level) at a temperature $T > 0$.

The work function Φ is defined as the energy required to remove an electron, which has an energy equal to the chemical potential (Fermi level), to the outside of the metal. The depth of the potential well in Figure 12.6(a) is equal to $[\mu(T) + \Phi]$. The chemical potential (Fermi level) of the free electrons in a metal is typically about 5 eV above the bottom of the potential well, and the work function is typically of the order of 3 eV, so that the depth of the potential well in Figure 12.6(a) is typically of the order of 8 eV in the case of the free electrons in a metal. According to equation (12.13), the pressure of the free electron gas at $T = 0$ is

$$p_0 = \tfrac{2}{5}(N/V)\varepsilon_f \ . \qquad (12.13)$$

In the case of the free electrons in copper, the concentration of free electrons (N/V) is equal to 8.5×10^{28} m^{-3} at 300K and the Fermi energy ε_f is 1.13×10^{-18} J. If the concentration remained equal to 8.5×10^{28} at the absolute zero, the value of p_0 would be 3.8×10^{10} N m^{-2}, which is 3.8×10^5 atmospheres. It can be seen from Figure 12.5(a) that at 300K the pressure of the free electron gas would be even higher. The free electrons are generally prevented from leaving the metal under this extreme pressure by the work function. At very high temperatures, some of the electrons can have sufficient energies to leave the metal. A reader interested in a full discussion of **thermionic emission** is referred to Gopal [4a].

12.2.2 The heat capacity of a free electron gas

It can be shown, see for example Huang [3a], that in the low temperature limit when $T \ll T_f$, which is true for the electrons in a metal at room temperatures, the energy of an ideal fermion gas is given approximately by

$$U = \tfrac{3}{5}N\varepsilon_f\{1 + (5\pi^2/12)(T/T_f)^2\} \ .$$

The heat capacity of the degenerate free electron gas, at constant volume, is

$$C_{\text{elec}} = (\partial U/\partial T)_V = (\pi^2/2)Nk(T/T_f) = 4.93Nk(T/T_f) \ . \qquad (12.27)$$

According to equation (12.27), when $T \ll T_f$, the contribution of the free electrons to the total heat capacity of a metal is proportional to the absolute temperature T. The reader can check that, near the origin where $T \ll T_f$, the slope of the curve relating C_V to T for an ideal fermion gas in Figure 12.5(b) is linear.

The linear dependence of the heat capacity of an ideal fermion gas on absolute temperature, when $T \ll T_f$, will be illustrated by a simple order of magnitude estimate. Assume that, for the free electrons in a typical metal, $T = 0.01T_f$ so that $kT = 0.01\varepsilon_f$. According to equation (12.6), when $T = 0.01T_f$, the chemical potential μ is equal to $0.999918\varepsilon_f$. The values of $f(\varepsilon)$, calculated using equation (12.2) for various values of ε, are given in Table 12.1. The numerical values for $f(\varepsilon)$ show that, in such a metal, $f(\varepsilon)$ is greater than 0.993 for all values of ε from zero up to $0.950\varepsilon_f$. Since $kT = 0.01\varepsilon_f$, this corresponds to an energy of $5kT$ below ε_f. When $\varepsilon = 0.990\varepsilon_f$, which is kT below ε_f, $f(\varepsilon)$ is down to 0.729. These results show that the mean occupancy of the single particle states, only drops by more than 25% in an energy range of the order of $kT = 0.01\varepsilon_f$ immediately below the Fermi energy. At a temperature T, which is very much less than the Fermi temperature T_f, it is mainly electrons, which at $T = 0$ would have been in single particle states having energy eigenvalues within an energy range of about kT immediately below the Fermi energy, which are found in excited states having energy eigenvalues above the Fermi energy, mainly in the energy range ε_f to $(\varepsilon_f + kT)$, which is from ε_f to $1.010\varepsilon_f$ in the present example. As a rough first approximation, it will be assumed that all the

Table 12.1—The Fermi–Dirac distribution function for $T = 0.01T_f$ and $\mu = 0.999\,918\,\varepsilon_f$

$\varepsilon/\varepsilon_f$	0.900	0.950	0.980	0.990	0.995	0.999	1.000	1.010	1.050
$f(\varepsilon)$	0.99995	0.993	0.880	0.729	0.621	0.523	0.498	0.267	0.007

electrons, which at $T = 0$ would have been in single particle states in the energy range $(\varepsilon_f - kT)$ to ε_f are, at an absolute temperature T, in excited single particle states in the energy range ε_f to $(\varepsilon_f + kT)$. It will be assumed that there are no changes outside these energy ranges. The number of single particle states in the small energy range $\Delta\varepsilon = kT$ immediately below the Fermi energy is approximately equal to the product of the density of states $D(\varepsilon)$ at the Fermi energy and $\Delta\varepsilon = kT$. Using equation (12.1), with $\Delta\varepsilon$ equal to kT, we find that, according to our approximate model, the number of electrons N_{ex} in excited states at an absolute temperature T is

$$N_{ex} \sim [(gV/4\pi^2)(2m/\hbar^2)^{3/2}\varepsilon_f^{1/2}]kT \quad . \tag{12.28}$$

Rearranging equation (12.10), we have

$$(gV/\pi^2)(2m/\hbar^2)^{3/2} = 6N/\varepsilon_f^{3/2} \quad .$$

Substituting in equation (12.28), we have

$$N_{ex} \sim \tfrac{3}{2}N(kT/\varepsilon_f) = \tfrac{3}{2}N(T/T_f) \quad .$$

The energy gained by each of these electrons is roughly of the order of kT. Hence at an absolute temperature T, a rough order of magnitude for the excitation energy of the electrons, which is the excess of their total energy over their total energy at $T = 0$, is given very approximately by

$$U_{ex} \sim 3NkT^2/2T_f \quad .$$

Hence a rough order of magnitude for the heat capacity of the electron gas is

$$C_{elec} = (\partial U_{ex}/\partial T) \sim 3Nk(T/T_f) \quad .$$

This result confirms that C_{elec} is proportional to T, but the numerical factor is 3 not the more correct value of 4.93 given by equation (12.27).

The total heat capacity of a metal is the sum of the contributions of the free electrons and the lattice vibrations. Since $T \sim 0.01T_f$ for a typical metal at room temperatures, according to equation (12.27) the heat capacity of the free electrons at room temperatures is $\sim 0.05Nk$, where N is the total number of free electrons. It was shown in Chapter 10 that the contribution of the lattice vibrations to the heat capacity of a solid at room temperatures is generally of the order of $3N'k$, where N' is the total number of atoms, which in a metal is generally equal to N, the number of free electrons, to within a factor of two. Hence at room temperatures, the contribution of the free electrons to the heat capacity of a metal is swamped by the contribution of the lattice vibrations. However, according to Debye's theory of heat capacities, developed in Section 10.3 of Chapter 10, at low temperatures the contribution of the lattice vibrations to the heat capacity is proportional to T^3, whereas the contribution of the free electrons given by equation (12.27) is proportional to T. The latter decreases more slowly than the contribution of the lattice vibrations as the temperature is reduced towards the absolute zero, so that the contribution of the free electrons is observable at low temperatures. At low temperatures when the Debye T^3 law is valid, one would expect the total variation of the total heat capacity of a metal to be given by

$$C_V = \gamma T + AT^3 \ , \qquad (12.29)$$

where γ and A are constants. The experimental results for silver and copper are shown in Figure 12.7, where C_V/T is plotted against T^2. The linear variation of C_V/T with T^2 confirms equation (12.29) at low temperatures. However, γ_{exp}, the experimental value of γ does not agree with γ_{theor}, the theoretical value of γ given by equation (12.27). The ratio of γ_{exp} to γ_{theor} is generally expressed in terms of a thermal effective mass m^* defined by the equation

$$m^*/m = \gamma_{exp}/\gamma_{theor}$$

The experimental value of m^*/m for potassium is 1.25.

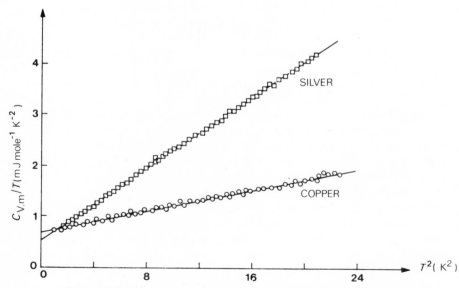

Figure 12.7—Low temperature molar heat capacity of metals; $C_{V,m}/T$ is plotted against T^2. The results are consistent with equation (12.29), according to which $C_{V,m}$ is equal to $\gamma T + AT^3$. The intercepts give the values of γ. The values of A can be determined from the slopes. (Reproduced with permission from *Statistical Mechanics and Properties of Matter* by E. S. R. Gopal, Ellis Horwood Ltd., Chichester.)

12.2.3 Discussion

The free electron theory of metals has been given as an example of an ideal fermion gas and as an introduction to some of the ideas used in solid state physics. The conduction electrons in a metal do not move in a uniform potential of the type shown in Figure 12.6(a), but in a periodic potential of the type sketched in Figure 12.6(b). The effect of the periodic potential is to introduce a band structure into the energy eigenvalue diagram. The variation of the density of states $D(\varepsilon)$ with ε is not given by equation (12.1), but there is a gap between the valence and conduction bands. An introduction to band theory is given by Gopal [4b]. For more comprehensive accounts, the reader is referred to text books on solid state physics, such as Kittel [5]. Many of the properties of metals and semiconductors arise from their band structures. The free electron theory of metals, outlined in this Section, is no more than a rough first approximation to the theory of metals.

12.3 THE ATOMIC NUCLEUS AS AN IDEAL FERMION GAS

Inside an atomic nucleus of atomic weight A and atomic number Z, there are Z protons and $N = (A - Z)$ neutrons. The term nucleon is applied collectively to both protons and neutrons, when we do not wish to distinguish between them, so that one would say that there are A nucleons in the nucleus. Both protons and neutrons are fermions of spin $s = \frac{1}{2}$. There are very strong attractive forces between the nucleons inside the nucleus. The mean binding energy of a nucleon in a nucleus is typically about 8 MeV, whereas the binding energy of an electron in an atom is only in the eV to keV range. Due to the strong interactions between nucleons, the protons and neutrons in an atomic nucleus do not constitute two non-interacting ideal fermion gases. However, it is an interesting application of the theory of Section 12.1 to use the ideal fermion gas model to determine a rough order of magnitude for the energies one might expect for the neutrons and the protons inside an atomic nucleus, assuming that they were two independent ideal fermion gases.

Experiments on the scattering of high energy atomic particles have shown that the radius R of a nucleus of atomic weight A is given approximately by $R = 1.3 \times 10^{-15} A^{1/3}$ metres. The concentration of neutrons in the nucleus is

$$\frac{N}{V} = \frac{(A - Z)}{V} = \frac{(A - Z)}{\frac{4}{3}\pi R^3} = 1.09 \times 10^{44} \left(\frac{A - Z}{A}\right) \text{ metre}^{-3} \ .$$

Using the values $m = 1.67 \times 10^{-27}$ kg for the mass of a neutron, $g = (2s + 1) = 2$ and $\hbar = 1.05 \times 10^{-34}$ J s in equation (12.10) and assuming that $(A - Z)/A = N/A$ is equal to 0.5 for a light nucleus, we find that the

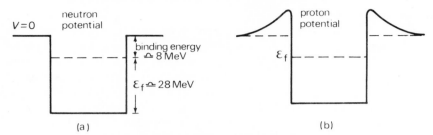

Figure 12.8—(a) Potential well for the neutrons in a typical atomic nucleus. (b) The potential well for the protons in a typical heavy nucleus. The potential well for the protons is not as deep as for the neutrons in a nucleus. In the case of protons, there is also a repulsive Coulomb potential barrier outside the nucleus, due to the electric charges on the protons in the nucleus.

Fermi energy of an ideal gas of neutrons in a volume equal to the volume of the atomic nucleus is $\varepsilon_f = 4.5 \times 10^{-12}$ J $\equiv 28$ MeV, as illustrated in Figure 12.8(a). According to equation (12.3) the corresponding Fermi temperature is 3.3×10^{11}K, showing that the ideal neutron gas is degenerate. The binding energy of each neutron in an atomic nucleus is about 8 MeV, so that, on the ideal fermion gas model, the total depth of the potential well for the neutrons inside the nucleus should be about 36 MeV, as shown in Figure 12.8(a). On the ideal fermion gas model the kinetic energies of the neutrons should range up to about 28 MeV. These rough orders of magnitude are confirmed by scattering experiments using high energy nucleons.

For $A > 100$, there are significantly more neutrons than protons in a nucleus, so that for a heavy nucleus the concentration and hence the Fermi energy is less for the protons than for the neutrons. Hence, the total depth of the potential well is less for protons than for neutrons, as shown in Figure 12.8(b). When a proton approaches the nucleus from outside the nucleus, there is an electric force of repulsion due to the positive electric charges of the approaching proton and the protons inside the atomic nucleus. Hence, in the case of protons there is a repulsive potential outside the nucleus, as shown in Figure 12.8(b).

12.4* WHITE DWARF AND NEUTRON STARS*

12.4.1* White dwarf stars*

Our approach is based primarily on the articles by Orear and Salpeter [6] and van Horn [7]. The main source of energy in main sequence stars, such as the Sun, is believed to be thermonuclear reactions near the centres of the stars, which convert hydrogen into helium. At the temperature near the centre of a star, which is typically of the order of 10^7K, the atoms are completely ionised. The pressure due to the thermal motions of the electrons and ions near the centres of the stars is sufficient in young stars to withstand the weight of the material above the centre of the star and prevent the gravitational collapse of the star. When the hydrogen near the centre of the star runs out, if the centre of the star begins to cool, the thermal motions of the ions and electrons near the centre of the star will not be sufficient to prevent the partial gravitational collapse of the star. According to the virial theorem, when the star undergoes partial gravitational collapse, part of the loss of gravitational potential energy goes into increasing the mean kinetic energies of the ions and the electrons. The temperature near the centre of the star is increased and further types of thermonuclear reactions, such as the conversion of three α particles (helium nuclei) into a $^{12}_{6}C$ nucleus, can take place. In some stars, the gravitational collapse of the core of the star is prevented by the pressure of a degenerate electron gas in the stars. In these cases the stars become white dwarfs.

Sirius B is a typical white dwarf star [7]. Its estimated mass is 2.09×10^{30} kg, which is 1.05 times the mass of the Sun. The estimated radius of Sirius B is $\simeq 5.57 \times 10^3$ km and its volume is $\simeq 7.23 \times 10^{20}$ m^3. For comparison, the radii of the Earth and the Sun are 6.4×10^3 km and 7.0×10^5 km respectively. A typical white dwarf star has about the mass of the Sun and about the size of the Earth.

A simplified model of a white dwarf star will be used. It will be assumed that all the hydrogen in the star has been converted into helium, and that the helium is completely ionised. Each helium nucleus (α-particle) consists of four nucleons, namely two protons and two neutrons, each of mass M_p approximately equal to 1.67×10^{-27} kg. (The difference between the masses of the proton and the neutron will be neglected in our approximate calculations.) Each helium atom loses two electrons, so that, on average, there are 0.5 free electrons per nucleon. The total number N of nucleons in the star is approximately equal to the ratio of the mass of the star to the mass of a nucleon. (This ignores the mass of the electrons in the star, but this is sufficiently accurate for our approximate calculations.) In the case of Sirius B, we have

$$N = (2.09 \times 10^{30})/(1.67 \times 10^{-27}) = 1.25 \times 10^{57} \ . \qquad (12.30)$$

Let N_e denote the total number of free electrons in the white dwarf star. Let x stand for the ratio N_e/N, where N is the total number of nucleons in the star. For completely ionised helium, $x = 0.5$. (If the white dwarf consisted of completely ionised $^{12}_{6}C$, there would be 6 free electrons for each 12 nucleons in a $^{12}_{6}C$ nucleus and x would again be equal to 0.5.) According to equation (12.10), which is based on the *non-relativistic* theory of a degenerate electron gas, the Fermi energy is

$$\varepsilon_f = (\hbar^2/2m)(6\pi^2 N_e/gV)^{2/3} = (\hbar^2/2m)(3\pi^2 xN/V)^{2/3} , \qquad (12.31)$$

where m is the mass of an electron and $g = (2s + 1) = 2$ for an electron of spin $\frac{1}{2}$. For Sirius B, using $x = 0.5$, $N = 1.25 \times 10^{57}$, $V = 7.23 \times 10^{20}$ m^3, $m = 9.108 \times 10^{-31}$ kg and $\hbar = 1.055 \times 10^{-34}$ J s, we find that

$$\varepsilon_f = 5.3 \times 10^{-14} \text{ J} \equiv 0.33 \text{ MeV} . \qquad (12.32)$$

This value for the Fermi energy of the free electron gas is getting close to the rest mass energy of an electron, which is $mc^2 = 0.5$ MeV. For larger values of (N/V), the full relativistic theory must be used. Using equation (12.3), we find that the Fermi temperature T_f for Sirius B is equal to 3.8 $\times$ 10^{9}K. The estimated temperature at the centre of Sirius B is ~2 $\times$ 10^{7}K, so that $T \ll T_f$ and the electron gas is degenerate. As a rough approximation, we shall use the equations valid for a completely degenerate, non-relativistic, ideal fermion gas at $T = 0$, developed in Section 12.1.6. According to equation (12.13) the pressure of an ideal fermion gas at $T = 0$ is

$$p_0 = \tfrac{2}{5}(N_e/V)\varepsilon_f = \tfrac{2}{5}(xN/V)\varepsilon_f \qquad (12.33)$$

For Sirius B, this pressure is equal to 1.8×10^{22} N m^{-2}, which is approximately 1.8×10^{17} atmospheres. This is sufficient to prevent the further gravitational collapse of a white dwarf star such as Sirius B.

12.4.2* The stability of white dwarf stars*

According to equation (12.12), the total energy of the N_e electrons in an electron gas at $T = 0$ is

$$U_{\text{elec}} = N_e(\tfrac{3}{5}\varepsilon_f) = \tfrac{3}{5}(xN)(\hbar^2/2m)(6\pi^2 xN/gV)^{2/3}$$

where N is the total number of nucleons in the white dwarf star, $g = 2$ for an electron and x is the ratio of the number of electrons to nucleons in the star. Since V is equal to $4\pi R^3/3$, where R is the radius of the white dwarf star, we have

$$U_{\text{elec}} = (3xN/5R^2)(\hbar^2/2m)(9\pi xN/4)^{2/3} . \qquad (12.34)$$

If the white dwarf star has uniform mass density, its gravitational potential energy is[†]

$$U_{\text{grav}} = -3GM^2/5R = -3GN^2 M_p^2/5R \qquad (12.35)$$

where $M = NM_p$ is the mass of the white dwarf star.

---

†Consider a sphere of uniform mass density ρ, of radius r and mass $4\pi r^3\rho/3$. The gravitational potential at its surface is $-G(4\pi r^3\rho/3)/r$. The work done in bringing an extra shell of thickness dr and mass $4\pi r^2\rho$ dr from infinity is $-16G\pi^2\rho^2 r^4$ d$r/3$. Integrating from $r = 0$ to $r = R$ and using $M = 4\pi R^3\rho/3$, we obtain equation (12.35).

The energy of the white dwarf star is

$$U = U_{elec} + U_{grav} = \frac{3xN}{5R^2}\left(\frac{\hbar^2}{2m}\right)\left(\frac{9\pi xN}{4}\right)^{2/3} - \frac{3GN^2M_p^2}{5R} \ . \quad (12.36)$$

This equation has the mathematical form

$$y = (a/R^2) - (b/R) \ .$$

The minimum value of y is when

$$dy/dR = -(2a/R^3) + (b/R^2) = 0 \ ,$$

which is when R is equal to $(2a/b)$. Hence the value of R for which $U(R)$ is a minimum is

$$R = (x/GNM_p^2)(\hbar^2/m)(9\pi xN/4)^{2/3} \ . \quad (12.37)$$

A white dwarf will continue to contract under the influence of gravitational forces until $U(R)$ is a minimum, when, according to our simple model, the radius of the star is given by equation (12.37). Substituting the following values for Sirius B, $x = 0.5$, $N = 1.25 \times 10^{57}$, $G = 6.67 \times 10^{-11}$ N m^2 kg^{-2}, $M_p = 1.67 \times 10^{-27}$ kg, $m = 9.108 \times 10^{-31}$ kg, $\hbar = 1.055 \times 10^{-34}$ J s, we find that the calculated radius Sirius B will have when $U(R)$ is a minimum is

$$R = 7 \times 10^6 \, \text{m} \ .$$

The estimated radius of Sirius B is 5.57×10^6 m, which is 0.008 solar radii. This suggests that Sirius B has probably contracted down to the radius, at which its energy $U(R)$ is a minimum.

In our simple example, it was assumed that the white dwarf star consisted only of fully ionised helium and that all the electron gas was degenerate. It is now believed that, in a typical white dwarf star, a degenerate core of the electron gas is surrounded by a non-degenerate outer layer [7]. It is believed that the nuclear reactions have stopped in white dwarf stars, possibly with some earlier burning of helium to give carbon and oxygen. White dwarf stars are now cooling slowly. The time scale before a white dwarf star cools to an invisible dwarf star is believed to be of the order of 10^{10} years. If the age of the Universe is 1.5×10^{10} years, very few white dwarf stars will have had time to evolve to the invisible dwarf stage. Recently some interesting calculations [7], suggest that the ions in white dwarf stars crystallise in a similar manner to a metal, which also has free electrons and positive ions in a crystal lattice.

12.4.3* The limiting mass of white dwarf stars*

So far, only the non-relativistic theory of the ideal fermion gas has been developed. According to equation (12.31), if (N_e/V) is increased, the Fermi energy ε_f increases, and the full relativistic theory must be used. The derivation in Appendix 6 of the equation

$$N(\lambda)\, d\lambda = gV4\pi\, d\lambda/\lambda^4 \qquad (A6.10)$$

giving the number of stationary wave solutions (normal modes) in the wavelength range λ to $(\lambda + d\lambda)$ depends only on the existence of stationary wave solutions satisfying the boundary conditions in a box. It holds for all types of wave motion. Since the de Broglie relation, $\lambda = h/p$, holds for relativistic particles, equation (A6.17) of Appendix 6, which is

$$N(p)\, dp = gVp^2\, dp/2\pi^2\hbar^3 \;, \qquad (12.38)$$

also holds in the relativistic limit.

At the absolute zero of temperature, all the single particle states (normal modes) up to the Fermi energy ε_f are occupied. Let the corresponding Fermi momentum be denoted by p_f. Integrating equation (12.38) up to the Fermi momentum and putting the spin degeneracy $g = 2$ for electrons, we have

$$N_e = \frac{V}{\pi^2\hbar^3} \int_0^{p_f} p^2\, dp = \frac{Vp_f^3}{3\pi^2\hbar^3} \;.$$

Hence in the relativistic limit

$$p_f = (3\pi^2\hbar^3 N_e/V)^{1/3} \;. \qquad (12.39)$$

According to relativistic mechanics, $E^2 = c^2p^2 + m^2c^4$, where m is the rest mass of the electron, c is the velocity of light in empty space and E is the total energy of the electron, which is equal to the sum of its kinetic energy and mc^2. In the *extreme relativistic limit*, when $E \gg mc^2$

$$\varepsilon_f \simeq cp_f = c(3\pi^2\hbar^3 N_e/V)^{1/3} \;. \qquad (12.40)$$

Since all the single particle states up to the Fermi energy are filled at $T = 0$, the mean momentum of one of the electrons in the electron gas at $T = 0$ is

$$\bar{p} = \frac{1}{N_e} \int_0^{p_f} pN(p)\, dp = \frac{1}{N_e} \int_0^{p_f} \frac{pVp^2\, dp}{\pi^2\hbar^3} = \frac{1}{N_e}\left(\frac{V}{\pi^2\hbar^3}\right)\left(\frac{p_f^4}{4}\right) \;.$$

Substituting for p_f^3 from equation (12.39), we find that $\bar{p}$ is equal to $3p_f/4$. Hence, in the extreme relativistic limit, the mean total energy of an electron at $T = 0$ is

$$\bar{\varepsilon} \simeq c\bar{p} = \tfrac{3}{4}cp_f \;. \qquad (12.41)$$

The mean energy of the electron gas is

$$U_{\text{elec}} \simeq N_e(\tfrac{3}{4}cp_f) = \tfrac{3}{4}N_e c(3\pi^2\hbar^3 N_e/V)^{1/3} \;. \qquad (12.42)$$

If the star is of uniform density, the gravitational potential energy of the star is given by equation (12.35). Using $N_e = xN$ and $V = 4\pi R^3/3$, we have

$$U = U_{\text{elec}} + U_{\text{grav}} = \frac{3xNc}{4R}\left(\frac{9\pi\hbar^3 xN}{4}\right)^{1/3} - \frac{3GN^2 M_p^2}{5R} . \quad (12.43)$$

In the extreme relativistic limit, both the terms on the right hand side of equation (12.43) are proportional to R^{-1}. The first term is proportional to $N^{4/3}$ whereas the second term is proportional to N^2, where N is the total number of nucleons in the star. In sufficiently massive stars the second term, which represents the gravitational potential energy, predominates and according to equation (12.43) the energy U continues to decrease with decreasing radius. The electron degeneracy pressure is not big enough to prevent the further gravitational collapse of such massive stars.

A rough order of magnitude for the critical number of nucleons N_{crit}, above which the star undergoes continual gravitational collapse, can be obtained by equating the two terms on the right hand side of equation (12.43). This leads to the relation

$$N_{\text{crit}} = (3\pi^{1/2}x^2/16)(5\hbar c/GM_p^2)^{3/2} . \quad (12.44)$$

Putting $x = 0.5$ for a white dwarf star, we find that $N_{\text{crit}} \simeq 2 \times 10^{57}$ nucleons. The critical mass is given by

$$M_{\text{crit}} = N_{\text{crit}}M_p \simeq 2 \times 10^{57} \times 1.67 \times 10^{-27} = 3.4 \times 10^{30} \text{ kg} .$$

For comparison, the mass of the Sun is equal to 2×10^{30} kg. Our rough estimate of the critical mass of a white dwarf star is approximately 1.7 times the mass of the Sun. More exact calculations give a critical mass of 1.44 times the mass of the Sun. This is known as the **Chandrasekhar limiting mass**. The electron degeneracy pressure is not sufficient to prevent the gravitational collapse of stars more massive than the critical mass.

12.4.4* Neutron stars*

If the kinetic energies of the electrons inside a star are high enough, the electrons can interact with the protons inside the star to produce neutrons by the inverse β-decay process. Such a process may take place following an implosion of the core of a supernova. The neutrons form a fermion gas. To simplify the discussion, it will be assumed that the star consists entirely of neutrons. Equation (12.37) is valid for the neutron gas, provided m, the mass of the electron, is changed to the nucleon mass M_p and x is put equal to unity. For a neutron star, equation (12.37) becomes

$$R_{\text{neut}} = (\hbar^2/GNM_p^3)(9\pi N/4)^{2/3} .$$

Putting $G = 6.67 \times 10^{-11}$ N m^2 kg^{-2}, $M_p = 1.67 \times 10^{-27}$ kg and $N = 1.2 \times 10^{57}$, corresponding to a neutron star of mass equal to the Sun, we find that the radius of the neutron star, when its energy $U(R)$ is a

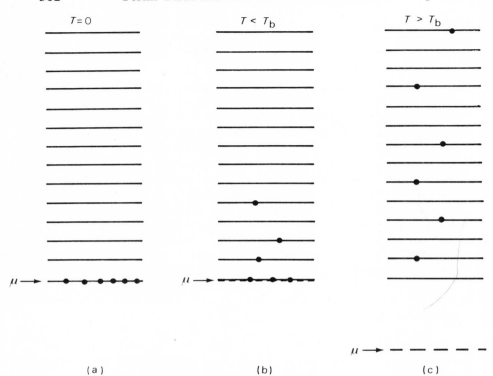

Figure 12.9.—A simple system consisting of 6 identical bosons of spin zero. (a) At $T = 0$ all the bosons are in the single particle ground state. (b) When $T > 0$ but $T < T_b$, where the Bose temperature T_b is given by equation (12.55), some of the bosons are in excited single particle states, but a significant fraction of the bosons are still in the single particle ground state. (c) When $T > T_b$, there are very few, if any, bosons in the single particle ground state. The case when $b(\varepsilon) \ll 1$ for all the single particle states corresponds to the classical limit.

The value of the chemical potential μ is shown in each case: (a) At $T = 0$, μ is equal to the energy of the lowest energy single particle state. In the text, this energy is chosen as the zero of the energy scale, in which case $\mu = 0$ at $T = 0$. (b) When $T < T_b$, μ is only fractionally less than the energy of the lowest energy single particle state. (c) As T increases above T_b, μ gets more and more negative.

minimum, is approximately $R_{neut} = 12.4$ km. The mass density of a neutron star of this radius and total mass equal to the mass of the Sun is 2.5×10^{17} kg m^{-3}. This is comparable with the density of nuclear matter. According to equation (12.31), using $g = (2s + 1) = 2$ for a neutron gas, the corresponding Fermi energy is

$$\varepsilon_f = (\hbar^2/2M_p)(3\pi^2 N/V)^{2/3} \simeq 9 \times 10^{-12} \text{ J} \equiv 56 \text{ MeV} \ .$$

This value for the Fermi energy is non-relativistic. For more massive neutron stars, the Fermi energy ε_f increases and the full relativistic theory must be used. Equation (12.44) can be used to make a rough order of magnitude calculation for the critical mass. Putting $x = 1$, we have

$$M_{\text{crit}} = N_{\text{crit}} M_p = (3\pi^{1/2}/16)(5\hbar c/GM_p^2)^{3/2} M_p \ . \qquad (12.45)$$

Substituting numerical values for the constants, we find that the critical mass given by equation (12.45) is 6.8 solar masses. To obtain a better estimate for the critical mass, one must use the theory of general relativity and allow for the effects of nuclear forces. These more comprehensive estimates of the critical mass vary from about 0.7 to 4.0 solar masses.

Only a simplified model of a homogeneous neutron star has been given. Theoretical studies suggest that, near the surface of a neutron star, the material consists mainly of iron nuclei and electrons. As the pressure increases with depth, the heavy nuclei break up and the star consists mainly of neutrons with a small proportion of protons and electrons. These neutrons may form a superfluid. Near the centre of a neutron star there should be some hyperons and mesons. (Reference: Shklovskii [8].) Experimental evidence for neutron stars has come from pulsars, which are believed to be rotating neutron stars.

Stars more massive than the critical masses of white dwarf and neutron stars probably end up as supernovae. In these stellar explosions, some of the outer material of the stars is blown off and the remnant may be small enough to end up as a neutron star. This is believed to be the case in the Crab Nebula. Stars more massive than about eight solar masses are probably too massive to end up as either white dwarf or neutron stars. Such stars probably end up as black holes.

12.5 THE IDEAL BOSON GAS

12.5.1 General introductory survey

Bosons are particles of integral spin (0, 1, 2 . . .) which obey **Bose–Einstein statistics**. There is no limit to the number of bosons that can be in any single particle state. [Reference: Section 2.1.5 of Chapter 2 and Section 11.4 of Chapter 11.] Consider an ideal boson gas consisting of N identical bosons, each of mass m in a box of volume V kept at an absolute temperature T, of the type shown in Figure 11.2 of Chapter 11. It will be assumed that there are no interactions between the bosons and that the bosons occupy single particle states (orbitals). These single particle states (orbitals) can be determined by solving Schrödinger's time independent equation in the way described for a particle in a cubical box in Appendix 3. The density of states $D(\varepsilon)$ is given by equation (12.1).

According to equation (11.20), the mean number of bosons in a *single* particle state, which has (single particle) energy eigenvalue ε, is given by the Bose–Einstein distribution function:

$$b(\varepsilon) = \frac{1}{[\exp\{(\varepsilon - \mu)/kT\} - 1]} = \frac{1}{[\exp\{\beta(\varepsilon - \mu)\} - 1]} \qquad (12.46)$$

In equation (12.46), T is the absolute temperature of the heat reservoir, and μ is the chemical potential of the particle reservoir, which in the present case is the bosons in all the other single particle states. Since there is no limit to the number of bosons that can be in any single particle state, at the absolute zero of temperature, all the bosons will be in the single particle ground state. Hence at $T = 0$, $b(\varepsilon)$ is equal to N for the single particle ground state, and $b(\varepsilon)$ is zero for all the excited single particle states. This is illustrated for the case of a system composed of six identical bosons of spin zero in Figure 12.9(a). When the temperature of the heat reservoir is raised above the absolute zero, some of the bosons occupy excited single particle states, though at sufficiently low temperatures a significant proportion of the bosons may still be in the single particle ground state, as shown for the case of a system composed of six identical bosons of spin zero in Figure 12.9(b). At high temperatures, the average number of bosons in any single particle state is much less than unity, as illustrated qualitatively for the case of a system composed of six identical bosons of spin zero in Figure 12.9(c). The case when $b(\varepsilon)$ is very much less than unity for all the single particle states corresponds to the classical limit.

To obtain the energy spectrum of the ideal boson gas in a box, the number of single particle states $D(\varepsilon)\,d\varepsilon$, which have (single particle) energy eigenvalues between ε and $(\varepsilon + d\varepsilon)$, must be multiplied by $b(\varepsilon)$, the mean number of bosons in each of these states. Using equations (12.1) and (12.46), we have

$$n(\varepsilon)\,d\varepsilon = \frac{gV}{4\pi^2}\left(\frac{2m}{\hbar^2}\right)^{3/2}\frac{\varepsilon^{1/2}\,d\varepsilon}{[\exp\{(\varepsilon - \mu)/kT\} - 1]}\ . \qquad (12.47)$$

This energy distribution differs from the Fermi–Dirac energy distribution, given by equation (12.4), by virtue of a negative rather than a positive sign inside the bracket in the denominator of equation (12.47). Typical velocity distributions for an ideal boson gas are shown in Figure 12.10.

Integrating equation (12.47) for an ideal boson gas consisting of N bosons in a box, we have

$$N = \frac{gV}{4\pi^2}\left(\frac{2m}{\hbar^2}\right)^{3/2}\int_0^\infty\frac{\varepsilon^{1/2}\,d\varepsilon}{[\exp\{(\varepsilon - \mu)/kT\} - 1]}\ . \qquad (12.48)$$

This is an integral equation which gives the chemical potential μ in terms of T, V and N. The variation of the chemical potential μ with temperature,

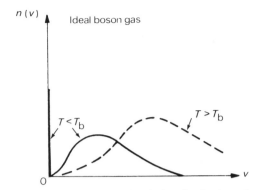

Figure 12.10—Sketches of typical velocity distributions for an ideal
boson gas. The case when $T > T_b$, where T_b is the Bose temperature
given by equation (12.55), tends to the Maxwell–Boltzmann velocity
distribution, but when $T < T_b$ there are significant differences. Notice
when $T < T_b$ there is a thick line just to the right of the ordinate. This
corresponds to the bosons in the single particle ground state.
(Reproduced with permission from *Statistical Mechanics and Properties
of Matter* by E. S. R. Gopal, Ellis Horwood Ltd., Chichester.)

calculated from equation (12.48), is shown in Figure 12.11. (Reference:
Turner and Betts [1a].) Since the number of bosons in any single particle
state cannot be negative, the denominator in equation (12.46) must always
be positive, so that, for an ideal boson gas, $\exp[(\varepsilon - \mu)/kT]$ must always be
greater than unity. This implies that $\varepsilon > \mu$. Hence the chemical potential of
an ideal boson gas must always be less than the energy of the lowest energy
single particle state. In our discussion of the ideal boson gas, we shall adopt
the convention of choosing the energy of the lowest energy *single* particle
state (the *single* particle ground state) as the zero of our energy scale.[†] The
chemical potential μ of the ideal boson gas is then always negative, as
shown in Figure 12.11. It will be shown in Section 12.5.2* that the Bose
temperature T_b is given by equation (12.55), which is

$$T_b = \frac{3.31\,\hbar^2}{mk}\left(\frac{N}{gV}\right)^{2/3}$$

where m is the mass of a boson and g is the spin degeneracy.

[†]It was shown in Section 2.1.4 of Chapter 2 using equation (2.7), that, for a $_2^4$He gas atom in a
cubical box of volume 22.4×10^{-3} m^3, the value of $\varepsilon_{1,1,1}$, the energy eigenvalue of the lowest
single particle state, is equal to 3.12×10^{-40} J, if it is assumed that the potential energy of the
$_2^4$He atom is zero inside the box. This value for $\varepsilon_{1,1,1}$ is very much less than the value of
$kT = 1.38 \times 10^{-23}T$ joules at any attainable low temperature. Hence the difference between the
energy scale in which $\varepsilon_{1,1,1}$ is zero and the energy scale in which the potential energy of the
particle is zero inside the box is generally negligible in practice.

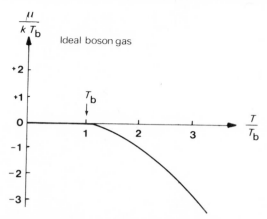

Figure 12.11—Variation of the chemical potential μ of an ideal boson gas with temperature. The Bose temperature T_b is given by equation (12.55). The chemical potential μ is always negative. It is virtually zero up to the Bose temperature. As T increases above T_b, μ gets more and more negative. [Adapted with permission from *Introductory Statistical Mechanics*, by R. E. Turner and D. S. Betts.]

It can be seen from Figure 12.11 that at $T = 0$, $\mu = 0$. [This is illustrated for the simple case of six identical bosons of spin zero by the position of μ in Figure 12.9(a).] At low temperature, when $T < T_b$, the chemical potential μ has a fairly constant numerical value just below zero. [This is illustrated for the case of six identical bosons of spin zero by the position of μ in Figure 12.9(b).] When T increases above T_b, the chemical potential μ gets more and more negative. [This is illustrated for the case of six identical bosons of spin zero by the position of μ in Figure 12.9(c).] Changes in V and N at a fixed value of T lead to a change in the value of μ through their effects on the value of T_b to be used in Figure 12.11.

It was shown in Section 5.9.8 of Chapter 5, that, in the high temperature classical limit, the chemical potential μ of an ideal monatomic gas is given by

$$\exp(-\mu/kT) = (gV/N)(mkT/2\pi\hbar^2)^{3/2} \gg 1 \ . \qquad (5.84)$$

For example, for a mole of $^4_2$He gas (a boson gas) at STP, $\exp(-\mu/kT)$ is equal to 2.56×10^5. [Reference: Section 5.9.8 of Chapter 5.] The corresponding value of (μ/kT) is (-12.45). Using the value of 2.56×10^5 for $\exp(-\mu/kT)$ in the expression for $b(\varepsilon)$ given by equation (12.46), the reader can show that, for a mole of $^4_2$He gas at STP, $b(\varepsilon)$ is equal to 8.7×10^{-7} for a $^4_2$He atom, which has an energy of $3kT/2$. (This is the mean energy of an atom of an ideal monatomic gas atom in the classical limit.) This result confirms the rough estimate of less than 3×10^{-6} made in

Section 4.9 and shows that at STP, the mean occupancies of the single particle states are very small. Due to the large value of $\exp(-\mu/kT)$ in the classical limit, $\exp[(\varepsilon - \mu)/kT] \gg 1$ and the expression for $b(\varepsilon)$ given by equation (12.46) reduces, in the classical limit, to

$$b(\varepsilon) = e^{\mu/kT}e^{-\varepsilon/kT} = \lambda e^{-\varepsilon/kT}$$

where λ is the absolute activity. This is the classical Boltzmann distribution function. Using equation (5.84) and putting $\varepsilon = \frac{1}{2}mv^2$, it can be shown that, in the classical limit, equation (12.47) goes over to the Maxwell–Boltzmann velocity distribution, given by equation (4.55) of Section 4.9, just as the ideal fermion gas did. [Reference: Section 12.1.7.]

Using equation (12.47) we find that the total energy U of the ideal boson gas is given by

$$U = \int_0^\infty \varepsilon\, n(\varepsilon)\, d\varepsilon = \frac{gV}{4\pi^2}\left(\frac{2m}{\hbar^2}\right)^{3/2} \int_0^\infty \frac{\varepsilon^{3/2}\, d\varepsilon}{[\exp\{(\varepsilon - \mu)/kT\} - 1]} \quad . (12.49)$$

The variation of U with T, at constant volume V is sketched in Figure 12.12(a). According to equation (5.74) of Section 5.9.4, U is equal to $3NkT/2$ for an ideal monatomic gas in the classical limit. This relation is shown by the dotted line in Figure 12.12(a). It can be seen that the total energy of an ideal boson gas is always less than the energy an ideal monatomic classical gas would have at the same values of V, T and N. It can be seen from Figure 12.12(a), that in the high temperature limit, the energy of the ideal boson gas approaches the ideal monatomic classical gas value. When $T = 0$, all the bosons are in the single particle ground state, which we have chosen as the zero of our energy scale. Hence for an ideal boson gas, $U = 0$ at $T = 0$, as shown in Figure 12.12(a). Contrast this with the behaviour of the ideal fermion gas illustrated in Figure 12.5(a). At $T = 0$ the fermions fill all the single particle states up to the Fermi energy, and U is not zero at $T = 0$.

According to equation (12.8) the pressure p of the ideal boson gas is equal to $(2U/3V)$, so that the variation of p with T at constant volume has the same form as the variation of U with T. It can be seen from Figure 12.12(a) that the pressure of an ideal boson gas is always less than the pressure an ideal monatomic classical gas would have, at the same values of V, T and N. In the case of a slightly degenerate ideal boson gas, the pressure is given by equation (12.18), provided the positive sign inside the bracket is changed to a negative sign.

The molar heat capacity $C_{V,m}$ of the ideal boson gas can be determined from the variation of U with T, using equation (10.1). The variation of $C_{V,m}$ with T is sketched in Figure 12.12(b). The curve has a cusp at the Bose temperature. [The expression for the Bose temperature will be developed in Section 12.5.2*, and is given by equation (12.55).] At the Bose tempera-

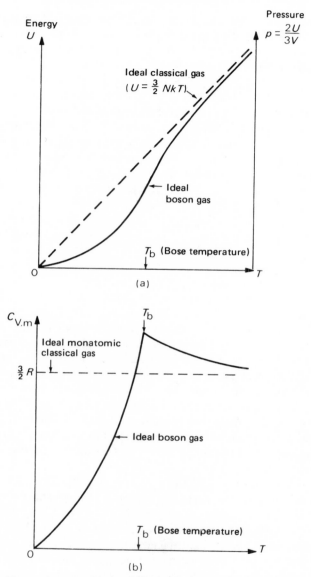

Figure 12.12—Behaviour of an ideal Boson gas. (a) The variations of the energy U and the pressure p with absolute temperature T are shown. (b) The variation of the molar heat capacity $C_{V,m}$ with absolute temperature T is shown. There is a cusp in the curve at the Bose temperature T_b, where T_b is given by equation (12.55). At high temperatures $C_{V,m}$ tends to the classical value of $3R/2$. (Adapted with permission from *Statistical Mechanics and Properties of Matter* by E. S. R. Gopal, Ellis Horwood Ltd., Chichester.)

ture the molar heat capacity of an ideal boson gas is greater than the value of $(3R/2)$, which is valid for an ideal monatomic gas in the classical limit.

12.5.2* The Bose–Einstein condensation*

There is no restriction on the number of bosons which can be in any single particle state. At the absolute zero of temperature, all the bosons should be in the single particle ground state, as shown in Figure 12.9(a). If we choose the single particle ground state as the zero of our energy scale, the $\varepsilon^{1/2}$ term in equation (12.48) is zero for the single particle ground state, so that the single particle ground state is not included in the integration in equation (12.48), which only gives N_{ex}, the number of bosons in excited states. Hence

$$N_{ex} = \frac{gV}{4\pi^2}\left(\frac{2m}{\hbar^2}\right)^{3/2}\int_0^\infty \frac{\varepsilon^{1/2}\,d\varepsilon}{[\exp(\varepsilon/kT)\exp(-\mu/kT) - 1]}\ . \quad (12.50)$$

If N is the total number of bosons, the number of bosons N_0 in the single particle ground state is

$$N_0 = N - N_{ex}\ . \quad (12.51)$$

At high temperatures $N_0 \ll N$ and equation (12.48) is a satisfactory approximation.

Near the absolute zero of temperature, all the N bosons tend to go into the single particle ground state, so that in the limit as T tends to zero, the Bose–Einstein distribution function $b(\varepsilon)$, which gives the mean number of bosons in a single particle state having energy eigenvalue ε, tends to N for the single particle ground state. If we put $\varepsilon = 0$ in equation (12.46), to allow for the fact that we have chosen the single particle ground state as our zero of energy, in the limit as T tends to zero, for the ground state we have

$$b = [\exp(-\mu/kT) - 1]^{-1} \simeq N\ .$$

Rearranging and taking the natural logarithm, we have

$$-\mu/kT \simeq \log[1 + 1/N]\ .$$

If $x \ll 1$, $\log(1 + x)$ is approximately equal to x. Hence for large N, at temperatures very close to the absolute zero, we have

$$-\mu/kT \simeq 1/N\ . \quad (12.52)$$

According to equation (12.52), the chemical potential of an ideal boson gas is zero at $T = 0$. For a small macroscopic system, consisting of $N = 10^{18}$ bosons, the value of $(-\mu/kT)$ given by equation (12.52) is approximately equal to 10^{-18}, which is extremely small. Hence at temperatures close to the absolute zero, it is a good approximation to assume that the

$\exp(-\mu/kT) = \exp(10^{-18})$ term is equal to unity in equation (12.50). Using the change of variable $x = \varepsilon/kT$, equation (12.50) then becomes

$$N_{ex} = \frac{gV}{4\pi^2} \left(\frac{2mkT}{\hbar^2}\right)^{3/2} \int_0^\infty \frac{x^{1/2}\,dx}{(e^x - 1)} \ .$$

The definite integral is equal to the product of the gamma function $\Gamma(\frac{3}{2}) = \pi^{1/2}/2$ and the Riemann zeta function $\zeta(\frac{3}{2}) = 2.612$. (Reference: Landau and Lifshitz [2d].) Hence, near the absolute zero, the number of bosons in excited states is

$$N_{ex} = 2.612gV(mkT/2\pi\hbar^2)^{3/2} \ . \tag{12.53}$$

The Bose temperature T_b, which is sometimes called the Bose–Einstein condensation temperature, is equal to that temperature at which, according to equation (12.53), all the bosons should be in excited states. Putting N_{ex} equal to N and $T = T_b$ in equation (12.53), we have

$$N = 2.612gV(mkT_b/2\pi\hbar^2)^{3/2} \ . \tag{12.54}$$

Rearranging, we have

$$T_b = \frac{2\pi\hbar^2}{mk}\left(\frac{N}{2.612gV}\right)^{2/3} = \frac{3.31\,\hbar^2}{mk}\left(\frac{N}{gV}\right)^{2/3} \ . \tag{12.55}$$

For example, for a boson gas composed of 6.02×10^{23} bosons ($_2^4$He atoms), each of mass 6.65×10^{-27} kg and spin zero, in a box of volume 22.4×10^{-3} m^3, the Bose temperature given by equation (12.55) is 0.036K. If there were no interactions between the bosons, the ideal boson gas would not liquefy. In practice, due to the interactions between the bosons, all real boson gases liquefy before their temperatures are reduced to their Bose temperatures. In the case of helium, the gas liquefies at a temperature of 4.21K at atmospheric pressure. This is long before the temperature gets down to 0.036K. This liquefaction increases the concentration (N/V). (If *liquid* helium were treated as an ideal boson gas, the Bose temperature given by equation (12.55) would be 3.1K.)

It can be seen from Figure 12.11 that the chemical potential of an ideal boson gas is virtually zero right up to the Bose temperature, so that the assumption made in equation (12.52) that $\exp(-\mu/kT)$ is equal to unity at very low temperatures is a good approximation right up to the Bose temperatures. Dividing equations (12.53) and (12.54), to a good approximation we have, when $T < T_b$,

$$N_{ex} = N(T/T_b)^{3/2} \ . \tag{12.56}$$

Substituting in equation (12.51) we find that, for temperatures below the Bose temperature, the number of bosons in the single particle ground state is given, to a good approximation, by

$$N_0 = N[1 - (T/T_b)^{3/2}] \ . \tag{12.57}$$

The variation of (N_0/N) with absolute temperature T is shown in Figure 12.13. For temperatures well below the Bose temperature, a significant fraction of the bosons are in the single particle ground state. [This is illustrated for the case of six identical bosons of spin zero in Figure 12.9(b).] Above the Bose temperature, the number of bosons, if any, in the single particle ground state is very much less than the total number of bosons. [This is illustrated for the case of six identical bosons of spin zero in Figure 12.9(c).]

The *velocity* distribution of an ideal boson gas for $T < T_b$, is sketched in Figure 12.10. Notice the thick line just fractionally to the right of the ordinate. This thick line represents the bosons in the single particle ground state. [The reader should remember that according to the uncertainty principle, the velocity of the bosons in the single particle ground state cannot be zero, if the bosons are inside a box of finite dimensions.]

The 'collapse' of bosons into the single particle ground state at very low temperatures is called the Bose–Einstein condensation. It is not a condensation in space, but a condensation into the same single particle ground state. It could be called a condensation in momentum space. The large differences between the predictions of the Bose–Einstein distribution and the Boltzmann distribution, near $T = 0$, are illustrated in Problem 12.11*.

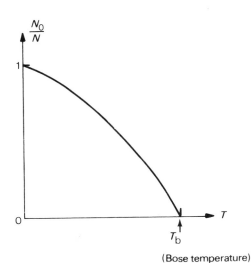

(Bose temperature)

Figure 12.13—The variation with absolute temperature T of the ratio of the number N_0 of bosons in the single particle ground state to N, the total number of bosons, for temperatures below the Bose temperature.

12.5.3* The thermodynamics of the ideal boson gas*

Consider the case of N non-interacting identical bosons inside a box. The expression for the grand potential Ω follows from equations (11.32), (12.8) and (12.49). We have

$$\Omega = -pV = -\frac{2U}{3} = -\frac{gV}{6\pi^2}\left(\frac{2m}{\hbar^2}\right)^{3/2}\int_0^\infty \frac{\varepsilon^{3/2}d\varepsilon}{[\exp\{\beta(\varepsilon - \mu)\} - 1]} .$$

(12.58)

The pressure p and the energy U can be determined directly from equation (12.58), once the value of the chemical potential μ of the boson gas has been calculated using equation (12.48). The entropy S can be calculated from the grand potential Ω using equation (11.31).

For the benefit of advanced readers, equation (12.58) will now be developed by calculating the grand partition function Ξ. According to equation (11.9), the grand partition function for the N boson system is, for fixed N,

$$\Xi = \sum_i \exp[\beta\{\mu N - E_i(V, N)\}] ,$$

where $E_i(V, N)$ is the energy eigenvalue of the ith N particle state of the N boson system, V is the volume, β is the temperature parameter of the heat reservoir and μ is the chemical potential of the particle reservoir. The summation is over all the possible N particle states of the N boson system, for fixed N. For the ith N particle state, we have

$$N = n_1 + n_2 + \ldots + n_j + \ldots$$
$$E_i = n_1\varepsilon_1 + n_2\varepsilon_2 + \ldots n_j\varepsilon_j + \ldots$$

where n_j is the number of bosons in the jth single particle state, which has (single particle) energy eigenvalue ε_j, when the N boson system is in its ith N particle state. Hence

$$\Xi = \sum \exp[\beta\mu(n_1 + n_2 + \ldots)]\exp[-\beta(n_1\varepsilon_1 + n_2\varepsilon_2 + \ldots)] .$$

Consider single particle state number one. To sum over all the possible N particle states, every possible value of n_1, which can vary from 0 to ∞ must be combined with every possible combination of $n_2, n_3, \ldots$. Hence Ξ can be factorised to give

$$\Xi = \{\exp(0) + \exp[\beta(\mu - \varepsilon_1)] + \exp[\beta(2\mu - 2\varepsilon_1)] + \ldots\} \times$$
$$\sum \exp[\beta\mu(n_2 + n_3 + \ldots)]\exp[-\beta(n_2\varepsilon_2 + n_3\varepsilon_3 + \ldots)] .$$

Since $(1 + x + x^2 + \ldots)$ is equal to $(1 - x)^{-1}$, putting x equal to $\exp\{\beta(\mu - \varepsilon_1)\}$, we have

$$\Xi = \{1 - \exp[\beta(\mu - \varepsilon_1)]\}^{-1}\sum \exp[\beta\mu(n_2 + n_3 + \ldots)]\exp[-\beta(n_2\varepsilon_2 + n_3\varepsilon_3 + \ldots)] .$$

Applying the above argument to each single particle state in turn, we have

$$\Xi = \{1 - \exp[\beta(\mu - \varepsilon_1)]\}^{-1}\{1 - \exp[\beta(\mu - \varepsilon_2)]\}^{-1} \dots . \quad (12.59)$$

Using equation (11.28) we have

$$\Omega = -kT \log \Xi = kT \sum_j \log\{1 - \exp[\beta(\mu - \varepsilon_j)]\} . \quad (12.60)$$

The summation in equation (12.60) is over all the single particle states of the system. For the case of an ideal boson gas in a box, the density of states $D(\varepsilon)$ is given by equation (12.1). Replacing the sum in equation (12.60) by an integration over the single particle states, we obtain equation (12.58). [Compare Section 12.1.8*.]

It can be seen from Figure 12.11 that, if the energy eigenvalue of the single particle ground state is chosen as the zero of the energy scale, the chemical potential μ of an ideal boson gas is very close to zero for all temperatures below the Bose temperature. In this temperature range, putting $\mu = 0$ and making the substitution $x = \varepsilon/kT$, equation (12.49) approximates to

$$U = \frac{gV}{4\pi^2}\left(\frac{2m}{\hbar^2}\right)^{3/2}(kT)^{5/2}\int_0^\infty \frac{x^{3/2}dx}{(e^x - 1)} . \quad (12.61)$$

The definite integral in equation (12.61) is equal to the product of the gamma function $\Gamma(\tfrac{5}{2}) = 3\pi^{1/2}/4$ and the Riemann zeta function $\zeta(\tfrac{5}{2}) = 1.34$. (Reference: Landau and Lifshitz [2d].) Using equation (12.55), we find finally that,

$$U = 0.770NkT(T/T_b)^{3/2} \qquad (T < T_b) . \quad (12.62)$$

The Bose temperature T_b is given by equation (12.55).

Using equations (12.8), (12.62) and (11.32), for $T < T_b$ we have

$$pV = \tfrac{2}{3}U = 0.513NkT(T/T_b)^{3/2} = -\Omega \qquad (T < T_b) . \quad (12.63)$$

When $T < T_b$, both U and p are proportional to $T^{5/2}$, if V is constant. The heat capacity for $T < T_b$ is

$$C_V = (\partial U/\partial T)_V = 1.93Nk(T/T_b)^{3/2} \qquad (T < T_b) . \quad (12.64)$$

Notice that C_V is proportional to $T^{3/2}$. Using equations (11.31) and (12.63), we have

$$S = -(\partial\Omega/\partial T)_{\mu,V} = 1.28Nk(T/T_b)^{3/2} \qquad (T < T_b) . \quad (12.65)$$

Notice that the entropy S is zero when $T = 0$, in agreement with the third law of thermodynamics. A reader interested in a fuller discussion of the thermodynamics of an ideal boson gas is referred to Huang [3b], Landau and Lifshitz [2e] and to Turner and Betts [1e].

12.5.4* The photon gas*

Consider the photon gas inside the cavity in the heat reservoir in Figure 9.1. The photon is a boson of spin 1. In this Section we shall apply the theory of the ideal boson gas to the photon gas. According to equation (A6.19) of Appendix 6, the number of single particle (photon) states per unit volume in the angular frequency range ω to $(\omega + d\omega)$ inside the cavity in Figure 9.1 is

$$D(\omega)\,d\omega = \omega^2\,d\omega/\pi^2 c^3 \ . \tag{12.66}$$

The photons in these single particle states have energies between $\hbar\omega$ and $\hbar(\omega + d\omega)$. Since the number of photons in each of these states can vary, at first sight it might appear that the grand canonical distribution, given by equation (11.7), should be used to calculate the mean number of photons in each single particle state, in the way described for bosons in Section 11.4. The question that arises is, what value should be used, in the case of photons, for the chemical potential μ of the particle reservoir in equation (11.7)? In an ideal boson gas, such as a gas of $^4_2$He atoms in a box, it is the bosons ($^4_2$He atoms) in the other single particle states, which act as the particle reservoir for the single particle state chosen for the application of equation (11.7). The chemical potential of an ideal boson gas, such as $^4_2$He, is given by equation (12.48). In the case of the photon gas, the total number of photons is not constant and equation (12.48) clearly cannot be applied. Furthermore, the photons cannot retain their identity and go directly from one single particle state to another by giving energy to, or receiving energy from the heat reservoir in Figure 9.1. A photon in the first single particle state must first be absorbed by the walls of the cavity in Figure 9.1, putting up the energy of the heat reservoir. Another photon may be emitted from the walls of the cavity into the second single particle state of different energy at a later time, but this is not the same photon as was absorbed earlier. The photons in the single particle state chosen for the application of (11.7) only interact with the walls of the cavity inside the heat reservoir in Figure 9.1. The photons in the other single particle (photon) states do not form part of the particle (photon) reservoir and have no effect on the chosen single particle state. When the number of photons in the chosen single particle state varies, due to the absorption or emission of a photon of the appropriate energy by the walls of the cavity in Figure 9.1, it is only the total energy of the heat reservoir in Figure 9.1 which changes. All the constituents making up the heat reservoir in Figure 9.1 are unchanged. Hence the number N_R of photons, if any, inside the material of the heat reservoir is unchanged, when a photon is absorbed by or emitted from the walls of the cavity. The numbers of all the other atomic constituents of the heat reservoir are also unchanged. In these conditions, equation (11.3) giving the probability that there are N_i photons each of energy $\hbar\omega_i$ in the single particle (photon) state of (single particle) energy eigenvalue $\hbar\omega_i$ is

$$P_{N_i, i} = C'g_R(U^* - \varepsilon_i, N_R, V_R)$$

where $\varepsilon_i = N_i\hbar\omega_i$ is the total energy of the N_i photons in the chosen single particle (photon) state, and both N_R, the number of photons, if any, in the reservoir and V_R, the volume of the heat reservoir (the material making up the surroundings of the cavity in Figure 9.1) are constant. Under these conditions equation (11.3) is the same as equation (4.2), and the grand canonical distribution given by equation (11.7) reduces to the Boltzmann distribution given by equation (4.5). This justifies the use of the Boltzmann distribution in Section 9.4 to develop equation (9.15), which gives the number of photons in a single particle (photon) state. It also justifies the use of the partition function approach, or alternatively, an approach based on the Helmholtz free energy F, to determine the properties of the photon gas in Section 9.5*. Comparing the Boltzmann distribution, equation (4.5), with the grand canonical distribution given by equation (11.7), it can be seen that, in the case of the photon gas, the chemical potential μ of the particle (photon) reservoir in Figure 9.1 must be put equal to zero. According to equation (3.80), this would suggest that the chemical potential of the photon gas itself is zero, when a state of thermodynamic equilibrium is reached in the example shown in Figure 9.1. This result is in agreement with equation (9.23).

One often finds the statement that the total number N of photons inside the cavity in Figure 9.1 adjusts itself so as to minimise the Helmoholtz free energy F. If μ were finite and kept constant by a particle (photon) reservoir, for constant V and T, but variable N it is the grand potential Ω, which is equal to $(U - TS - \mu N)$, which is a minimum at thermodynamic equilibrium. [Reference: Problem 11.2 of Chapter 11.] If μ is zero, Ω is equal to $(U - TS)$, which is the Helmholtz free energy F.

Putting μ equal to zero in equation (12.46), we find that the ensemble average number of photons in a single particle (photon) state having (single particle) energy eigenvalue $\varepsilon = \hbar\omega$ is

$$\bar{n} = 1/[\exp(\hbar\omega/kT) - 1] \ . \qquad (12.67)$$

This is the same as equation (9.15). To obtain Planck's law for the total energy of the photons per unit volume, which have energies between $\hbar\omega$ and $\hbar(\omega + d\omega)$, corresponding to angular frequencies in the range ω to $(\omega + d\omega)$, we must multiply the number of single particle states per unit volume in this angular frequency range, given by equation (12.66), by the mean number of photons in each of these states, given by equation (12.67), and by $\hbar\omega$, the mean energy of each of these photons. This leads to equation (9.3).

2.5.5* Liquid helium*

Helium gas consists mainly of the isotope $^4_2$He, which has a resultant spin of zero and is a boson. The isotope $^3_2$He has a resultant spin of $\frac{1}{2}$ and is a fermion. Since $^3_2$He forms only 1.3 parts per million of naturally occurring helium, the properties of liquid helium are dominated by the properties of the isotope $^4_2$He. [The properties of liquid $^3_2$He have been determined using separated isotopes. They differ markedly from the properties of liquid $^4_2$He.]

The critical temperature of helium is 5.2K, so that helium cannot exist as a liquid above 5.2K, whatever the external pressure. At normal atmospheric pressure (760 mm of mercury), the boiling point of liquid $^4_2$He is 4.21K. The interatomic (van der Waals) forces are too weak for liquid $^4_2$He to solidify at atmospheric pressure, even at the lowest temperatures that have been attained. The phase diagram for $^4_2$He is shown in Figure 12.14. As the temperature of liquid helium is reduced, there is a large anomaly in the heat capacity, which, for liquid helium under its own saturation vapour pressure, is at 2.17K, as shown in Figure 12.15. This is called the λ-point. Liquid $^4_2$He at temperatures above the λ-point is called liquid He I. Below

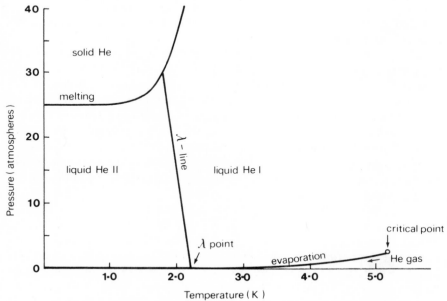

Figure 12.14—The phase diagram for liquid helium. The solid phase cannot exist as T tends to zero, unless a pressure of about 25 atmospheres is applied. There is a transition from liquid He I to liquid He II at the λ point. The co-existence curve for liquid He I and liquid He II is called the λ-line.

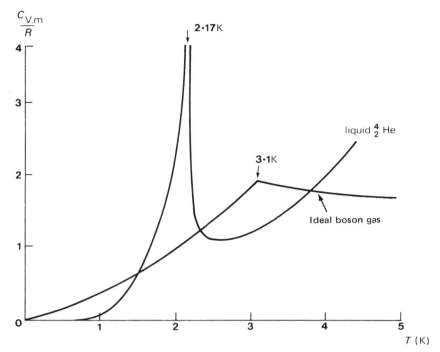

Figure 12.15—Experimental molar heat capacity $C_{V,m}$ of liquid $^4_2$He is compared with the predictions for an ideal boson gas of equal number density. (Reproduced with permission from *Introductory Statistical Mechanics* by R. E. Turner and D. S. Betts.)

the λ-point it is called liquid He II. Apart from its very low mass density, liquid He I is a fairly normal liquid. On the other hand, liquid He II has some remarkable properties. For example, the viscosity of liquid He II, measured by the rate of flow of the liquid through a capillary tube, is effectively zero. (A full account of the superfluid properties of liquid He II is given by Zemansky [9].)

In the two fluid theory of liquid He II, it is assumed that, at temperatures below the λ-point, liquid He II is a mixture of two fluids, namely a normal fluid, which behaves like a normal liquid having viscosity etc., and a superfluid, which has no viscosity and which gives rise to the superfluid properties of liquid He II. It is assumed that the proportion of the superfluid component increases from zero at the λ-point to unity at the absolute zero of temperature.

The coexistence curve for liquid He I and liquid He II is shown in Figure 12.14. It is called the λ-line, and joins the liquid He I, liquid He II, $^4_2$He vapour triple point at 2.17K to another triple point at 1.74K at which liquid He I, liquid He II and solid $_2^4$He coexist. [By contrast, liquid $_2^3$He does not have a λ-point, and does not have the same superfluid properties as liquid He II.[†] The differences between the behaviour of liquid $_2^4$He and liquid $_2^3$He are probably due, to some extent at least, to the fact that the $_2^4$He atom is a boson, whereas the $_2^3$He atom is a fermion.]

Since the interatomic (van der Waals forces) are weak in liquid helium, as a first approximation, it will be assumed that liquid $_2^4$He is an ideal boson gas. The volume of a mole of liquid $_2^4$He is 27×10^{-6} m^3, so that the concentration (N/V) is $2.2. \times 10^{28}$ m^{-3}. Using the value of 6.65×10^{-27} kg for the mass of a $_2^4$He atom in equation (12.54), we find that the Bose temperature T_b is 3.1 K. According to the ideal boson gas model, a significant proportion of the $_2^4$He atoms should be in the single particle ground state below the Bose temperature. These could correspond to the superfluid component of liquid He II. The $_2^4$He atoms in excited single particle states could correspond to the normal component of liquid He II. Thus the superfluid properties of liquid He II could be due, in part at least, to a Bose–Einstein type condensation at the λ point, but this cannot be the whole story. It can be seen from Figure 12.15 that there are very significant differences between the properties of liquid $_2^4$He and the ideal boson gas model. For example, the ideal boson gas model predicts a cusp in the variation of the molar heat capacity of liquid $_2^4$He at the Bose temperature, whereas the λ point is a much sharper peak. According to equation (12.64), the heat capacity of an ideal boson gas below the Bose temperature should be proportional to $T^{3/2}$, whereas below 0.6 K the experimentally determined heat capacity of liquid He II obeys the Debye T^3 law. This experimental result suggests that, even though liquid He II is a liquid, most of its internal energy below 0.6 K is in its collective longitudinal vibrational modes (longitudinal phonons) rather than in the motions of individual $_2^4$He atoms. The small interactions between $_2^4$He atoms are important at these very low temperatures, and for this reason the ideal boson gas model is inadequate. It is now believed that the superfluid properties of liquid He II are associated with the difficulty of producing excitations in liquid $_2^4$He, due to the form of the dispersion curve relating the energy and the momentum of the excitations in liquid He II. (References: Kittel and Kroemer [11] and Turner and Betts [1f].)

REFERENCES

[1] Turner, R. E. and Betts, D. S. *Introductory Statistical Mechanics*, (Sussex University Press, 1974), (a) pages 125 and 200, (b) page 202, (c) page 122, (d) page 123, (e) page 129, (f) page 142.

†Liquid $_2^3$He does show superfluid properties below 0.003 K, but these are associated with the formation of Cooper type pairs of $_2^3$He atoms. These pairs behave as bosons. (Reference: Hook [10].)

[2] Landau, L. D. and Lifshitz, E. M. *Statistical Physics* (Pergamon Press, 1969), (a) page 150, (b) page 151, (c) page 148, (d) page 155, (e) page 160.

[3] Huang, K. *Statistical Mechanics* (John Wiley, 1963), (a) Chapter 11, (b) Chapter 12.

[4] Gopal, E. S. R. *Statistical Mechanics and Properties of Matter*, (Ellis Horwood Ltd., Chichester, 1974), (a) page 163, (b) page 167.

[5] Kittel, C. *Introduction to Solid State Physics*, Fourth Edition, (John Wiley, 1971), Chapters 7, 8, 9 and 10.

[6] Orear, J. and Salpeter, E. E. *Amer. Journ. Phys.* **41**, 1131 (1973).

[7] Van Horn, H. M. *Physics Today, 32*, No. 1, page 23 (1979).

[8] Shklovskii, I. S. *Stars, Their Birth, Life and Death*, (W. H. Freeman, 1978), page 349.

[9] Zemansky, M. W. *Heat and Thermodynamics*, Fifth Edition (McGraw-Hill, 1968), Chapter 15.

[10] Hook, J. R. *Physics Bulletin*, **29**, page 513, (1978).

[11] Kittel, C. and Kroemer, H. *Thermal Physics*, (2nd Edition, W. H. Freeman, 1980), page 212.

Problems

Boltzmann's constant $k = 1.380 \times 10^{-23}$ J K^{-1}
Planck's constant $h = 6.625 \times 10^{-34}$ J s
$\hbar = 1.055 \times 10^{-34}$ J s
Mass of an electron $m = 9.10 \times 10^{-31}$ kg
1 electron volt 1 eV $= 1.602 \times 10^{-19}$ J
Avogadro's constant $N_A = 6.022 \times 10^{23}$ mole^{-1}

Problem 12.1
Assume that the temperature $T = 0.1T_f$ is low enough for the chemical potential to be given by equation (12.6). Calculate the value of the mean fermion occupancy $f(\varepsilon)$ for values of ε equal to $0.1\varepsilon_f$, $0.4\varepsilon_f$, $0.6\varepsilon_f$, $0.8\varepsilon_f$, $0.85\varepsilon_f$, $0.9\varepsilon_f$, $0.95\varepsilon_f$, $0.99\varepsilon_f$, $1.00\varepsilon_f$, $1.05\varepsilon_f$, and $1.10\varepsilon_f$, where ε_f is the Fermi energy and T_f is the Fermi temperature. Comment on the variation of $f(\varepsilon)$ with ε for the energy range within $\pm kT (= \pm 0.1kT_f)$ of the Fermi energy.

Problem 12.2
Assume that the temperature $T = 0.05T_f$ is low enough for the chemical potential to be given by equation (12.6). Plot the variation of the Fermi–Dirac distribution function $f(\varepsilon)$ with the ratio $\varepsilon/\varepsilon_f$ in the range 0 to 3. Make a large scale enlargement of the region, where $\varepsilon/\varepsilon_f$ is in the range 0.9 to 1.1.

Problem 12.3
Assume that for $T = 3T_f$ the value of the chemical potential is $-5.6\varepsilon_f$. Calculate the value of the Fermi–Dirac distribution function $f(\varepsilon)$ at the temperature T, for values of $\varepsilon/\varepsilon_f$ of (a) 0, (b) 0.2, (c) 0.4, (d) 0.8, (e) 1.0, (f) 1.5, (g) 2.0, (h) 3.0.

Problem 12.4

Use equation (12.6) to show that, if the temperature of an ideal fermion gas is equal to $0.1T_f°$, the chemical potential is equal to $0.9918\varepsilon_f°$, where $T_f°$ and $\varepsilon_f°$ are the (initial) values of the Fermi temperature and Fermi energy respectively. Use equation (12.2) to show that $f(\varepsilon)$, the mean occupancy of a single particle state having energy eigenvalue $\varepsilon = 1.6\,\varepsilon_f°$ is equal to 2.28×10^{-3}. The concentration (N/V) of the fermions is increased by a factor of eight, by increasing the number of fermions at a fixed volume V. Use equation (12.10) to show that the new values ε_f' of the Fermi energy and T_f' of the Fermi temperature are equal to $4\varepsilon_f°$ and $4T_f°$ respectively. The temperature of the fermion gas is kept at $T = 0.1T_f° = 0.025T_f'$. Use equation (12.6) to show that the new value of chemical potential is equal to $0.99949\varepsilon_f'$ which is equal to $3.99794\varepsilon_f°$ or 4.03 times the initial value of chemical potential. Show that the value of $f(\varepsilon)$ for $\varepsilon = 1.6\varepsilon_f° = 0.4\varepsilon_f'$ is now extremely close to unity. (This example illustrates how increasing the concentration of the fermions increases the chemical potential and changes the mean occupancies of the single particle states. The reader may find it helpful to draw an energy eigenvalue diagram showing ε and the initial and final values of the Fermi energy and chemical potential.)

Problem 12.5

The number of free electrons per metre^3 in silver is 6×10^{28}. Determine the Fermi energy. Express your answer in both joules and electron volts. What is the Fermi temperature? What is the velocity of an electron having kinetic energy equal to the Fermi energy? Calculate the pressure of the degenerate electron gas assuming that equation (12.13) is valid. Express your answer in atmospheres. (1 atmosphere $= 101325$ N m^{-2}.)

Problem 12.6

Consider an ideal fermion gas at $T = 0$. Convert the energy distribution given by equation (12.4) for $T = 0$ into a velocity distribution. Show that the mean speed of the fermions is equal to $\frac{3}{4}v_f$, where v_f is the speed of a fermion which has kinetic energy equal to the Fermi energy.

Problem 12.7

Consider a mole of ^3_2He gas atoms in a volume of 0.0224 m^3 at 273K. Use equation (5.84) to determine $\exp(-\mu/kT)$. Determine $f(\varepsilon)$, the mean occupancy of a single particle state which has energy eigenvalue equal to $3kT/2$. What is the Fermi temperature of the gas? The mass of a ^3_2He atom is 5.11×10^{-27} kg and the spin degeneracy $g = 2$.

Problem 12.8

It was shown in Section 5.9.8 that the chemical potential of a mole of ^4_2He gas at STP is $(-4.69 \times 10^{-20}$ J). Use equation (12.46) to show that the mean occupancy of a single particle state having energy eigenvalue $3kT/2$ is 8.7×10^{-7} at STP.

Problem 12.9
Assume that the chemical potential of an ideal boson gas at a temperature equal to $3T_b$ is $(-2.6kT_b)$. Determine the mean number of bosons $b(\varepsilon)$ in single particle states having energy eigenvalues of (a) 0; (b) $0.5kT_b$; (c) $1.0kT_b$; (d) $1.5kT_b$; (e) $2.0kT_b$; (f) $3.0kT_b$.

Problem 12.10* Consider an ideal boson gas consisting of bosons of mass 6.65×10^{-27} kg and spin zero. The concentration of the bosons is 10^{26} m^{-3}. Use equation (12.55) to determine the Bose temperature. Use equation (12.57) to determine the fraction of the bosons in the single particle ground state at a temperature of $0.2T_b$.

Problem 12.11*
Consider a mole of an ideal boson gas consisting of $^4_2$He atoms, each of mass 6.65×10^{-27} kg and spin zero inside a cubical box of volume 0.0224 m^3. Calculate the Bose temperature. Use equation (2.5) to check the values for the energy eigenvalues of the single particle states having quantum numbers $n_x = n_y = n_z = 1$ and $n_x = 2, n_y = n_z = 1$ given in Section 2.1.4. Change the energy scale so that the energy of the 1, 1, 1 state is zero. Use equation (12.52) to determine the chemical potential of the boson gas at a temperature of 10^{-4}K. Calculate the ratio of the mean number of bosons in these two single particle states at 10^{-4}K using equation (12.46). What would the ratio be if the Boltzmann distribution were used?

Problem 12.12*
Apply equation (11.31) to the expression for the grand potential Ω of an ideal boson gas, given by equation (12.60) to show that the entropy of an ideal boson gas is

$$S = -k \sum \{b(\varepsilon_j)\log b(\varepsilon_j) - [1 + b(\varepsilon_j)]\log[1 + b(\varepsilon_j)]\}$$

where $b(\varepsilon_j)$ is given by equation (12.46), and the summation is over all the *single* particle states of the system.
Hint: it follows from equation (12.46) that

$$\exp[(\mu - \varepsilon_j)/kT] = b(\varepsilon_j)/[1 + b(\varepsilon_j)] \ .$$

Taking the natural logarithm of both sides we have

$$(\mu - \varepsilon_j)/kT = \log\{b(\varepsilon_j)/[1 + b(\varepsilon_j)]\}.$$

Appendix 1

Mathematical Appendix

A1.1 LOGARITHMIC AND EXPONENTIAL FUNCTIONS

It is assumed that the reader is familiar with differential and integral calculus, and with the logarithmic and exponential functions. The natural logarithm of x, that is the logarithm of x to the base e is usually denoted by either $\log_e x$ or $\ln x$. Since we shall only use the natural logarithm of x to the base e in the text, the algebraic expressions are simplified by denoting the natural logarithm of x simply as $\log x$.

If $y = \log x$, then $x = e^y$. Occasionally, it will be more convenient to rewrite e^y as $\exp(y)$.

If $y = 0$, $e^y = 1$, so that $\log 1 = 0$.

The reader should be thoroughly familiar with the following types of operations:

$$\log A + \log B = \log(AB)$$
$$\log A - \log C = \log(A/C)$$
$$\log A^n \qquad = \log A + \log A + \ldots = n \log A \ .$$

The reader should be able to recognise the following type of step, when it is carried out in either direction:

$$\log[A^3 B^2 C^5/D^4] = 3 \log A + 2 \log B + 5 \log C - 4 \log D \ .$$

If
$$\log A \qquad = \log B + x$$
$$\log(A/B) = x \ .$$

Taking exponentials we have

$$A = Be^x \ .$$

In going in the reverse direction, the reader should remember that $\log(e^x)$ is equal to x.

A1.2 PARTIAL DIFFERENTIATION

The expression $f(x, y, z)$ means that the function f is a function of the independent variables x, y and z. For example, the expression $S(U, V, N)$

means that the entropy S is a function of the energy U, the volume V and the number of particles N. Another typical example used in the text is the expression $g_1(U^* - U_2)$, which means that g_1 is a function of the variable $(U^* - U_2)$.

The differentiation of $f(x, y, z)$ with respect to x keeping y and z constant is called the partial differential coefficient of f with respect to x, and is denoted $\partial f/\partial x$. For example, if

$$f(x, y, z) = x^2yz^2 + 3xy^2z^4$$
$$\partial f/\partial x = 2xyz^2 + 3y^2z^4 \ .$$

Similarly $\partial f/\partial y$ is the differential coefficient of f with respect to y keeping x and z constant. We have

$$\partial f/\partial y = x^2z^2 + 6xyz^4 \ .$$

Similarly, $$\partial f/\partial z = 2x^2yz + 12xy^2z^3 \ .$$

Differentiating $\partial f/\partial x$ partially with respect to x, we have

$$\frac{\partial^2 f}{\partial x^2} = \frac{\partial}{\partial x}\left(\frac{\partial f}{\partial x}\right) = 2yz^2 \ .$$

Differentiating $\partial f/\partial x$ partially with respect to y, we have

$$\frac{\partial^2 f}{\partial y \, \partial x} = \frac{\partial}{\partial y}\left(\frac{\partial f}{\partial x}\right) = 2xz^2 + 6yz^4 \ .$$

Differentiating $\partial f/\partial y$ partially with respect to x, we have

$$\frac{\partial^2 f}{\partial x \, \partial y} = \frac{\partial}{\partial x}\left(\frac{\partial f}{\partial y}\right) = 2xz^2 + 6yz^4 \ .$$

Notice that $$\frac{\partial^2 f}{\partial x \, \partial y} = \frac{\partial^2 f}{\partial y \, \partial x} \ . \tag{A1.1}$$

It is not always obvious in thermodynamics and statistical physics which variables are being kept constant. The convention adopted is to place the variables kept constant outside brackets. For example, $(\partial S/\partial U)_{V,N}$ means that the entropy S is differentiated partially with respect to the energy U, keeping the volume V and the number of particles N constant.

A1.3 SOME STATISTICAL FORMULAE

Consider three particles coloured red, white and blue and denoted R, W and B respectively. The three particles can be arranged in the following six ways: R, W, B; R, B, W; W, R, B; W, B, R; B, R, W; and B, W, R. The first particle can be chosen in 3 ways namely R or W or B. That leaves 2 choices for the second particle for each of the three first choices. The first two particles chosen can be arranged in 3×2 ways namely R, W; R, B; W, R; W, B; B, R and B, W. There is only one choice left for the third particle in each of these cases. Hence the total number of ways of arranging the three particles is $3 \times 2 \times 1$, which is three factorial and is denoted 3! In the general case, if there are N distinguishable objects, the first can be chosen in N ways. Having chosen one object, there are $(N - 1)$ left, so that there are $(N - 1)$ ways of choosing the second object, for each of the N possibilities for the first choice. Hence the first two objects can be chosen in $N(N - 1)$ different ways. Having chosen two objects, the third can be chosen in $(N - 2)$ ways and so on. Hence the total number of ways of arranging the N distinguishable objects is $N(N - 1)(N - 2) \dots$ which is $N!$

It will now be assumed that there are N distinguishable objects to be placed into two distinguishable boxes. Let M denote the number of ways we can select n_1 distinguishable objects for box number 1 and $n_2 = (N - n_1)$ distinguishable objects for box number 2. It does not matter which order the objects are put into boxes 1 and 2, what matters is which objects are in which box. As an illustrative example consider 5 particles, coloured red(R), white (W), blue(B), green(G) and yellow(Y). We shall consider the M ways we can select 3 particles for box 1 and 2 particles for box 2. One possible selection would be to put R, W and B in box 1 and G and Y in box 2. Having made this selection, the R, W and B particles are taken from box 1. These 3 particles can be arranged in 3! ways, namely R, W, B; R, B, W; W, R, B; W, B, R; B, R, W and B, W, R. The two particles are then taken from box 2 and placed after the particles from box 1. The two particles from box 2 can be arranged in 2! ways, namely G, Y and Y, G. Each of the 3! ways of arranging the 3 particles from box 1 can be combined with each of the 2! ways of arranging the two particles from box 2 to give $3! \times 2!$ arrangements of the type B, W, R, Y, G for the selection of R, W and B for box 1 and G and Y for box 2. Each of the M different selections of three distinguishable particles for box 1 and two distinguishable particles for box 2 can be arranged with the particles from box 1 first followed by the particles from box 2 in $3! \times 2!$ ways. The $M \times 3! \times 2!$ arrangements obtained in this way give the 5! arrangements[†] of R, W, B, G and Y. Hence

$$M\,3!\,2! = 5!$$
$$M = 5!/3!\,2! = 10 \ .$$

[†]The reader could check this point by writing out the 10 selections and then writing out the 12 arrangements for each selection to show that this gives all the 5! arrangements. It will be a lot quicker for the reader to do it for 3 distinguishable particles, with 2 particles in box 1 and 1 particle in box 2, showing that this gives the 3! arrangements.

In the general case, if we have N distinguishable objects, the number of ways M of selecting n_1 objects for box 1 and $n_2 = (N - n_1)$ objects for box 2 is given by

$$Mn_1! \, n_2! = N!$$

or,
$$M = N!/n_1! n_2! = N!/n_1! \, (N - n_1)! \, . \qquad (A1.2)$$

The number of ways n objects can be selected from a set of N different objects, which is equivalent to putting n objects into box 1 and $(N - n)$ objects into box 2, is called the number of combinations of N objects taken n at a time, and is generally denoted either by $_N C_n$ or $\binom{N}{n}$. From equation (A1.2), we have

$$_N C_n = \binom{N}{n} = \frac{N!}{n! \, (N - n)!} \, . \qquad (A1.3)$$

If we have r distinguishable boxes with n_1 distinguishable objects in box 1, n_2 distinguishable objects in box 2, etc., generalising equation (A1.2) we have

$$M = \frac{N!}{n_1! \, n_2! \, n_3! \ldots n_r!} \, , \qquad (A1.4)$$

where N is the total number of distinguishable objects. In applying equations (A1.2), (A1.3) and (A1.4) the reader should assume that 0! is equal to unity. Equation (A1.4) is applied in Section 2.2.1 of Chapter 2 to calculate how many ways N distinguishable particles can be arranged with n_1 particles in single particle state 1, n_2 particles in single particle state 2, etc.

Another formula used in Section 2.2.1 is equivalent to calculating the number of ways one can put X *indistinguishable* apples into N *distinguishable* boxes, without any restriction on how many of the N apples are in each box. Obviously, if all the apples are to be in one or other of the boxes, we must start by picking a box. This can be done in N ways leaving $(N - 1)$ boxes and X apples. We can then either choose another box, which would leave the first box empty, or we can pick an apple to put into the first box. If the apples are distinguishable this choice can be made in $(N + X - 1)$ ways. The next choice can be made in $(N + X - 2)$ ways and so on until we end up with either an apple to put into the last box or with a box which remains empty. Hence the total number of ways we can put X distinguishable apples into N distinguishable boxes is $N(N + X - 1)!$ For *each* choice of the order of the boxes, there are $X!$ ways of selecting the order in which the X distinguishable apples are put into the boxes. If the apples are indistinguishable, the order they are put into the boxes does not matter, and we must divide $N(N + X - 1)!$ by $X!$ Since there are $N!$ ways of selecting N distinguishable boxes, if we are not interested in the order in which we pick the boxes, but only in how many indistinguishable apples are in each box, we must also divide by $N!$ Hence the number of ways of putting X indistinguishable apples into N distinguishable boxes is

$$g = \frac{N(N + X - 1)!}{N!\,X!} = \frac{(N + X - 1)!}{(N - 1)!\,X!}\ . \tag{A1.5}$$

For example, if there are $X = 3$ apples for $N = 2$ boxes,

$$g = 4!/1!\,3! = 4\ .$$

These are 3, 0; 0, 3; 2, 1 and 1, 2, where the first and second numbers are the numbers of apples in boxes 1 and 2 respectively. Equation (A1.5) is used in Section 2.2.1 to calculate how many ways we can give $X = U$ indistinguishable energy units (which are equivalent to the apples) to N distinguishable particles (which are equivalent to the distinguishable boxes).

A1.4 AVERAGE VALUES AND THE STANDARD DEVIATION

The average value (or mean value) of a set of N values $x_1, x_2, \ldots x_N$ of x is denoted by either $\bar{x}$ or $\langle x \rangle$, and is given by

$$\bar{x} = \langle x \rangle = \frac{x_1 + x_2 + \ldots + x_N}{N} = \frac{1}{N} \sum_{j=1}^{N} x_j\ . \tag{A1.6}$$

The summation is over all the N values of x_j. For example, if the values of x_j are 6, 7, 6, 7, 7, 8, 9, 7, 5, 8, the average value of x is

$$\bar{x} = \langle x \rangle = (6 + 7 + 6 + 7 + 7 + 8 + 9 + 7 + 5 + 8)/10$$
$$\bar{x} = 7\ .$$

Since there are one five, two sixes, four sevens, two eights and one nine, the expression for $\bar{x}$ can be written in the form

$$\bar{x} = (1 \times 5 + 2 \times 6 + 4 \times 7 + 2 \times 8 + 1 \times 9)/10\ .$$

This can be rewritten in the form

$$\bar{x} = \frac{1}{10} \times 5 + \frac{2}{10} \times 6 + \frac{4}{10} \times 7 + \frac{2}{10} \times 8 + \frac{1}{10} \times 9\ .$$

For the numerical values given, the probability of getting a 5 is $1/10$, the probability of getting a 6 is $2/10$ etc. (This probability distribution is shown in Figure A1.1(a)). Hence in the general case, equation (A1.6) can be rewritten in the form

$$\bar{x} = \langle x \rangle = \Sigma x_i P_i \tag{A1.7}$$

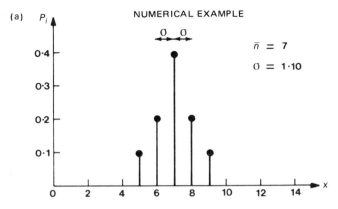

(a) P_i NUMERICAL EXAMPLE

$\bar{n} = 7$

$\sigma = 1\cdot10$

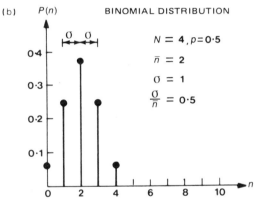

(b) $P(n)$ BINOMIAL DISTRIBUTION

$N = 4, p = 0\cdot5$

$\bar{n} = 2$

$\sigma = 1$

$\dfrac{\sigma}{\bar{n}} = 0\cdot5$

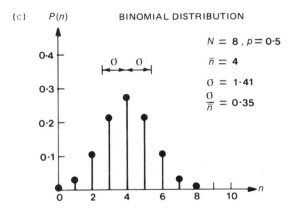

(c) $P(n)$ BINOMIAL DISTRIBUTION

$N = 8, p = 0\cdot5$

$\bar{n} = 4$

$\sigma = 1\cdot41$

$\dfrac{\sigma}{\bar{n}} = 0\cdot35$

Figure A1.1—(a) The probability distribution for the values of x given in Section A1.4. (b) The binomial distribution, given by equation (A1.10), for the case when $N = 4$, $p = 0.5$. (c) The binomial distribution, given by equation (A1.10), for the case when $N = 8$, $p = 0.5$.

where P_i is the probability of getting the value x_i. The summation is over all possible values of x_i. In our numerical example, all the P_i are zero, except when x_i is equal to 5, 6, 7, 8 or 9, in which case the values of P_i are 0.1, 0.2, 0.4, 0.2 and 0.1 respectively. According to equation (A1.7) the value of $\bar{x}$ is equal to $(0.1 \times 5 + 0.2 \times 6 + 0.4 \times 7 + 0.2 \times 8 + 0.1 \times 9)$, which is equal to 7. The standard deviation σ is defined by the equation

$$\sigma^2 = \langle (x_j - \bar{x})^2 \rangle = \frac{1}{N} \sum_{j=1}^{N} (x_j - \bar{x})^2 \ . \tag{A1.8}$$

The summation is over all the N given values of x_j. The reader can check that for our numerical example $\sigma = 1.10$. The quantity σ^2 is sometimes called the variance. Alternatively, by comparison with equation (A1.7),

$$\sigma^2 = \Sigma (x_i - \bar{x})^2 P_i \ . \tag{A1.9}$$

where P_i is the probability of getting the value x_i, and the summation is over all possible values of x_i. For our numerical example, since $\bar{x} = 7$, using equation (A1.9), we have

$$\sigma^2 = [0.1(5-7)^2 + 0.2(6-7)^2 + 0.2(8-7)^2 + 0.1(9-7)^2] = 1.2 \ .$$

Hence, $\sigma = 1.10$.

The standard deviation σ is a good measure of the width of the probability distribution. In the numerical example illustrated in Figure A1.1(a), 8 of the 10 values of x lie between $(\bar{x} - \sigma)$ and $(\bar{x} + \sigma)$. Two further examples, based on the binomial distribution, are given in Figures A1.1(b) and A1.1(c). These will be discussed in the next Section. In the case of the Gaussian distribution, developed later in Section A1.10, 68% of the values of x are between $(\bar{x} - \sigma)$ and $(\bar{x} + \sigma)$, and 99.7% of the values of x lie between $(\bar{x} - 3\sigma)$ and $(\bar{x} + 3\sigma)$.

A1.5 THE BINOMIAL DISTRIBUTION

According to the binomial distribution, the probability $P(n)$ of getting n successes in N attempts is

$$P(n) = \frac{N!}{n! \, (N - n)!} p^n q^{(N-n)} \ , \tag{A1.10}$$

where p is the probability of a success per attempt and $q = (1 - p)$ is the probability of a failure per attempt. For example, in the case of an unloaded die, the probability of throwing a six is $1/6$ and the probability of not throwing a six is $5/6$. As an introductory example, consider the probability of throwing $n = 3$ sixes using $N = 5$ dice. First of all we shall calculate the number of ways we can select 3 dice from 5. According to equation (A1.2) this can be done in $M = 5!/3! \, 2! = 10$ ways. Having selected three dice, we must calculate the probability of getting a six with each of the three chosen dice and the probability of not throwing a six with each of the other two dice. The probability of getting three sixes with the three chosen dice is $(1/6)^3$, and the probability of not getting a six with the other two dice is $(5/6)^2$. Hence the probability of getting 3 sixes with the 3 chosen dice and no six with the other two dice is $(1/6)^3(5/6)^2 = 0.0032$. This is true for each of the $M = 10$ ways of selecting 3 dice from 5. Hence the total probability of obtaining exactly 3 sixes, when we throw 5 dice is $10 \times 0.0032 = 0.032$. The reader can check this result using equation (A1.10).

In the general case, according to equation (A1.2) the number of ways of selecting n dice from a total of N dice is

$$M = N!/n! \, (N - n)! \ .$$

The probability of getting exactly n successes with each of the n chosen dice and failures with the other $(N - n)$ dice is $p^n q^{(N-n)}$. Hence the total probability of getting exactly n successes from N attempts is given by equation (A1.10).

The probability distributions given by equation (A1.10) for the cases when $N = 4$, $p = 0.5$ and $N = 8$, $p = 0.5$ are shown in Figure A1.1(b) and A1.1(c) respectively.

Using the binomial theorem, which is equation (A1.25) of Section A1.7, we have

$$(p + q)^N = q^N \left(1 + \frac{p}{q} \right)^N$$

$$= q^N \left[1 + \frac{Np}{q} + \frac{N(N - 1)}{2!} \left(\frac{p}{q} \right)^2 + \cdots + \frac{N!}{n!(N - n)!} \left(\frac{p}{q} \right)^n + \cdots \right]$$

Hence $\qquad (p + q)^N = \sum_{n=0}^{N} \frac{N!}{n!(N - n)!} p^n q^{(N-n)} = \sum_{n=0}^{N} P(n) \ ,$

where $P(n)$ is given by equation (A1.10). It can be seen that the term in p^n in the expansion of $(p + q)^N$ is equal to the expression for $P(n)$ given by equation (A1.10). Since $(p + q)$ is equal to unity, $\Sigma P(n)$ is equal to unity, confirming that equation (A1.10) is normalised. The mean value of n, which is denoted by $\bar{n}$ or by $\langle n \rangle$, is

$$\bar{n} = \langle n \rangle = \sum_{n=0}^{N} nP(n) = \sum_{n=0}^{N} \frac{N!}{n!(N - n)!} np^n q^{(N-n)} \ . \qquad (A1.11)$$

Since $np^n = p(\partial/\partial p)(p^n)$, the expression for $\bar{n}$ can be rewritten in the form

$$\bar{n} = \sum_{n=0}^{N} \frac{N!}{n! \, (N - n)!} \left[p \, \frac{\partial}{\partial p} \, p^n \right] q^{(N-n)} \ . \qquad (A1.12)$$

Interchanging the order of summation and differentiation, we have

$$\bar{n} = p \frac{\partial}{\partial p} \sum_{n=0}^{N} P(n) = p \frac{\partial}{\partial p} (p + q)^N = pN(p + q)^{(N-1)} .$$

Since $(p + q)$ is equal to unity,

$$\bar{n} = \langle n \rangle = Np . \tag{A1.13}$$

According to equation (A1.8), the standard deviation σ is given by

$$\sigma^2 = \langle (n - \bar{n})^2 \rangle = \overline{n^2} - 2(\bar{n})(\bar{n}) + (\bar{n})^2$$
$$\sigma^2 = \overline{n^2} - (\bar{n})^2 . \tag{A1.14}$$

Now $\quad \overline{n^2} = \langle n^2 \rangle = \sum_{n=0}^{N} n^2 P(n) = \sum_{n=0}^{N} n^2 \frac{N!}{n!\,(N-n)!} p^n q^{(N-n)} .$

The reader can check that $n^2 p^n = [p(\partial/\partial p)]^2 p^n$.

Hence, $\quad \overline{n^2} = \sum_{n=0}^{N} \frac{N!}{n!\,(N-n)!} \left(p \frac{\partial}{\partial p} \right)^2 p^n q^{N-n} .$

Interchanging the order of summation and differentiation, we have

$$\overline{n^2} = \left(p \frac{\partial}{\partial p} \right)^2 \sum_{n=0}^{N} P(n) = \left(p \frac{\partial}{\partial p} \right) \left(p \frac{\partial}{\partial p} \right) (p + q)^N$$
$$= p \frac{\partial}{\partial p} [pN(p + q)^{(N-1)}]$$
$$= p[N(p + q)^{(N-1)} + pN(N - 1)(p + q)^{(N-2)}] .$$

Putting $(p + q)$ equal to unity, we have

$$\overline{n^2} = Np + Np^2(N - 1) . \tag{A1.15}$$

According to equation (A1.13),

$$(\bar{n})^2 = N^2 p^2 . \tag{A1.16}$$

Substituting from equations (A1.15) and (A1.16) into equation (A1.14) we obtain

$$\sigma^2 = Np(1 - p) = Npq . \tag{A1.17}$$

[Equations (A1.13) and (A1.17) are also valid for the Gaussian and the Poisson distributions, which are special cases of the binomial distribution.]

In the example of the binomial distribution, illustrated in Figure A1.1(b), $N = 4$, $p = q = 0.5$ so that, according to equations (A1.13) and (A1.17), $\bar{n} = 2$ and $\sigma = 1$. The ratio of the standard deviation σ to the mean value $\bar{n}$ is 0.5. In this example, the probability that n is in the range

$(\bar{n} - \sigma)$ to $(\bar{n} + \sigma)$ is 0.875. In the example illustrated in Figure A1.1(c), $N = 8$ and $p = q = 0.5$. In this case, $\bar{n} = 4$, $\sigma = 1.41$ and $\sigma/\bar{n}$ is equal to 0.35. The probability that n lies between $(\bar{n} - \sigma)$ and $(\bar{n} + \sigma)$ is 0.71. It can be seen that for a binomial distribution, most of the values of n are within a range of $\pm\sigma$ of the mean value. The standard deviation is a good measure of the width of the probability distribution. It can be seen that as N increases, the ratio of the standard deviation to mean value decreases. According to equations (A1.13) and (A1.17), $\sigma/\bar{n}$ is equal to $(q/Np)^{1/2}$. For $N = 10^{18}$ and $p = q = 0.5$, the ratio $\sigma/\bar{n}$ is equal to 10^{-9}.

A1.6 STIRLING'S APPROXIMATION

Factorial N is given by

$$N! = 1 \times 2 \times 3 \times \ldots \times N \ .$$

Hence $\log N! = \log 1 + \log 2 + \log 3 + \ldots + \log N \ .$ (A1.18)

The right hand side of equation (A1.18) is equal to the sum of the areas of the rectangles in Figure A1.2 up to $x = N$. In Figure A1.2, we have also plotted $\log x$ against x. It can be seen that for large N, the area under the curve of $\log x$ plotted against x approximates closely to the sum of the areas of the rectangles. Hence for large N, $\log N!$ to the base e is given by

$$\log N! \simeq \int_1^N \log x \ dx = [x \log x - x]_1^N = [N \log N - N + 1] \ .$$

Hence, for $N \gg 1$, $\log N! \simeq N \log N - N \ .$ (A1.19)

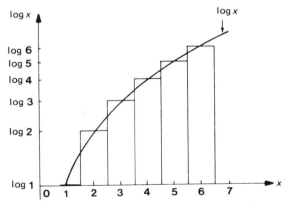

Figure A1.2—Derivation of Stirling's approximation: $\log N!$ is equal to the sum of the areas of the rectangles. For large N this area can be approximated by the area under the curve of $\log x$.

For example, the reader can check using a calculator that 50! is equal to 3.0414×10^{64}, and that log 50! is equal to 148.478. The value of [50 log 50 − 50] is equal to 145.601, which is correct to within about 2%. A better approximation is given by

$$\log N! = N \log N - N + \tfrac{1}{2} \log(2\pi N) \ . \tag{A1.20}$$

For $N = 50$, the term $\tfrac{1}{2} \log(2\pi N)$ is equal to 2.875. Applying this correction to [50 log 50 − 50] gives a value of 148.476 as an approximation for log 50! This is correct to within about 1 part in 10^5. If N is equal to 10^{18}, [$N \log N - N$] is equal to 4.0446×10^{19} whereas the $\tfrac{1}{2} \log(2\pi N)$ term is only equal to 21.64 and is negligible. For very large N, equation (A1.19) is a satisfactory approximation for log N!

A1.7 THE TAYLOR SERIES

Assume that the function $f(x)$ can be expanded in the power series

$$f(x) = a_0 + a_1 x + a_2 x^2 + a_3 x^3 + \ldots \tag{A1.21}$$

Differentiating this series term by term, we have

$$f'(x) = df/dx = a_1 + 2a_2 x + 3a_3 x^2 + \ldots \tag{A1.22}$$

$$f''(x) = d^2 f/dx^2 = 2a_2 + 6a_3 x + \ldots \ . \tag{A1.23}$$

Putting $x = 0$ in equations (A1.21), (A1.22) and (A1.23) etc., we have

$$a_0 = f(0); \qquad a_1 = f'(0); \qquad a_2 = f''(0)/2! \quad \text{etc.}$$

Substituting in equation (A1.21), in the range where the series is convergent, we have

$$f(x) = f(0) + xf'(0) + (x^2/2!)f''(0) + \ldots + (x^n/n!)f^{(n)}(0) + \ldots \tag{A1.24}$$

This is **Maclaurin's theorem**. For example, if $f(x)$ is equal to $(1 + x)^n$

$$f(0) = 1; \qquad f'(0) = n; \qquad f''(0) = n(n - 1) \quad \text{etc.}$$

Substituting in equation (A1.24), we have

$$(1 + x)^n = 1 + nx + n(n - 1)x^2/2! + \ldots \tag{A1.25}$$

This is the **binomial theorem**.

A function $y = f(x)$ is plotted against x in Figure A1.3. The value f_C of the function $f(x)$ at the point C, having abscissa x, will be expressed in terms of the values of the differential coefficients f', f'', ... evaluated at the point B in Figure A1.3, which has an abscissa $x = a$. Choose a new origin at the point O', which is on the x axis at $x = a$ in Figure A1.3. The ordinate through O' goes through B. Introduce a new variable h which is equal to

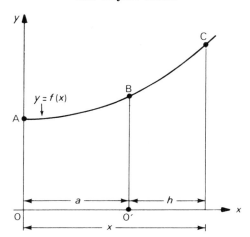

Figure A1.3—Derivation of the Taylor series.

$(x - a)$. The abscissa of the point C in the new coordinate system is $h = (x - a)$, and the abscissa of the point B is $h = 0$. Applying Maclaurin's theorem, equation (A1.24), using h as variable instead of x, we have for the value of the function f at the point C

$$f_C = f(h = 0) + h(df/dh) + (h^2/2!)(d^2f/dh^2) + \ldots . \quad \text{(A1.26)}$$

The differential coefficients must be evaluated at $h = 0$. Since they are both equal to the slope of the curve at the point B in Figure A1.3, the value of df/dh evaluated at $h = 0$ is equal to the value of df/dx evaluated at $x = a$. The latter will be denoted by $f'(a)$. Similarly d^2f/dh^2 is equal to $f''(a)$ etc. Since h is equal to $(x - a)$ and $f(h = 0)$ is equal to $f(a)$, equation (A1.26), giving the value of f at the point C, which has abscissa x in the coordinate system having O as origin, can be rewritten in the form

$$f_C = f(x) = f(a) + (x - a)f'(a) + [(x - a)^2/2!]f''(a) + \ldots . \quad \text{(A1.27)}$$

The differential coefficients in equation (A1.27) are evaluated at the point B which has an abscissa $x = a$ in the coordinate system having O as origin. Alternatively, with $x = (a + h)$, where a is a constant and h is a variable

$$f(a + h) = f(a) + hf'(a) + \frac{h^2}{2!}f''(a) + \ldots$$

$$= \sum_{n=0}^{\infty} \frac{h^n}{n!} f^{(n)}(a) . \quad \text{(A1.28)}$$

The differential coefficients $f^{(n)}(a)$ are evaluated at the point $h = 0$.
Equations (A1.27) and (A1.28) are two forms of the **Taylor series**.

A1.8 THE TOTAL DIFFERENTIAL

Let f be a function of the two independent variables x and y. Let x be changed to $(x + \delta x)$ and let y be changed to $(y + \delta y)$. First we shall expand the function $f(x + \delta x, y + \delta y)$ in a Taylor series about the point x, $(y + \delta y)$ keeping $(y + \delta y)$ constant. Putting $a = x$ and $h = \delta x$ in equation (A1.28) we have

$$f(x + \delta x, y + \delta y) = f(x, y + \delta y) + \delta x \cdot (\partial f / \partial x) + \ldots .$$

(A1.29)

The partial differential coefficients are evaluated at the point $(x, y + \delta y)$. Expanding $f(x, y + \delta y)$ as a Taylor series with $a = y$ and $h = \delta y$ in equation (A1.28) and keeping x constant, we have

$$f(x, y + \delta y) = f(x, y) + \delta y (\partial f / \partial y) + \ldots ,$$

where the partial differential coefficients are evaluated at the point (x, y). If the $\partial f / \partial x$ term in equation (A1.29), which is a function of x and $(y + \delta y)$, is expanded as a Taylor series about the point (x, y) it is equal to the value of $\partial f / \partial x$ evaluated at (x, y) plus terms in ∂y, $(\partial y)^2$ etc. Substituting in equation (A1.29) for $f(x, y + \delta y)$ and for $\partial f / \partial x$ and assuming that both δx and δy are infinitesimal, so that second order terms can be ignored, with $\delta x = dx$ and $\delta y = dy$, we have

$$f(x + dx, y + dy) = f(x, y) + (\partial f / \partial x)dx + (\partial f / \partial y)dy . \quad (A1.30)$$

Hence $df = f(x + dx, y + dy) - f(x, y) = (\partial f / \partial x)dx + (\partial f / \partial y)dy$.

The partial differential coefficients are evaluated at the point (x, y). It is straightforward to extend the method to the general case when $f(x, y, z, \ldots)$ is a function of the independent variables $x, y, z, \ldots$. The total differential of $f(x, y, z, \ldots)$ is

$$df = \left(\frac{\partial f}{\partial x}\right) dx + \left(\frac{\partial f}{\partial y}\right) dy + \left(\frac{\partial f}{\partial z}\right) dz + \ldots . \quad (A1.31)$$

For example, if the entropy $S(U, V, N)$ is a function of the energy U, the volume V and the number of particles N, using equation (A1.31), we have

$$dS = \left(\frac{\partial S}{\partial U}\right)_{V,N} dU + \left(\frac{\partial S}{\partial V}\right)_{U,N} dV + \left(\frac{\partial S}{\partial N}\right)_{U,V} dN . \quad (A1.32)$$

A1.9 GAUSSIAN INTEGRALS

To evaluate the integral $I = \displaystyle\int_{-\infty}^{+\infty} \exp(-x^2) \, dx$, multiply by I to give

$$I^2 = \int_{-\infty}^{+\infty} \exp(-x^2)\, dx \int_{-\infty}^{+\infty} \exp(-y^2)\, dy = \int_{-\infty}^{+\infty}\int_{-\infty}^{+\infty} \exp[-(x^2 + y^2)]\, dx\, dy \ .$$

(A1.33)

This double integral extends over the whole of the xy plane. Introduce polar coordinates r and θ, where $r^2 = (x^2 + y^2)$. The element of area in polar coordinates is $r\, dr\, d\theta$. Hence equation (A1.33) can be rewritten in the form

$$I^2 = \int_0^\infty\int_0^{2\pi} \exp(-r^2)\, r\, dr\, d\theta = 2\pi \int_0^\infty \exp(-r^2)\, r\, dr \ .$$

If $r^2 = u$, $du = 2r\, dr$, and

$$I^2 = \pi \int_0^\infty e^{-u}\, du = \pi[-e^{-u}]_0^\infty = \pi \ .$$

Hence,

$$I = \int_{-\infty}^{+\infty} \exp(-x^2)\, dx = \pi^{1/2} \ .$$

(A1.34)

Since the function $\exp(-x^2)$ is symmetrical about $x = 0$, it follows that

$$\int_0^\infty \exp(-x^2)\, dx = \frac{\pi^{1/2}}{2} \ .$$

(A1.35)

To evaluate the integral $I_0 = \int_0^\infty \exp(-\alpha v^2)\, dv$,

let $w = \alpha^{1/2} v$, so that $dw = \alpha^{1/2}\, dv$. Using equation (A1.35), we have

$$I_0 = \frac{1}{\alpha^{1/2}} \int_0^\infty \exp(-w^2)\, dw = \frac{1}{2} \left(\frac{\pi}{\alpha}\right)^{1/2} \ .$$

(A1.36)

To evaluate

$$I_1 = \int_0^\infty v \exp(-\alpha v^2)\, dv \ ,$$

let $z = \alpha v^2$, so that $dz = 2\alpha v\, dv$. We have

$$I_1 = \frac{1}{2\alpha} \int_0^\infty e^{-z}\, dz = \frac{1}{2\alpha}[-e^{-z}]_0^\infty = \frac{1}{2\alpha} \ .$$

(A1.37)

To evaluate the integral

$$I_n = \int_0^\infty v^n \exp(-\alpha v^2)\, dv = -\frac{1}{2\alpha} \int_0^\infty v^{(n-1)} \left[\frac{d[\exp(-\alpha v^2)]}{dv}\right] dv$$

where n is an integer, integrate by parts to give

$$I_n = -\frac{1}{2\alpha}\left[[v^{(n-1)}\exp(-\alpha v^2)]_0^\infty - \int_0^\infty \exp(-\alpha v^2)\left[\frac{dv^{(n-1)}}{dv}\right]dv\right] .$$

Since the integrated term vanishes at both limits, we have

$$I_n = \frac{1}{2\alpha}\int_0^\infty (n-1)v^{(n-2)}\exp(-\alpha v^2)\,dv = \left(\frac{n-1}{2\alpha}\right)I_{(n-2)} . \quad (A1.38)$$

This is a recurrence relation which enables us to determine any of the I_n in terms of either I_0 or I_1. Summarising, we have

$$I_0 = \int_0^\infty \exp(-\alpha v^2)\,dv = \frac{1}{2}\left(\frac{\pi}{\alpha}\right)^{1/2} \qquad\qquad (A1.36)$$

$$I_1 = \int_0^\infty v\exp(-\alpha v^2)\,dv = \frac{1}{2\alpha} \qquad\qquad (A1.37)$$

$$I_2 = \int_0^\infty v^2\exp(-\alpha v^2)\,dv = \frac{\pi^{1/2}}{4\alpha^{3/2}} \qquad\qquad (A1.39)$$

$$I_3 = \int_0^\infty v^3\exp(-\alpha v^2)\,dv = \frac{1}{2\alpha^2} \qquad\qquad (A1.40)$$

$$I_4 = \int_0^\infty v^4\exp(-\alpha v^2)\,dv = \frac{3\pi^{1/2}}{8\alpha^{5/2}} \qquad\qquad (A1.41)$$

$$I_5 = \int_0^\infty v^5\exp(-\alpha v^2)\,dv = \frac{1}{\alpha^3} \qquad\qquad (A1.42)$$

$$I_6 = \int_0^\infty v^6\exp(-\alpha v^2)\,dv = \frac{15\pi^{1/2}}{16\alpha^{7/2}} . \qquad\qquad (A1.43)$$

A1.10 THE GAUSSIAN DISTRIBUTION

The Gaussian distribution will be developed from the binomial distribution, namely equation (A1.10), for the special case when N and n are both very large numbers. Taking the natural logarithm of equation (A1.10), we have

$$\log P(n) = \log N! - \log n! - \log(N-n)! + n\log p + (N-n)\log q .$$

Applying Stirling's theorem, in the form of equation (A1.19), we obtain

$$\log P(n) = N\log N - n\log n - (N-n)\log(N-n)$$
$$+ n\log p + (N-n)\log q .$$

Differentiating with respect to n, keeping N, p and q constant, we have

$$d(\log P(n))/dn = -\log n + \log(N - n) + \log p - \log q$$
$$= \log\{(N - n)p/nq\} \ . \tag{A1.44}$$

The maximum of $\log P(n)$ is when

$$d(\log P)/dn = \log\{(N - n)p/nq\} = 0 \ . \tag{A1.45}$$

which is when

$$(N - n)p/nq = (N - n)p/n(1 - p) = 1 \ .$$

Rearranging, $n = Np$

Hence the most probable value of n, which is denoted by $\tilde{n}$ and is the value of n at which $\log P(n)$ and hence $P(n)$ is a maximum, is given by

$$\tilde{n} = Np \ .$$

Since the Gaussian distribution is a special case of the binomial distribution, it follows from equation (A1.13), that Np is equal to the mean value $\bar{n}$. Hence, for a Gaussian distribution

$$\tilde{n} = \bar{n} = Np \ . \tag{A1.46}$$

Let $n = \bar{n} + x$, so that x is the deviation from the mean value $\bar{n}$. To expand $\log[P(\bar{n} + x)]$ in a Taylor series, let f stand for $\log P$, let a stand for $\bar{n}$ and let h stand for x in equation (A1.28). We have

$$\log[P(\bar{n} + x)] = \log[P(\bar{n})] + x \left(\frac{d \log P}{dn}\right) + \frac{x^2}{2!}\left(\frac{d^2 \log P}{dn^2}\right) + \dots \ .$$
$$\tag{A1.47}$$

The differential coefficients are evaluated at the point where $x = 0$. It follows from equation (A1.45) that, since $n = \bar{n} = \tilde{n}$ at the point $x = 0$, $d(\log P)/dn$, evaluated at $x = 0$, is zero in equation (A1.47). Differentiating equation (A1.44) with respect to n keeping N, p and q constant we have

$$\frac{d^2 \log P}{dn^2} = -\frac{1}{n} - \frac{1}{(N - n)} = -\frac{N}{n(N - n)} \ .$$

At $x = 0$, $n = \bar{n}$, where according to equation (A1.46) $\bar{n} = Np$. Substituting Np for n, at the point $x = 0$ we obtain

$$d^2(\log P)/dn^2 = -1/Np(1 - p) \ .$$

Since we started with the binomial distribution, according to equation

(A1.17), $Np(1 - p)$ is equal to σ^2, where σ is the standard deviation. Substituting in equation (A1.47) and ignoring terms of order x^3 and above, we have

$$\log[P(\bar{n} + x)] = \log[P(\bar{n})] - x^2/2\sigma^2 .$$

Taking exponentials, we obtain

$$P(n) = P(\bar{n} + x) = P(\bar{n})\exp(-x^2/2\sigma^2) . \qquad (A1.48)$$

Since $\Sigma P(n)$ is equal to unity, we have

$$P(\bar{n})\int_{-\infty}^{+\infty} \exp(-x^2/2\sigma^2)\,dx = 1 .$$

According to equation (A1.36), if $\alpha = 1/2\sigma^2$ we have

$$\int_{-\infty}^{+\infty} \exp(-\alpha x^2)\,dx = 2I_0 = \left(\frac{\pi}{\alpha}\right)^{1/2} = (2\pi\sigma^2)^{1/2} .$$

Hence $$P(\bar{n}) = (2\pi\sigma^2)^{-1/2} .$$

Substituting in equation (A1.48) we have

$$P(n) = \frac{1}{(2\pi\sigma^2)^{1/2}} \exp\left[-\frac{(n - \bar{n})^2}{2\sigma^2} \right] , \qquad (A1.49)$$

where $$\sigma^2 = Np(1 - p) = Npq , \qquad (A1.50)$$

and $$\bar{n} = \tilde{n} = Np . \qquad (A1.46)$$

Equation (A1.49) is the expression for the Gaussian distribution. It can be shown that, for a Gaussian distribution, the probability of getting a value of n between $(\bar{n} - \sigma)$ and $(\bar{n} + \sigma)$ is 68%. The probability of obtaining a value of n between $(\bar{n} - 3\sigma)$ and $(\bar{n} + 3\sigma)$ is 99.7%. [See also Problem 2.11.]

Appendix 2

The Carnot Cycle

A2.1 THE CARNOT CYCLE

Consider a gas contained in a cylinder by a frictionless piston, as shown in Figure A2.1(a). The cylinder is placed on a heat reservoir, which has an empirical temperature θ_1, as shown in Figure A2.1(a). By adjusting the external pressure appropriately, for example by changing the weights in Figure A2.1(a), the gas is allowed to expand reversibly and isothermally from the initial equilibrium state A to equilibrium state B, as illustrated on the pV indicator diagram in Figure A2.2. (For the special case of an ideal gas, the curve AB would be given by pV is a constant.) During this isothermal expansion, the gas system receives a quantity of heat Q_1 from the heat reservoir. According to the Kelvin–Planck statement of the second law of thermodynamics, we cannot convert all of this heat Q_1 into work in a cyclic process, which brings the gas back to its initial state A in Figure A2.2.

In the Carnot cycle, the cylinder is then placed on an insulating stand, as shown in Figure A2.1(b). The gas is allowed to expand reversibly and adiabatically, by adjusting the external pressure appropriately, until the gas reaches equilibrium state C in Figure A2.2. The empirical temperature of the gas drops from θ_1 to θ_2, when the gas goes from state B to state C. No heat enters or leaves the gas in this part of the cycle. (For an ideal gas, the curve BC is given by pV^γ equals a constant, where γ is the ratio of the heat capacities of the gas at constant pressure and constant volume.) According to equation (1.8), the work done *by* the gas in the transition from state A to state C is

$$W_1' = \int_A^C p \, dV \ .$$

This is equal to the area ABCSR between the curve ABC and the abscissa in Figure A2.2. (The work done *on* the gas is equal to $-W_1'$.) If the engine is to work in a cyclic process, the working substance must be brought back to the initial state A in Figure A2.2. In the Carnot cycle, the gas system is

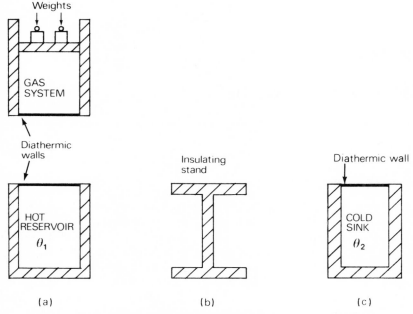

Figure A2.1—Apparatus for taking a gas system around a Carnot cycle.

then placed on a cold sink, which is at an empirical temperature θ_2, as shown in Figure A2.1(c). The gas is compressed reversibly and isothermally so that it proceeds along the path CD in Figure A2.2. During this process, the gas gives up heat Q_2' to the cold sink. (The heat gained by the gas is equal to $-Q_2'$.) One cannot avoid some loss of heat by the gas, when the system is taken along any path from the adiabatic CB to join the adiabatic DA, which passes through the initial state A. Thus not all the heat Q_1 absorbed by the gas in the section AB in Figure A2.2 can be converted into work, if the gas is to return to its initial state A. When the gas reaches equilibrium state D in Figure A2.2, the cylinder is placed on the insulating stand again, and the gas is compressed reversibly and adiabatically until it reaches the initial state A. The empirical temperature of the gas increases from θ_2 to θ_1 in this adiabatic process. The work W_2 done *on* the gas in going from C to A via D is equal to the area CDARS in Figure A2.2.

At the point A, the gas (working substance) is back in its initial state. According to the first law of thermodynamics, the total change ΔU in the internal energy of the gas in one complete cycle is zero. The net work done *on* the gas per cycle is

$$\Delta W = -W_1' + W_2 \ .$$

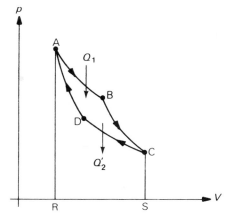

Figure A2.2—Carnot cycle pV indicator diagram.

[The net work done *by* the gas per cycle is $(W_1' - W_2)$.] The net heat supplied to the gas per cycle is

$$\Delta Q = Q_1 - Q_2' \ .$$

Substituting in equation (1.6), we have

$$\Delta U = \Delta W + \Delta Q = -W_1' + W_2 + Q_1 - Q_2' = 0 \ .$$

Rearranging, $$W_1' - W_2 = Q_1 - Q_2' \ . \qquad\qquad (A2.1)$$

The net work $(W_1' - W_2)$ done *by* the gas in one complete cycle is equal to the area ABCD in Figure A2.2. The efficiency η of the engine is defined as the ratio of the work done by the engine per cycle to the heat Q_1 taken from the heat reservoir of empirical temperature θ_1 per cycle. Using equation (A2.1), we have

$$\eta = (W_1' - W_2)/Q_1 = (Q_1 - Q_2')/Q_1 \ . \qquad\qquad (A2.2)$$

The Carnot cycle is reversible and the system can be taken from A to D to C to B to A. In this reverse cycle the engine acts as a refrigerator.

A2.2 CARNOT'S THEOREMS

Consider a reversible engine R and another hypothetical engine S, which is more efficient than R. When the two engines are working in the normal way, as shown in Figure A2.3(a), in each cycle the reversible engine R takes heat Q_1 from the hot reservoir, does work W and gives up heat Q_2 to the sink. If in each cycle, the more efficient engine S takes the same heat Q_1 from the hot reservoir, it will do more mechanical work $(W + \Delta Q)$ than R and give up less heat $(Q_2 - \Delta Q)$ to the cold sink.

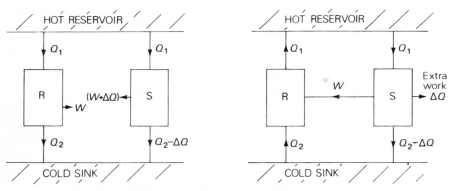

Figure A2.3—Proof of Carnot's theorems: (a) The reversible engine R and the engine S, which is assumed to be more efficient than R, are both working in the forward direction. (b) The more efficient engine S is used to drive the reversible engine R in the reverse direction. The net result contravenes the second law of thermodynamics.

In Figure A2.3(b) the engines are connected in such a way that the engine S drives the reversible engine R backwards. When R is working in reverse, in each cycle R takes heat Q_2 from the cold sink, work equal to W is done on R, and R gives up heat Q_1 to the hot reservoir. In each cycle, let S still take heat Q_1 from the hot reservoir, so that there is no net change in the internal energy of the hot reservoir. In each cycle the more efficient engine S is able to provide the work W necessary to drive R in the reverse direction, and still do an extra amount of mechanical work equal to ΔQ. Since R takes heat Q_2 from the cold sink but S only gives up heat $(Q_2 - \Delta Q)$ to the cold sink, the internal energy of the cold sink goes down by ΔQ in each cycle. The net result is that in each cycle, the composite engine takes in heat ΔQ from the cold sink and converts *all* of this heat into external work. This contravenes the Kelvin–Planck statement of the second law of thermodynamics. Hence to be consistent with the second law of thermodynamics, we must conclude that the engine S cannot be more efficient than a reversible engine working between the same limits of temperature.

Consider a hypothetical *reversible* engine S, which is more efficient than the reversible engine R. The above argument can be used to show that, on the basis of the second law of thermodynamics, we must conclude that S cannot be more efficient than R. Similarly, the reversible engine R cannot be more efficient than the reversible engine S. Hence, according to the second law of thermodynamics, all reversible engines working between the same limits of temperature have the same efficiency, irrespective of the nature of the working substance.

Appendix 3

Energy Levels for a Particle in a Box

A3.1 ENERGY LEVELS IN A ONE DIMENSIONAL POTENTIAL WELL

Consider a particle which is free to move in one dimension along the x axis between the points $x = 0$ and $x = L$. According to Schrödinger's time independent equation, for motion in one dimension, we have

$$\frac{\hbar^2}{2m} \frac{d^2\psi}{dx^2} + (\varepsilon - V_0)\psi = 0 \ . \tag{A3.1}$$

In equation (A3.1), $\psi(x)$ is the wave function, $\hbar$ is Planck's constant h divided by 2π, m is the mass of the particle, V_0 is the potential energy of the particle and ε is the total energy (potential energy plus kinetic energy) of the particle. It will be assumed that the potential energy V_0 is infinite for $x \geqslant L$ and $x \leqslant 0$ and that V_0 is zero for $L > x > 0$, as illustrated in Figure A3.1. In the region $L > x > 0$, where V_0 is zero, equation (A3.1) can be rewritten in the form

$$d^2\psi/dx^2 = -k^2\psi \ , \tag{A3.2}$$

where

$$k^2 = 2m\varepsilon/\hbar^2 \ . \tag{A3.3}$$

The solution of equation (A3.2) can be expressed in the form

$$\psi = A \sin kx + B \cos kx \ , \tag{A3.4}$$

where A and B are constants. Since V_0 is infinite at $x = 0$ and $x = L$, the wave function ψ must be zero at $x = 0$ and $x = L$. If $\psi = 0$ at $x = 0$, since $\sin 0$ is zero and $\cos 0$ is unity, at $x = 0$ we have

$$\psi = A \sin 0 + B \cos 0 = B = 0 \ .$$

Hence equation (A3.4) becomes

$$\psi = A \sin kx \ . \tag{A3.5}$$

To satisfy the boundary condition that $\psi = 0$ at $x = L$, we must have

$$A \sin kL = 0 \ . \tag{A3.6}$$

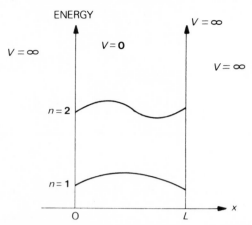

Figure A3.1—The standing wave solutions for a particle in a one dimensional infinite potential well.

If A were zero, according to equation (A3.5) ψ would be zero for all x, and the probability $\psi^*\psi \, dx$ that the particle was between x and $(x + dx)$ would be zero everywhere. This solution would correspond to the case when there was no particle in the potential well. Hence, to satisfy equation (A3.6) with a particle in the box, sin kL must be zero, so that kL must be equal to $n\pi$, that is

$$k = n\pi/L , \qquad (A3.7)$$

where $n = 1, 2, 3, \ldots$. The term $n = 0$ is absent, as this would make ψ zero everywhere. The solution of equation (A3.2) corresponding to the quantum number n is

$$\psi_n = A_n \sin[(n\pi/L)x] . \qquad (A3.8)$$

Substituting for k from equation (A3.7) into equation (A3.3), we have

$$\varepsilon_n = n^2\hbar^2\pi^2/2mL^2 , \qquad (A3.9)$$

where n is a positive integer greater than zero. A negative value of n would give the same value of ε_n and the same value for $\psi^*\psi$ everywhere and would be the same solution as for the positive value of n.

It is only for the values of energy given by equation (A3.9) that equation (A3.1) has a solution which satisfies the boundary conditions. The values of energy given by equation (A3.9) are the *energy eigenvalues*. One quantum number n suffices to specify the wave function ψ_n and the energy ε_n in the case of a particle in a one dimensional infinite potential well. The dependence of ε_n on L comes in via the boundary condition that ψ is zero when $x = L$.

A mechanical analogy of the above mathematical solution is a stretched string of length L fixed at both ends. The only solutions which satisfy the boundary conditions, that the displacement of the string is zero at $x = 0$ and at $x = L$, are the fundamental and its harmonics. The fundamental has a wavelength $\lambda_1 = 2L$ and a wave number $k_1 = 2\pi/\lambda_1 = \pi/L$ in agreement with equation (A3.7).

A3.2 SINGLE PARTICLE IN A CUBICAL BOX

Consider a particle of mass m and spin zero inside a cubical box of side L. Choose axes as shown in Figure A3.2. In this case the potential is zero inside the box and infinite on the inside surface of the box, so that ψ is zero, whenever $x = 0, y = 0, z = 0, x = L, y = L$, or $z = L$. For a particle free to move in three dimensions inside the box, Schrödinger's equation becomes

$$\partial^2\psi/\partial x^2 + \partial^2\psi/\partial y^2 + \partial^2\psi/\partial z^2 = -2m\varepsilon\psi/\hbar^2 \ . \qquad (A3.10)$$

The wave function ψ is now a function of x, y and z. By analogy with equation (A3.5), a solution of equation (A3.10) which satisfies the boundary condition that ψ is zero whenever $x = 0, y = 0$ or $z = 0$ is

$$\psi = A\,(\sin k_x x)(\sin k_y y)(\sin k_z z) \ . \qquad (A3.11)$$

To satisfy the boundary condition that ψ is zero whenever $x = L, y = L$ or $z = L$, we must have

$$k_x = n_x\pi/L \, ; \qquad k_y = n_y\pi/L \, ; \qquad k_z = n_z\pi/L \ , \qquad (A3.12)$$

where n_x, n_y and n_z are positive integers greater than zero.

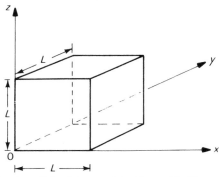

Figure A3.2—A particle of mass m is inside a cubical box of side L. The standing wave solutions of Schrödinger's equation are developed, using the boundary condition that the wave function is zero everywhere on the inside surface of the box.

Substituting for ψ from equation (A3.11) into equation (A3.10), carrying out the partial differentiations and dividing both sides by ψ, we obtain

$$\varepsilon = (\hbar^2/2m)(k_x^2 + k_y^2 + k_z^2) \ .$$

Using equations (A3.12) and remembering that L^3 is equal to V, the volume of the box, we have

$$\varepsilon = \frac{\hbar^2\pi^2}{2mV^{2/3}}(n_x^2 + n_y^2 + n_z^2) = \frac{\hbar^2\pi^2}{2mV^{2/3}}n^2 \qquad (A3.13)$$

where $$n^2 = n_x^2 + n_y^2 + n_z^2 \ . \qquad (A3.14)$$

It is only for values of ε for which n_x, n_y and n_z are all positive integers greater than zero that solutions of equation (A3.10) exist which satisfy the boundary conditions. The dependence of ε on V comes in via the boundary conditions used to solve equation (A3.10). The energies given by equation (A3.13) are the possible **energy eigenvalues**. To specify the wave function $\psi(x, y, z)$ we need to know the three quantum numbers n_x, n_y and n_z. The lowest energy eigenvalue is when $n_x = 1, n_y = 1$ and $n_z = 1$. Each different set of quantum numbers n_x, n_y and n_z represents a different solution of equation (A3.10) and corresponds to a different eigenstate (microstate) of the system which is composed of a particle in a cubical box.

A3.3 DENSITY OF STATES

Equation (A3.13) can be rewritten in the form

$$n_x^2 + n_y^2 + n_z^2 = 2mV^{2/3}\varepsilon/\hbar^2\pi^2 = R^2 \ , \qquad (A3.15)$$

where R is given by the equation

$$R^2 = 2mV^{2/3}\varepsilon/\hbar^2\pi^2 \ . \qquad (A3.16)$$

It is possible to represent the various possible eigenstates (microstates) of the system by dots in a number space using n_x, n_y and n_z as coordinates, as shown for the $n_z = 0$ plane only in Figure A3.3. Each dot in Figure A3.3, represents an independent solution of equation (A3.10), which satisfies the boundary conditions and corresponds to one possible eigenstate (microstate) of the system. The quantum numbers n_x, n_y and n_z have positive integer values. There are no dots on the axes in Figure A3.3.

According to equation (A3.15), the possible combinations of n_x, n_y and n_z, which give the same value of ε, lie on the surface of a sphere of radius R in number space. Since n_x, n_y and n_z must be positive, the dots corresponding to the same value of ε lie on one octant of the sphere. Each dot in number space is separated from its nearest neighbours by $\Delta n_x = \pm1$, or $\Delta n_y = \pm1$, or $\Delta n_z = \pm1$. There is one dot per unit volume of number

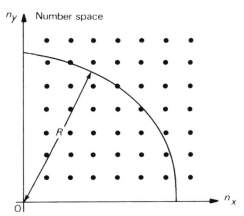

Figure A3.3—The quantum numbers of each of the various eigenstates (standing wave solutions) of Schrödinger's equation for a particle in a cubical box can be represented by a dot having coordinates n_x, n_y and n_z in a three dimensional number space. Only the $n_z = 0$ plane is shown in the figure.

space. The total number of eigenstates (microstates), denoted by Φ, having energy eigenvalues less than or equal to ε, is equal to the number of dots in one octant of the sphere of radius R, that is in a volume $(4\pi R^3/3)/8 = \pi R^3/6$ of number space. Since there is one dot per unit volume, the number of eigenstates (microstates) having energy eigenvalues less than or equal to ε is

$$\Phi(\leq\varepsilon) = \frac{\pi R^3}{6} = \frac{V}{6\pi^2}\left(\frac{2m}{\hbar^2}\right)^{3/2}\varepsilon^{3/2} \ . \tag{A3.17}$$

Let Φ increase to $(\Phi + d\Phi)$, when ε is increased to $(\varepsilon + d\varepsilon)$. If $D(\varepsilon)$ is the number of microstates per unit energy range in the vicinity of ε, $d\Phi$ is equal to $D(\varepsilon)\,d\varepsilon$. Differentiating equation (A3.17), we have

$$D(\varepsilon)\,d\varepsilon = d\Phi = \frac{d\Phi}{d\varepsilon}\,d\varepsilon = \frac{V}{4\pi^2}\left(\frac{2m}{\hbar^2}\right)^{3/2}\varepsilon^{1/2}\,d\varepsilon \ . \tag{A3.18}$$

The quantity $D(\varepsilon)$ is generally called the *density of states*. Equation (A3.18) is the same as equation (A6.18) of Appendix 6.

The results derived in this Appendix are only valid for particles of spin zero. For particles having spin quantum number s, the right hand sides of equations (A3.17) and (A3.18) must be multiplied by the spin degeneracy $g = (2s + 1)$.

Appendix 4*

Two Macroscopic Systems in Thermal Contact*

A4.1* FLUCTUATIONS IN ENERGY*

Consider the two macroscopic systems shown in Figure 3.1(b) of Section 3.2. The two subsystems are allowed to reach thermal equilibrium. The total energy is given by

$$U_1 + U_2 = U^* \ .$$

The volumes V_1 and V_2 and the numbers of particles N_1 and N_2 in the two subsystems are kept constant. Let $\tilde{U}_2$ be the value of the energy of subsystem 2 at which the probability $P_2(U_2)$, that subsystem 2 has energy U_2, is a maximum. It is shown in Section 3.2 that $P_2(U_2)$ is a maximum and $U_2 = \tilde{U}_2$ when

$$(\partial \log g_1/\partial U_1)_{V_1,N_1} = (\partial \log g_2/\partial U_2)_{V_2,N_2} \ , \tag{3.13}$$

or

$$\beta_1 = \beta_2 \ , \tag{3.15}$$

where the temperature parameter β is defined by equation (3.14).

According to equation (3.5) the probability that, at thermal equilibrium, subsystem 2 has energy U_2 equal to $(\tilde{U}_2 + x)$ and subsystem 1 has energy $(\tilde{U}_1 - x)$ is

$$P_2(U_2) = P_2(\tilde{U}_2 + x) = g_2(\tilde{U}_2 + x)g_1(\tilde{U}_1 - x)/g_T \ . \tag{A4.1}$$

According to Taylor's theorem, equation (A1.28) of Appendix 1,

$$f(a + h) = f(a) + hf'(a) + (h^2/2!)f''(a) + \ldots \ .$$

To expand $\log[g_2(\tilde{U}_2 + x)]$ in a Taylor series, let f stand for $\log g_2$, let $a = \tilde{U}_2$ and let $h = x$. Ignoring terms of order x^3 and remembering that the value of $(\partial \log g_2/\partial U_2)$ evaluated at $U_2 = \tilde{U}_2$ is β_2, we have

$$\log[g_2(\tilde{U}_2 + x)] \simeq \log[g_2(\tilde{U}_2)] + x\beta_2 + \frac{x^2}{2}\left(\frac{\partial \beta_2}{\partial U_2}\right) \ . \tag{A4.2}$$

Similarly, $\log[g_1(\tilde{U}_1 - x)] \simeq \log[g_1(\tilde{U}_1)] - x\beta_1 + \dfrac{x^2}{2}\left(\dfrac{\partial\beta_1}{\partial U_1}\right)$.   (A4.3)

Adding equations (A4.2) and (A4.3), and using $\beta_1 = \beta_2$, we have

$$\log[g_2(\tilde{U}_2 + x)g_1(\tilde{U}_1 - x)] \simeq \log[g_2(\tilde{U}_2)g_1(\tilde{U}_1)] + \dfrac{x^2}{2}\left[\dfrac{\partial\beta_2}{\partial U_2} + \dfrac{\partial\beta_1}{\partial U_1}\right] .$$

Taking exponentials, we obtain

$$g_2(\tilde{U}_2 + x)g_1(\tilde{U}_1 - x) \simeq g_2(\tilde{U}_2)g_1(\tilde{U}_1)\exp\left[\dfrac{x^2}{2}\left(\dfrac{\partial\beta_2}{\partial U_2} + \dfrac{\partial\beta_1}{\partial U_1}\right)\right] .$$

Substituting in equation (A4.1), we have

$$P_2(U_2) = P_2(\tilde{U}_2 + x) = \dfrac{g_2(\tilde{U}_2)g_1(\tilde{U}_1)}{g_T}\exp\left[\dfrac{x^2}{2}\left(\dfrac{\partial\beta_2}{\partial U_2} + \dfrac{\partial\beta_1}{\partial U_1}\right)\right] .\text{(A4.4)}$$

Comparing with equation (A1.48) of Appendix 1, it can be seen that equation (A4.4) is a Gaussian distribution of the form

$$P_2(U_2) = P_2(\tilde{U}_2 + x) \simeq (P_2)_{\max}\exp[-x^2/2\sigma_2{}^2] \qquad (A4.5)$$

where $(P_2)_{\max} = \dfrac{g_2(\tilde{U}_2)g_1(\tilde{U}_1)}{g_T}$ (A4.6)

is the maximum value of P_2, which is when $U_2 = \tilde{U}_2$ and $U_1 = \tilde{U}_1 = (U^* - \tilde{U}_2)$. The standard deviation σ_2 in equation (A4.5) is given by

$$1/\sigma_2{}^2 = -[(\partial\beta_2/\partial U_2) + (\partial\beta_1/\partial U_1)] . \qquad (A4.7)$$

The differential coefficients in equation (A4.7) are evaluated at $U_2 = \tilde{U}_2$ and $U_1 = \tilde{U}_1$ respectively. For a Gaussian distribution, the mean value $\bar{U}_2$ is equal to the most probable value $\tilde{U}_2$. Similarly $\tilde{U}_1 = \bar{U}_1$.

To illustrate the order of magnitude of σ_2, which is the standard deviation in the energy of subsystem 2 at thermal equilibrium, it will be assumed that subsystems 1 and 2 in Figure 3.1(b) are non-conducting solids. It is shown in Chapter 10 that, in the classical limit, the heat capacity of an insulating solid, consisting of N atoms, is equal to $3Nk$, and that, in the classical limit, the mean internal energies of the non-conducting solids are given, to a good approximation, by

$$U_1 = 3N_1kT_1; \qquad \beta_1 = 3N_1/U_1$$
$$U_2 = 3N_2kT_2; \qquad \beta_2 = 3N_2/U_2$$

where N_1 and N_2 are the numbers of atoms in subsystems 1 and 2 respectively. Differentiating and substituting in equation (A4.7), for $U_1 = \bar{U}_1$ and $U_2 = \bar{U}_2$, we have

$$1/\sigma_2{}^2 = [(3N_2/\bar{U}_2{}^2) + (3N_1/\bar{U}_1{}^2)] . \qquad (A4.8)$$

Case 1: Let $N_1 = N_2$ so that at thermal equilibrium $\bar{U}_1 = \bar{U}_2$. Equation (A4.8) then becomes

$$1/\sigma_2^2 = 6N_2/\bar{U}_2^2 \ .$$

Hence $\sigma_2/\bar{U}_2 = 1/(6N_2)^{1/2} \ .$ (A4.9)

Case 2: Let $N_2 \ll N_1$. (We shall generally adopt the convention of labelling the smaller of the two subsystems in Figure 3.1(b) subsystem 2.) In this case, since $U_2 = 3N_2kT$ and $U_1 = 3N_1kT$, $3N_2/\bar{U}_2^2 \gg 3N_1/\bar{U}_1^2$, so that equation (A4.8) gives

$$\sigma_2/\bar{U}_2 = 1/(3N_2)^{1/2} \ .$$ (A4.10)

Equation (A4.10) is also derived in Section 5.2, using the approach to statistical thermodynamics based on the partition function.

Summarising, if N_2 is approximately equal to N_1, or if $N_2 \ll N_1$ then $\sigma_2/\bar{U}_2$, the relative fluctuation in the energy of subsystem 2 at thermal equilibrium, is very roughly of the order of $1/N_2^{1/2}$. We shall therefore assume that $\sigma_2/\bar{U}_2 = \gamma/N_2^{1/2}$ where γ is of the order of magnitude of unity.

A4.2* THE HEAT CAPACITY AT CONSTANT VOLUME*

It was pointed out in Section 1.11 that, in the context of classical equilibrium thermodynamics, for the thermal equilibrium of two systems in thermal contact to be stable equilibrium, the heat capacity at constant volume must be positive. For the thermal equilibrium of subsystems 1 and 2 in Figure 3.1(b) to be stable, the probability distribution $P_2(U_2)$ given by equation (A4.5) must have a maximum not a minimum at $U_2 = \bar{U}_2$. For the probability distribution to approximate to a Gaussian, as shown in Figure 3.2(b), σ_2^2 in equation (A4.5) must be positive. If σ_2^2 were negative, the curve would have a minimum at $U_2 = \bar{U}_2$. According to equation (A4.7) σ_2^2 is positive if $(\partial\beta/\partial U)_V$ is negative for both subsystems 1 and 2, that is if

$$(\partial U/\partial\beta)_V < 0$$ (A4.11)

Since $\beta = 1/kT$, $\partial/\partial\beta = -kT^2\partial/\partial T$, so that equation (A4.11) can be rewritten in the form

$$kT^2(\partial U/\partial T)_V > 0 \ .$$ (A4.12)

Since Boltzmann's constant k, defined by equation (3.22), is positive and T^2 is positive, equation (A4.12) implies that, for the probability distribution given by equation (A4.5) to have a maximum, we must have

$$C_V = (\partial U/\partial T)_V > 0 \ .$$

Conversely, if C_V is positive, $(\partial U/\partial T)_V$ is positive, so that $(\partial U/\partial\beta)_V$ is negative, σ_2^2 is positive in equation (A4.5), and the thermal equilibrium of subsystems 1 and 2 in Figure 3.1(b) will be stable. [See also Sections 1.11 and 3.3.]

A4.3* ENTROPY OF THE COMPOSITE SYSTEM IN FIGURE 3.1(b)*

An introduction to the type of approximations to be made in this Section is given in Section 2.5.3.

Since at thermal equilibrium subsystem 2 in Figure 3.1(b) must have some value of energy,

$$\Sigma P_2(U_2) = 1 \ .$$

Using equations (A4.5) and (A4.6) we have

$$\{g_1(\tilde{U}_1)g_2(\tilde{U}_2)/g_T\}\Sigma \exp[-x^2/2\sigma_2^2] = 1 \ . \tag{A4.13}$$

In the simplified model adopted in Section 3.2, it was assumed that subsystem 2 had a series of energy levels, each of which had an enormous degeneracy of the order of $\exp(N_2)$. In order to convert the summation over the energy levels in equation (A4.13) to an integral, it will be assumed that the energy separation of the energy levels of subsystem 2 is δ. In the energy range U_2 to $(U_2 + dU_2)$ there are dU_2/δ energy levels. The probability that, at thermal equilibrium, subsystem 2 is in any particular one of these energy levels is given by equation (A4.5). Hence the probability $P'(U_2) dU_2$ that at thermal equilibrium subsystem 2 has an energy between U_2 and $(U_2 + dU_2)$ is

$$P'(U_2) \, dU_2 = \left(\frac{dU_2}{\delta}\right) \frac{g_1(\tilde{U}_1)g_2(\tilde{U}_2)}{g_T} \exp[-x^2/2\sigma_2^2] \ .$$

Normalising, we have $\displaystyle\int_0^{U^*} P'(U_2) \, dU_2 = 1 \ .$

Since the Gaussian distribution, given by equation (A4.5), is extremely narrow, no great error is involved, if we change the variable to $x = (U_2 - \tilde{U}_2)$ and integrate between the limits $x = -\infty$ to $x = +\infty$. In this case, since $dU_2 = dx$, we have

$$\frac{g_1(\tilde{U}_1)g_2(\tilde{U}_2)}{g_T\delta} \int_{-\infty}^{\infty} \exp(-x^2/2\sigma_2^2) \, dx = 1 \ . \tag{A4.14}$$

According to equation (A1.36) of Appendix 1

$$I_0 = \int_0^{\infty} \exp(-\alpha x^2) \, dx = \frac{\pi^{1/2}}{2\alpha^{1/2}} \ .$$

Putting $\alpha = 1/(2\sigma_2^2)$ we have

$$\int_{-\infty}^{+\infty} \exp(-x^2/2\sigma_2^2)\, dx = 2I_0 = (2\pi\sigma_2^2)^{1/2} \ .$$

Substituting in equation (A4.14) we have

$$g_1(\tilde{U}_1)g_2(\tilde{U}_2)(2\pi\sigma_2^2)^{1/2}/g_T\delta = 1 \ .$$

Rearranging and taking the natural logarithm, we obtain

$$\log g_T = \log g_1(\tilde{U}_1) + \log g_2(\tilde{U}_2) - \log \delta + \tfrac{1}{2}\log 2\pi + \log \sigma_2 \ .$$
$$\text{(A4.15)}$$

According to equation (2.42) of Section 2.5.2, for an isolated macroscopic system consisting of N particles, $\log g$ is of order αN, where α is generally in the range 0.5 to 100. Hence, $\log[g_1(\tilde{U}_1)]$ and $\log[g_2(\tilde{U}_2)]$ are of order N_1 and N_2 respectively. It was shown in Section A4.1 that σ_2 is of order $U_2/N_2^{1/2}$ which is of order $N_2^{1/2}$. Hence $\log \sigma_2$ is of order $\log N_2$. As an example of a model for subsystem 2 in Figure 3.1(b), consider a system of N particles of spin $\tfrac{1}{2}$ in a magnetic field of magnetic induction B. It is shown in Section 8.3* that, in this example, there are $(N + 1)$ energy levels of energy $U = (N - 2n)\mu B$ where n, the number of magnetic moments pointing in the direction of the magnetic field, can vary from $n = 0$ to $n = N$. [According to equation (8.3) of Chapter 8, the degeneracy of the levels is $N!/n!(N - n)!$.] In this example, the separation of the energy levels is $\delta = 2\mu B$. If μ is equal to one Bohr magneton (9.27×10^{-24} A m^2) and $B = 1$ tesla, $\delta = 2\mu B = 18.54 \times 10^{-24}$ and $\log \delta = -52.3$, which is negligible compared with N_1 and N_2. Hence to an excellent approximation equation (A4.15) can be written as

$$\log g_T = \log g_1(\tilde{U}_1) + \log g_2(\tilde{U}_2) \ . \tag{A4.16}$$

Multiplying by Boltzmann's constant k, and using equation (3.22), we find that the entropy S_f of the composite system in Figure 3.1(b), when subsystems 1 and 2 are in thermal equilibrium, is given, to an excellent approximation, by

$$S_f = k \log g_T = S_1(\tilde{U}_1) + S_2(\tilde{U}_2) \tag{A4.17}$$

where $S_1(\tilde{U}_1)$ and $S_2(\tilde{U}_2)$ are the entropies subsystems 1 and 2 would have, if they were isolated systems having energies $U_1 = \tilde{U}_1 = \bar{U}_1$ and $U_2 = \tilde{U}_2 = \bar{U}_2$ respectively.

A4.4* USE OF THE DENSITY OF STATES*

This section should be read in conjunction with Section 3.8*. As in Section 3.8*, it will be assumed that the degeneracies of the energy levels are split by the interactions between the particles, and we shall work with the density of states D defined by equation (3.89). According to equation (3.97)

$$P_2'(U_2)\, dU_2 = D_1 D_2\, \Delta U_1\, dU_2/g_T$$

where g_T, the total number of accessible microstates, is a constant and where ΔU_1 is a factor which allows for the small spread of ΔU^* in the total energy U^* of the composite system. (Reference: Section 3.8.3.*) Proceeding as in Section A4.1* it can be shown that

$$D_2(\tilde{U}_2 + x)D_1(\tilde{U}_1 - x) = D_2(\tilde{U}_2)D_1(\tilde{U}_1) \exp\left[\frac{x^2}{2}\left(\frac{\partial\beta_2}{\partial U_2} + \frac{\partial\beta_1}{\partial U_1}\right)\right] ,$$

so that

$$P_2'(U_2)\, dU_2 = \frac{D_1(\tilde{U}_1)D_2(\tilde{U}_2)\, \Delta U_1}{g_T}\, \exp[-x^2/2\sigma_2^2]\, dU_2 . \quad (A4.18)$$

This probability distribution is an extremely sharp Gaussian distribution. Normalising by integrating over U_2, we find that

$$g_T = D_1(\tilde{U}_1)D_2(\tilde{U}_2)\, \Delta U_1(2\pi\sigma_2^2)^{1/2} .$$

Taking logarithms, to an excellent approximation, we have

$$\log g_T \simeq \log D_1(\tilde{U}_1) + \log D_2(\tilde{U}_2) .$$

Multiplying by k and using equation (3.91), we again obtain

$$S_f \simeq S_1(\tilde{U}_1) + S_2(\tilde{U}_2) . \quad (A4.19)$$

Appendix 5

The Pressure of a Macroscopic System

A5.1 THE PRESSURE OF A SYSTEM

The pressure on a surface is defined in terms of mechanical quantities as the force per unit area perpendicular to the surface. Consider the system shown in Figure 1.2(a) of Chapter 1. A fluid, such as a gas, is contained inside rigid adiabatic walls by a piston of area A made from a rigid adiabatic material. The piston can move in and out without friction. It will be assumed that the system is in its ith microstate, which has energy eigenvalue ε_i. It will be assumed that in this microstate the pressure exerted by the system on the piston is p_i. To determine p_i we shall use the principle of virtual work. If the piston were given a virtual displacement dx inwards, the change in the volume of the system would be equal to $A\,dx$, and the work done on the system would be $p_i A\,dx = -p_i\,dV$, where dV would be the increase in the volume of the system. If all the work went into increasing the energy of the microstate from ε_i to $(\varepsilon_i + d\varepsilon_i)$, then

$$d\varepsilon_i = -p_i\,dV \ . \tag{A5.1}$$

Hence, when the system is in its ith microstate having energy eigenvalue ε_i, according to the principle of virtual work the pressure of the system is

$$p_i = -\partial\varepsilon_i/\partial V \ . \tag{A5.2}$$

A5.2 WORK DONE IN A REVERSIBLE ADIABATIC CHANGE

Consider a real change of volume. If the piston moves in extremely slowly so that the change is quasistatic, according to the time dependent perturbation theory of quantum mechanics, the probability that the system is in any particular microstate does not change during the quasi-static process. This is an example of the adiabatic approximation of quantum mechanics (Ehrenfest's principle). (References: Landau and Lifshitz [1] and Wannier [2].) If the probability that the system is in any particular microstate does not change during the quasi-static process, the change takes place at constant entropy.†

According to equation (2.27) the ensemble average energy is

$$U = \sum_{\text{(states)}} P_i \varepsilon_i \; . \tag{A5.3}$$

Notice that the probability that the system is in its ith microstate is denoted by a capital P_i to distinguish it from the symbol p_i which denotes the pressure of the system, when it is in its ith microstate. From equation (A5.3), we have

$$dU = \sum P_i \, d\varepsilon_i + \sum \varepsilon_i \, dP_i \; .$$

For a reversible adiabatic change, all the P_i are constant, so that in this case

$$dU = \sum P_i \, d\varepsilon_i \; .$$

From equation (A5.1)

$$d\varepsilon_i = -p_i \, dV \; .$$

Hence, for a reversible adiabatic change

$$dU = -\sum P_i p_i \, dV = -dV \sum P_i p_i \; .$$

According to equation (2.28) of Section 2.3.3, with $\alpha_{2i} = p_i$, the ensemble average pressure is

$$p = \sum_{\text{(states)}} P_i p_i \; . \tag{A5.4}$$

Hence, for an infinitesimal reversible adiabatic change

$$dU \left(= \sum P_i \, d\varepsilon_i \right) = -p \, dV \; . \tag{A5.5}$$

This is in agreement with equation (1.8), which was developed from classical equilibrium thermodynamics in Section 1.6 of Chapter 1.

A5.3 ENSEMBLE AVERAGE PRESSURE

Substituting for p_i from equation (A5.2) into equation (A5.4), we find that the ensemble average pressure is given by

$$p = \sum_{\text{(states)}} P_i \left(-\frac{\partial \varepsilon_i}{\partial V} \right) \; . \tag{A5.6}$$

---

†One way of seeing that, if the probability P_i that the system is in its ith microstate does not change, the entropy is constant is to use Boltzmann's definition of entropy, namely,

$$S = -k \sum P_i \log P_i \; .$$

If all the P_i are constant, the entropy is constant. A fuller discussion is given in Sections 7.1.2* and 7.1.5*.

If the entropy S is constant, the P_i do not change and terms such as $\partial P_i/\partial V$ are zero. Hence for constant S, equation (A5.6) can be rewritten as

$$p = -(\partial/\partial V)\sum P_i\varepsilon_i \qquad (S \text{ constant}) .$$

Using equation (A5.3) we obtain

$$p = -(\partial U/\partial V)_{S,N} . \qquad (A5.7)$$

Equation (A5.7) is the same as equation (1.21) which was developed from classical equilibrium thermodynamics in Sections 1.8 of Chapter 1.

Since S is a function of U, V and N, using equation (A1.32) of Appendix 1, for constant N we have

$$dS = (\partial S/\partial U)_{V,N}\, dU + (\partial S/\partial V)_{U,N}\, dV .$$

If S is constant, $dS = 0$. Dividing by dV, for constant S and N we have

$$0 = (\partial S/\partial U)_{V,N}(\partial U/\partial V)_{S,N} + (\partial S/\partial V)_{U,N} .$$

Substituting from equations (3.23) and (A5.7), we obtain

$$p = T(\partial S/\partial V)_{U,N} . \qquad (A5.8)$$

Equation (A5.8) is the same as equation (1.20) of classical equilibrium thermodynamics.

REFERENCES

[1] Landau, L. D. and Lifshitz, E. M., *Quantum Mechanics* (Pergamon Press, 1958), page 144.
[2] Wannier, G. H., *Statistical Physics* (Wiley, 1966), page 92.

Appendix 6

Density of States

A6.1 STATIONARY WAVE SOLUTIONS

A6.1.1 The general case of wave motion

The **general wave equation** is of the form

$$\nabla^2 \Phi - \frac{1}{c^2}\frac{\partial^2 \Phi}{\partial t^2} = 0 \ . \tag{A6.1}$$

where c is the velocity of the waves. To obtain stationary wave solutions of equation (A6.1), assume that

$$\Phi = \phi(\mathbf{r})T(t) \ ,$$

where $\phi(\mathbf{r})$ is a function of the position vector $\mathbf{r}$ only and $T(t)$ is a function of the time t only. Substituting in equation (A6.1) we have

$$\nabla^2(\phi T) - \frac{1}{c^2}\frac{\partial^2(\phi T)}{\partial t^2} = T\nabla^2\phi - \frac{\phi}{c^2}\frac{\partial^2 T}{\partial t^2} = 0 \ .$$

Dividing by ϕT, and rearranging, we have

$$\frac{1}{\phi}\nabla^2\phi = \frac{1}{c^2 T}\frac{\partial^2 T}{\partial t^2} \ . \tag{A6.2}$$

The left hand side of equation (A6.2) depends only on $\mathbf{r}$, whereas the right hand side depends only on t. If equation (A6.2) is to hold for independent variations in $\mathbf{r}$ and t, both the left hand side and the right hand side must be equal to the same constant. It will simplify the expressions for the solutions of the differential equations, if we put this constant equal to $(-k^2)$, so that

$$\frac{1}{\phi}\nabla^2\phi = -k^2 \tag{A6.3}$$

$$\frac{1}{c^2 T}\frac{\partial^2 T}{\partial t^2} = -k^2 \ . \tag{A6.4}$$

Equation (A6.3) can be rewritten in the form

$$\nabla^2\phi(\mathbf{r}) + k^2\phi(\mathbf{r}) = 0 \ . \tag{A6.5}$$

Equation (A6.5) for ϕ is the same as equation (A3.10) of Appendix 3, which was obtained from Schrödinger's time independent equation for a particle in a box, provided $k^2 = 2m\varepsilon/\hbar^2$. We shall consider standing wave solutions in the cubical box of side L and volume $V = L^3$, shown in Figure A6.1(a). Proceeding as in Section A3.2 of Appendix 3, it follows that the solutions of equation (A6.5) which satisfy the boundary condition that ϕ is zero on the inside surface of the box are

$$\phi = A \, \sin(k_x x)\sin(k_y y)\sin(k_z z) \ , \tag{A6.6}$$

where A is a constant,

$$k_x = n_x\pi/L; \qquad k_y = n_y\pi/L; \qquad k_z = n_z\pi/L \ . \tag{A6.7}$$

and n_x, n_y and n_z can have any positive integer value from 1 to infinity. These are the **standing wave solutions** of equation (A6.5). They are the **normal modes**. It can be shown that the total energy of each normal mode can vary independently.

The sets of values of k_x, k_y and k_z can be plotted in a k space, as shown in Figure A6.1(b). Each dot represents an independent standing wave solution of equation (A6.5). For a fixed value of k_y and k_z, if n_x is increased by 1, then $\Delta k_x = \pi/L$ so that the distance between neighbouring dots in Figure A6.1 (b) is π/L and there is one dot per volume of π^3/L^3 of k space. Let

$$k^2 = k_x^2 + k_y^2 + k_z^2 \ . \tag{A6.8}$$

The number of points having a value of k between k and $(k + dk)$ is the number of dots in k space in one octant of a spherical shell of inner radius k and outer radius $(k + dk)$, that is in a volume of $(4\pi k^2 \, dk)/8$ of k space. Since there is one dot per volume of π^3/L^3 of k space, the number of standing wave solutions (normal modes) having a value of wave vector k between k and $(k + dk)$ is

$$D(k) \, dk = (\pi k^2 \, dk/2) \div (\pi^3/L^3) = Vk^2 \, dk/2\pi^2 \ . \tag{A6.9}$$

The number of standing wave solutions (normal modes) per unit volume is equal to $(k^2 \, dk/2\pi^2)$. [It can be shown that this result is valid whatever the shape of the box provided the dimensions of the box are much bigger than the wavelength λ. (Reference: Courant and Hilbert [1]).] Equation (A6.9) does not depend on the type of wave motion, but only on fitting standing waves in a box, subject to the boundary condition that ϕ is zero everywhere on the inner surface of the box. Since $k = 2\pi/\lambda$,

$$dk = -2\pi \, d\lambda/\lambda^2$$

Hence the number of standing wave solutions (normal modes) $D(\lambda) \, d\lambda$

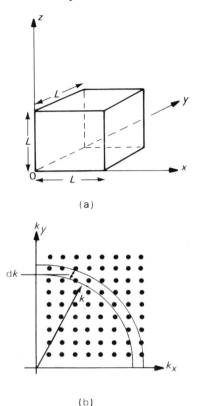

(a)

(b)

Figure A6.1—(a) The standing wave solutions of the general wave equation inside a cubical box are determined using the boundary condition that the amplitude is zero on the inner surface of the box. (b) The possible values of k_x, k_y and k_z are plotted in k-space. Only the $k_z = 0$ plane is shown.

having wavelengths between λ and $(\lambda + \mathrm{d}\lambda)$ is

$$D(\lambda)\, \mathrm{d}\lambda = V4\pi\, \mathrm{d}\lambda/\lambda^4 \ . \tag{A6.10}$$

The phase velocity c is equal to ω/k so that for a non-dispersive medium for which c is constant, $\mathrm{d}k = \mathrm{d}\omega/c$. Hence, for a non-dispersive medium, equation (A6.9) becomes

$$D(\omega)\, \mathrm{d}\omega = V\omega^2\, \mathrm{d}\omega/2\pi^2 c^3 \ , \tag{A6.11}$$

where $D(\omega)\, \mathrm{d}\omega$ is the number of stationary wave solutions (normal modes) which have angular frequencies in the range ω to $(\omega + \mathrm{d}\omega)$. Since $\omega = 2\pi\nu$,

$$D(\nu)\, \mathrm{d}\nu = V4\pi\nu^2\, \mathrm{d}\nu/c^3 \ , \tag{A6.12}$$

where $D(v)$ dv is the number of stationary wave solutions (normal modes) having frequencies in the range v to $(v + dv)$.

Multiplying equation (A6.4) by ϕ, for a fixed position **r**, we have $\partial^2\Phi/\partial t^2 = -k^2c^2\Phi = -\omega^2\Phi$, where k is given by equation (A6.8). Hence for any fixed value of **r**, the value of Φ for each normal mode obeys the same equation as a harmonic oscillator of angular frequency $\omega = ck$. [Notice the similarity between a normal mode and a harmonic oscillator.] To get Φ, we must multiply the expression for $\phi(\mathbf{r})$, given by equation (A.6.6) by $\sin(\omega t + \alpha)$, the solution of equation (A6.4). The general solution of equation (A6.1) is a linear combination of the normal modes.

Some special types of wave motion will now be considered.

A6.1.2 Electromagnetic waves

In the case of electromagnetic waves $\Phi(\mathbf{r}, t)$ could stand for either the electric field intensity **E** or the magnetising force **H**. Electromagnetic waves are transverse waves, which can have two independent directions of polarisation for each value of k. Hence for electromagnetic waves

$$D(\omega)\ d\omega = 2(V\omega^2\ d\omega/2\pi^2c^3) = V\omega^2\ d\omega/\pi^2c^3 \ . \qquad (A6.13)$$

A6.1.3 Elastic waves in solids (sound waves)

In a solid one can have both longitudinal waves of velocity c_L and transverse waves of velocity c_T. The velocity of longitudinal waves is equal to $[(K + 4G/3)/\rho]^{1/2}$ where ρ is the density, K is the bulk modulus and G is the rigidity (or shear) modulus. The velocity of transverse waves is equal to $(G/\rho)^{1/2}$. The transverse waves can have two independent directions of polarisation. Hence, according to equation (A6.11) the total number of stationary wave solutions (normal modes) in the angular frequency range ω to $(\omega + d\omega)$ is

$$D(\omega)\ d\omega = (V\omega^2\ d\omega/2\pi^2)[1/c_L^3 + 2/c_T^3] \ . \qquad (A6.14)$$

A6.1.4 Schrödinger's equation

Consider a particle of mass m and *spin zero* in the cubical box shown in Figure A6.1(a). (This is the same as the case discussed in Section A3.2 of Appendix 3.) According to the de Broglie relation, $\lambda = h/p$, where λ is the wavelength, p is the momentum of the particle and h is Planck's constant. Hence $p = h/\lambda = \hbar k$, where $k = 2\pi/\lambda$. Using $k = p/\hbar$ in equation (A6.9) we obtain

$$D(p)\ dp = Vp^2\ dp/2\pi^2\hbar^3 \ . \qquad (A6.15)$$

This expression depends only on fitting standing waves of wavelength $\lambda = h/p$ in a box. It holds in both the relativistic and non-relativistic regions. In the non-relativistic limit, when $p = mv$, and $\varepsilon = \frac{1}{2}mv^2 = p^2/2m$, equation (A6.15) gives

$$D(\varepsilon)\,d\varepsilon = \frac{V}{4\pi^2}\left(\frac{2m}{\hbar^2}\right)^{3/2}\varepsilon^{1/2}\,d\varepsilon \qquad \text{(non relativistic limit)} \ , \ (A6.16)$$

where $D(\varepsilon)\,d\varepsilon$ is the number of stationary wave solutions in which the particle in the box has kinetic energy between ε and $(\varepsilon + d\varepsilon)$. Equation (A6.16) is the same as equation (A3.18) of Appendix 3.

In the case of particles, the term single particle quantum state is used in the text rather than stationary wave solution or normal mode. The term orbital is often used, particularly in chemistry, for a single particle quantum state. If the particle has spin s, there are $g = (2s + 1)$ states associated with each set of values of n_x, n_y and n_z, and the right hand sides of equations (A6.15) and (A6.16) must be multiplied by the spin degeneracy g to give

$$D(p)\,dp = gVp^2\,dp/2\pi^2\hbar^3 \qquad (A6.17)$$

$$D(\varepsilon)\,d\varepsilon = \frac{gV}{4\pi^2}\left(\frac{2m}{\hbar^2}\right)^{3/2}\varepsilon^{1/2}\,d\varepsilon \qquad \text{(non-relativistic limit)} \ . \ (A6.18)$$

A6.1.5 Photons

Consider a single photon in the cubical box shown in Figure A6.1(a). The possible single particle (photon) states are the standing wave solutions (normal modes) obtained by solving the wave equation. This leads to the conditions given by equation (A6.7) and, ignoring the spin of the photon, leads up to equation (A6.11) for the number of single particle (photon) states in which the angular frequency of the photon is between ω and $(\omega + d\omega)$. Photons are bosons of spin $s = 1$. However, for photons, the number of spin states g is equal to 2 not $(2s + 1) = 3$. The degeneracy of 2 arises from the fact that photons travel at the speed of light. (Reference: Wigner [2].) Hence equation (A6.11) must be multiplied by 2 not 3, so that the number of single particle (photon) states, in which the photon has angular frequency between ω and $(\omega + d\omega)$, corresponding to a photon having energy between $\hbar\omega$ and $\hbar(\omega + d\omega)$, is

$$D(\omega)\,d\omega = V\omega^2\,d\omega/\pi^2c^3 \ . \qquad (A6.19)$$

A6.2* PROGRESSIVE WAVE SOLUTIONS OF SCHRÖDINGER'S EQUATION*

Progressive wave solutions of Schrödinger's equation for a particle can be obtained by introducing periodic boundary conditions. Consider a cube of side L. Let one corner of this cube be at the origin of the coordinate system, as shown in Figure A6.2(a). It will be assumed that the wave

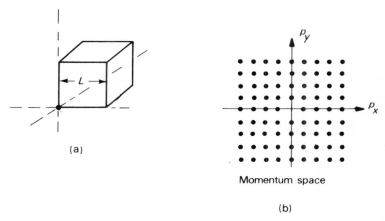

(a)

Momentum space

(b)

Figure A6.2—(a) Determination of progressive wave solutions of
Schrödinger's equation for a particle, using periodic boundary
conditions. (b) The allowed momenta of the progressive wave solutions
can be represented by dots in momentum space. Only the $p_z = 0$ plane
is shown.

function ψ has the same value at $x = L$ as at $x = 0$, the same value at $y = L$
as at $y = 0$ and the same value at $z = L$ as at $z = 0$. It will be found that
the number of progressive wave solutions per unit volume is independent
of the value of L. A progressive wave solution of Schrödinger's equation is
of the type

$$\psi = A \exp[i(k_x x + k_y y + k_z z - \omega t)] \qquad (A6.20)$$

where A is a constant. According to de Moivre's theorem,

$$\exp[i(\theta + 2\pi n)] = \cos(\theta + 2\pi n) + i \sin(\theta + 2\pi n) \ .$$

If n is a positive or negative integer or zero, both $\cos(\theta + 2\pi n)$ is equal to
$\cos \theta$ and $\sin(\theta + 2\pi n)$ is equal to $\sin \theta$, so that $\exp[i(\theta + 2\pi n)]$ is equal to
$\exp(i\theta)$. For ψ to have the same value at $x = 0$ and $x = L$ at all times t, $k_x L$
must be equal to $2\pi n_x$, where n_x is an integer. Hence for a wave travelling
in an arbitrary direction, in order to satisfy the periodic boundary conditions, we must have

$$k_x = 2\pi n_x/L; \qquad k_y = 2\pi n_y/L; \qquad k_z = 2\pi n_z/L \ .$$

According tò the de Broglie relation, $\lambda = h/p$ so that, since $k = 2\pi/\lambda$, then
$p = kh/2\pi$. Hence the possible values of the components of the momentum
p are

$$p_x = n_x h/L; \qquad p_y = n_y h/L; \qquad p_z = n_z h/L \ , \qquad (A6.21)$$

where p_x, p_y and p_z are the components of the momentum of the particle and n_x, n_y and n_z can be positive or negative integers or zero. Negative integer values of n_x, n_y and n_z correspond to progressive waves going in the negative x, y and z directions respectively.

Equation (A6.20) satisfies Schrödinger's equation, namely equation (A3.10) of Appendix 3, if

$$\varepsilon = (\hbar^2/2m)(k_x^2 + k_y^2 + k_z^2) = (1/2m)(p_x^2 + p_y^2 + p_z^2) \ .$$

The possible values of p_x, p_y and p_z given by equation (A6.21) can be plotted in momentum space, that is a space in which p_x, p_y and p_z are used as rectangular axes. The possible values of p_x, p_y and p_z are represented by dots, which are a distance h/L apart in Figure A6.2(b), so that there is one dot per volume of h^3/L^3 of momentum space. Consider the volume of momentum space between p_x and $(p_x + dp_x)$, p_y and $(p_y + dp_y)$ and p_z and $(p_z + dp_z)$. Let the volume element $dp_x \, dp_y \, dp_z$ of momentum space be represented by $d^3\mathbf{p}$. The position of the volume element in momentum space is given by the vector $\mathbf{p}$, which has components p_x, p_y, p_z. The number of dots in the volume element $d^3\mathbf{p}$ of momentum space is $d^3\mathbf{p} \div h^3/L^3$, which is equal to $L^3 d^3\mathbf{p}/h^3$. The number of progressive wave solutions satisfying periodic boundary conditions on the surface of the cube, of volume $V = L^3$, is

$$N(\mathbf{p})d^3\mathbf{p} = V \, dp_x \, dp_y \, dp_z/h^3 \qquad (A6.22)$$

If the volume V tends to infinity, the states get closer and closer together, and tend to a continuum. If the particle has a spin degeneracy g, the right hand side of equation (A6.22) must be multiplied by g.

REFERENCES

[1] Courant, R. and Hilbert, D. *Methods of Mathematical Physics*, (Interscience, 1953), Vol. 1, p. 429.
[2] Wigner, E. P. *Rev. Mod. Phys.*, **29**, 255 (1957).

Answers to Problems

1.1 602 km.

1.2 $+184 \text{ J K}^{-1}$.

2.1 The configurations are: 1, 1, 1; 2, 1, 1; 2, 2, 1; 3, 1, 1 and 2, 2, 2. The corresponding energy eigenvalues are 2.48×10^{-41}, 4.96 $\times 10^{-41}$, 7.44×10^{-41}, 9.09×10^{-41} and 9.92×10^{-41} J. The corresponding degeneracies are 1, 3, 3, 3 and 1 respectively. (a) All the energy eigenvalues are increased by a factor of 4. (b) All the energy eigenvalues are decreased by a factor of 4.

2.2 Check: In the cases of (a) three distinguishable particles, (b) three identical bosons and (c) three identical fermions the total numbers of accessible microstates are (a) 15, (b) 4 and (c) 14 respectively.

2.3 Check: In the cases of (a) four distinguishable particles, (b) four identical bosons and (c) four identical fermions, the total numbers of accessible microstates are (a) 20, (b) 3 and (c) 4, all from one configuration.

2.4 (ii) 20 accessible microstates. (iii) There are 3 configurations and 9 accessible microstates.

2.6 There are 84 accessible microstates. $\bar{U}_2 = 4.5$.

2.7 Initially system A has 1 accessible microstate and system B has 3 accessible microstates. After the systems are placed in thermal contact, the composite system has 25 accessible microstates. The increase in entropy is $k \log(25/3)$. $\bar{U}_A = 2.4$. $\bar{U}_B = 5.6$.

In the case of bosons, initially systems A and B have 1 and 2 accessible microstates respectively. After the systems are placed in thermal contact, the composite system has 15 accessible microstates. $\Delta S = k \log(15/2)$. $\bar{U}_A = 2\frac{2}{3}$. $\bar{U}_B = 5\frac{1}{3}$.

2.8 $\bar{U}_2 = 10$; $\sigma_2 = 4.3$.

2.10 (a) $N = 10^4$, (b) $N = 10^{12}$. The probability that all the molecules are in part A of the box is $1/2^N$. This gives (a) 10^{-3010}, (b) $10^{-3.01 \times 10^{11}}$

2.12 There are 26^{26} ways of typing the 26 letters of the alphabet. The average time taken would be 5.85×10^{30} years.

2.13 Fractional change $[\log N_2 - \log N_1]/\log N_1$ is 3.56×10^{-22}.
2.15 1; 10; 92378; 4.26×10^{12}; 2.88×10^{21}; 2.77×10^{30}; 2.76×10^{39}.
2.16 (a) $\exp(7.24 \times 10^{23})$, (b) $\exp(7.24 \times 10^{22})$, (c) $\exp(7.24 \times 10^{21})$, (d) $\exp(7.24 \times 10^{20})$, (e) $\exp(7.24 \times 10^{19})$.

3.1 $\exp(8.55 \times 10^{17})$.
3.2 $T = 2U/3Nk$; $p = 2U/3V$; $\mu = -kT \log[(gV/N^{5/2})(mU/3\pi\hbar^2)^{3/2}]$.
3.3 $T = (4/3C)(U/V)^{1/4}$.

4.1 (i) 0.020, (ii) 0.126, (iii) 0.239, (iv) 0.471, (v) 1056K, (vi) 2862K, (vii) 2.9×10^4K.
4.2 9.95K; 5.42×10^{-3}K.
4.3 2.2×10^{23}; 2.04 A m^2.
4.4 2.48×10^{-10}.
4.5 The degeneracies are 3 and 6.
4.6 The relative population is 1.224. The populations are equal at 8.54×10^4K.
4.7 The partition functions are (i) 1.000 045 402, (ii) 1.5032, (iii) 2.724. The probabilities P_0, P_1 and P_2 are (i) 0.999 954 698, 0.000 045 400 and 0.000 000 002. (ii) 0.665, 0.245 and 0.090. (iii) 0.367, 0.332, 0.301. The ensemble average energies are (i) 0.000 045ε, (ii) 0.425ε, (iii) 0.933ε.
4.8 The probabilities are 0.486, 0.295 and 0.219. $Z = 2.056$. $\bar{U} = 0.323\varepsilon$.
4.9 $Z = 2.78$. The relative populations are 1 : 1.10 : 0.68. The average energy is $264k$. The populations of the two levels are equal at 587K.
4.10 The rms, average and most probable speeds are 433, 399 and 353 m s^{-1} respectively.
4.11 $\exp[-2.367 \times 10^{21}]$. The scale height is 4.2×10^{-22} m.
4.12 5.81×10^{23} mole^{-1}.
4.15 4.06×10^{-4} m.

5.1 $Z = (1.553)^N$; $U = 0.507N\varepsilon$; $S = 0.948Nk$.
5.2 $Z = (2.476)^N$; $U = 0.976N\varepsilon$; $S = 1.882Nk$.
5.3 $U = 4kaT^5V$; $p = kAT^5$; $S = 5kaT^4V$.
5.4 $Z = (2.050)^N$, where $N = 10^{24}$; $U = -2.042$ J; $C_V = 0.658$ J K^{-1}.
5.5 $0.92Nk$.
5.6 $1.30 \times 10^{-11}Nk$; $0.0062Nk$; $0.926Nk$.
5.7 124 J K^{-1}.
5.8 $\mu = -5.99 \times 10^{-20}$ J.

6.1 $F = -0.823N\varepsilon$.

9.1 a : b : c = 1 : 10^8 : 10^{16}.

9.2 At 3K: $u = 6.124 \times 10^{-14}$ J m^{-3}; $\bar{N} = 5.45 \times 10^8$; $p = 2.06 \times 10^{-14}$ N m^{-2}. At 300K: $u = 6.124 \times 10^{-6}$ J m^{-3}; $\bar{N} = 5.45 \times 10^{14}$; $p = 2.06 \times 10^{-6}$ N m^{-2}. At 30 000K: $u = 6.124 \times 10^2$ J m^{-3}; $\bar{N} = 5.45 \times 10^{20}$; $p = 2.06 \times 10^2$ N m^{-2}.

9.4 2 900 nm.

9.5 1.76×10^{11} Hz.

12.1 0.9999; 0.997; 0.981; 0.872; 0.805; 0.715; 0.603; 0.504; 0.480; 0.358; 0.253.

12.3 (a) 0.134 (b) 0.126 (c) 0.119 (d) 0.106 (e) 0.100 (f) 0.086 (g) 0.074 (h) 0.054.

12.4 The new value of $f(\varepsilon)$ is $(1 - 3.85 \times 10^{-11})$.

12.5 The Fermi energy is 8.954×10^{-19} J, which is 5.59 eV. The Fermi temperature is 6.49×10^4K. The velocity of an electron having kinetic energy equal to the Fermi energy is 1.4×10^6 m s^{-1}. The pressure is 2.12×10^5 atmospheres.

12.7 $\exp(-\mu/kT) = 3.40 \times 10^5$; $f(\varepsilon) = 6.57 \times 10^{-7}$; $T_f = 0.068$K.

12.9 (a) 0.725 (b) 0.552 (c) 0.431 (d) 0.342 (e) 0.275 (f) 0.183.

12.10 $T_b = 0.087$K; 0.91.

12.11 $T_b = 0.036$K, $\mu = -2.29 \times 10^{-51}$ J. The ratio, calculated using equation (12.46), is 7.39×10^{-12}. If the Boltzmann distribution were used, the ratio would be 0.999 999 999 999 775.

Index